INTRODUCTION

TO

HIGHER ALGEBRA

BY

MAXIME BÔCHER

PROFESSOR OF MATHEMATICS IN HARVARD UNIVERSITY

PREPARED FOR PUBLICATION WITH THE COÖPERATION OF

E. P. R. DUVAL

INSTRUCTOR IN MATHEMATICS IN THE UNIVERSITY
OF WISCONSIN

DOVER PUBLICATIONS, INC.
Mineola, New York

DOVER PHOENIX EDITIONS

Bibliographical Note

This Dover edition, first published in 1964, and reissued in 2004, is an unabridged and unaltered republication of the work first published by The Macmillan Company in 1907.

Library of Congress Cataloging-in-Publication Data

Bôcher, Maxime, 1867–1918.
    Introduction to higher algebra / by Maxime Bôcher ; prepared for publication with the coöperation of E.P.R. Duval.
        p. cm. — (Dover phoenix editions)
    Originally published: New York : Macmillan Co., 1907.
    Includes index.
    ISBN 0-486-49570-1
     1. Algebra. I. Title. II. Series.

QA155.B66 2004
512—dc22

                                        2004043834

PREFACE

An American student approaching the higher parts of mathematics usually finds himself unfamiliar with most of the main facts of algebra, to say nothing of their proofs. Thus he has only a rudimentary knowledge of systems of linear equations, and he knows next to nothing about the subject of quadratic forms. Students in this condition, if they receive any algebraic instruction at all, are usually plunged into the detailed study of some special branch of algebra, such as the theory of equations or the theory of invariants, where their lack of real mastery of algebraic principles makes it almost inevitable that the work done should degenerate to the level of purely formal manipulations. It is the object of the present book to introduce the student to higher algebra in such a way that he shall, on the one hand, learn what is meant by a proof in algebra and acquaint himself with the proofs of the most fundamental facts, and, on the other, become familiar with many important results of algebra which are new to him.

The book being thus intended, not as a compendium, but really, as its title states, only as an *introduction* to higher algebra, the attempt has been made throughout to lay a sufficiently broad foundation to enable the reader to pursue his further studies intelligently, rather than to carry any single topic to logical completeness. No apology seems necessary for the omission of even such important subjects as Galois's Theory and a systematic treatment of invariants. A selection being necessary, those subjects have been chosen for treatment which have proved themselves of greatest importance in geometry and analysis, as well as in algebra, and the relations of the algebraic theories to geometry have been emphasized throughout. At the same time it must be borne in mind that the subject primarily treated is algebra, not analytic geometry, so that such geometric information as is given is necessarily of a fragmentary and somewhat accidental character.

No algebraic knowledge is presupposed beyond a familiarity with elementary algebra up to and including quadratic equations, and such a knowledge of determinants and the method of mathematical induction as may easily be acquired by a freshman in a week or two. Nevertheless, the book is not intended for wholly immature readers, but rather for students who have had two or three years' training in the elements of higher mathematics, particularly in analytic geometry and the calculus. In fact, a good elementary knowledge of analytic geometry is indispensable.

The exercises at the ends of the sections form an essential part of the book, not merely in giving the reader an opportunity to think for himself on the subjects treated, but also, in many cases, by supplying him with at least the outlines of important additional theories. As illustrations of this we may mention Sylvester's Law of Nullity (page 80), orthogonal transformations (page 154 and page 173), and the theory of the invariants of the biquadratic binary form (page 260).

On a first reading of Chapters I–VII, it may be found desirable to omit some or all of sections 10, 11, 18, 19, 20, 25, 27, 34, 35. The reader may then either take up the subject of quadratic forms (Chapters VIII–XIII), or, if he prefer, he may pass directly to the more general questions treated in Chapters XIV–XIX.

The chapters on Elementary Divisors (XX–XXII) form decidedly the most advanced and special portion of the book. A person wishing to read them without reading the rest of the book should first acquaint himself with the contents of sections 19 (omitting Theorem 1), 21–25, 36, 42, 43.

In a work of this kind, it has not seemed advisable to give many bibliographical references, nor would an acknowledgement at this point of the sources from which the material has been taken be feasible. The work of two mathematicians, however, Kronecker and Frobenius, has been of such decisive influence on the character of the book that it is fitting that their names receive special mention here. The author would also acknowledge his indebtedness to his colleague, Professor Osgood, for suggestions and criticisms relating to Chapters XIV–XVI.

This book has grown out of courses of lectures which have been delivered by the author at Harvard University during the last ten years. His thanks are due to Mr. Duval, one of his former pupils, without whose assistance the book would probably never have been **written.**

CONTENTS

CHAPTER I

POLYNOMIALS AND THEIR MOST FUNDAMENTAL PROPERTIES

CHAPTER II

A FEW PROPERTIES OF DETERMINANTS

CHAPTER III

THE THEORY OF LINEAR DEPENDENCE

CHAPTER IV

LINEAR EQUATIONS

CHAPTER V

SOME THEOREMS CONCERNING THE RANK OF A MATRIX

CHAPTER VI

LINEAR TRANSFORMATIONS AND THE COMBINATION OF MATRICES

CHAPTER VII

INVARIANTS. FIRST PRINCIPLES AND ILLUSTRATIONS

CHAPTER VIII

BILINEAR FORMS

CHAPTER IX

GEOMETRIC INTRODUCTION TO THE STUDY OF QUADRATIC FORMS

CHAPTER X

QUADRATIC FORMS

CHAPTER XI

REAL QUADRATIC FORMS

CHAPTER XII

THE SYSTEM OF A QUADRATIC FORM AND ONE OR MORE LINEAR FORMS

CHAPTER XIII

PAIRS OF QUADRATIC FORMS

CHAPTER XIV

SOME PROPERTIES OF POLYNOMIALS IN GENERAL

CHAPTER XIX

POLYNOMIALS SYMMETRIC IN PAIRS OF VARIABLES

CHAPTER XX

ELEMENTARY DIVISORS AND THE EQUIVALENCE OF λ-MATRICES

CHAPTER XXI

THE EQUIVALENCE AND CLASSIFICATION OF PAIRS OF BILINEAR FORMS AND OF COLLINEATIONS

CHAPTER XXII

THE EQUIVALENCE AND CLASSIFICATION OF PAIRS OF QUADRATIC FORMS

INTRODUCTION TO HIGHER ALGEBRA

CHAPTER I

POLYNOMIALS AND THEIR MOST FUNDAMENTAL PROPERTIES

1. Polynomials in One Variable. By an integral rational function of x, or, as we shall say for brevity, a polynomial in x, is meant a function of x determined by an expression of the form

$$(1) \qquad c_1 x^{a_1} + c_2 x^{a_2} + \cdots + c_k x^{a_k},$$

where the α's are integers positive or zero, while the c's are any constants, real or imaginary. We may without loss of generality assume that no two of the α's are equal. This being the case, the expressions $c_i x^{a_i}$ are called the *terms* of the polynomial, c_i is called the *coefficient* of this term, and α_i is called its *degree*. The highest degree of any term whose coefficient is not zero is called the *degree of the polynomial*.

It should be noticed that the conceptions just defined — terms, coefficients, degree — apply not to the polynomial itself, but to the particular *expression* (1) which we use to determine the polynomial, and it would be quite conceivable that one and the same function of x might be given by either one of two wholly different expressions of the form (1). We shall presently see (cf. Theorem 5 below) that this cannot be the case except for the obvious fact that we may insert in or remove from (1) any terms we please with zero coefficients.

By arranging the terms in (1) in the order of decreasing α's and supplying, if necessary, certain missing terms with zero coefficients, we may write the polynomial in the normal form

$$(2) \qquad a_0 x^n + a_1 x^{n-1} + \cdots + a_{n-1} x + a_n.$$

It should, however, constantly be borne in mind that a polynomial in this form is not necessarily of the nth degree; but will be of the nth degree when and only when $a_0 \neq 0$.

DEFINITION. *Two polynomials, $f_1(x)$ and $f_2(x)$, are said to be identically equal $(f_1 \equiv f_2)$ if they are equal for all values of x. A polynomial $f(x)$ is said to vanish identically $(f \equiv 0)$ if it vanishes for all values of x.*

We learn in elementary algebra how to add, subtract, and multiply * polynomials; that is, when two polynomials $f_1(x)$ and $f_2(x)$ are given, to form new polynomials equal to the sum, difference, and product of these two.

THEOREM 1. *If the polynomial*

$$f(x) \equiv a_0 x^n + a_1 x^{n-1} + \cdots + a_n$$

vanishes when $x = \alpha$, there exists another polynomial

$$\phi_1(x) \equiv a_0 x^{n-1} + a_1' x^{n-2} + \cdots + a_{n-1}',$$

such that
$$f(x) \equiv (x - \alpha)\phi_1(x).$$

For since by hypothesis $f(\alpha) = 0$, we have

$$f(x) \equiv f(x) - f(\alpha) \equiv a_0(x^n - \alpha^n) + a_1(x^{n-1} - \alpha^{n-1}) + \cdots + a_{n-1}(x - \alpha).$$

Now by the rule of elementary algebra for multiplying together two polynomials we have

$$x^k - \alpha^k \equiv (x - \alpha)(x^{k-1} + \alpha x^{k-2} + \cdots + \alpha^{k-1}).$$

Hence

$$f(x) \equiv (x - \alpha) \left[a_0(x^{n-1} + \alpha x^{n-2} + \cdots + \alpha^{n-1}) + a_1(x^{n-2} + \alpha x^{n-3} + \cdots + \alpha^{n-2}) + \cdots + a_{n-1} \right].$$

If we take as $\phi_1(x)$ the polynomial in brackets, our theorem is proved.

Suppose now that β is another value of x distinct from α for which $f(x)$ is zero. Then

$$f(\beta) = (\beta - \alpha)\phi_1(\beta) = 0;$$

---

* The question of division is somewhat more complicated and will be considered in § 63.

and since $\beta - \alpha \neq 0$, $\phi_1(\beta) = 0$. We can therefore apply the theorem just proved to the polynomial $\phi_1(x)$, thus getting a new polynomial

$$\phi_2(x) \equiv a_0 x^{n-2} + a_1'' x^{n-3} + \cdots + a_{n-2}'',$$

such that

$$\phi_1(x) \equiv (x - \beta)\phi_2(x),$$

and therefore

$$f(x) \equiv (x - \alpha)(x - \beta)\phi_2(x).$$

Proceeding in this way, we get the following general result:

THEOREM 2.   *If α_1, α_2, $\cdots \alpha_k$ are k distinct constants, and if*

$$f(x) \equiv a_0 x^n + a_1 x^{n-1} + \cdots + a_n \qquad (n \geqq k),$$

and

$$f(\alpha_1) = f(\alpha_2) = \cdots = f(\alpha_k) = 0,$$

then

$$f(x) \equiv (x - \alpha_1)(x - \alpha_2) \cdots (x - \alpha_k)\phi(x),$$

where

$$\phi(x) \equiv a_0 x^{n-k} + b_1 x^{n-k-1} + \cdots + b_{n-k}.$$

Applying this theorem in particular to the case $n = k$, we see that if the polynomial $f(x)$ vanishes for n distinct values α_1, α_2, $\cdots \alpha_n$ of x, then

$$f(x) \equiv a_0(x - \alpha_1)(x - \alpha_2) \cdots (x - \alpha_n).$$

Accordingly, if $a_0 \neq 0$, there can be no value of x other than $\alpha_1, \cdots \alpha_n$ for which $f(x) = 0$.   We have thus proved

THEOREM 3.   *A polynomial of the nth degree in x cannot vanish for more than n distinct values of x.*

Since the only polynomials which have no degree are those all of whose coefficients are zero, and since such polynomials obviously vanish identically, we get the fundamental result :

THEOREM 4.   *A necessary and sufficient condition that a polynomial in x vanish identically is that all its coefficients be zero.*

Since two polynomials in x are identically equal when and only when their difference vanishes identically, we have

THEOREM 5.   *A necessary and sufficient condition that two polynomials in x be identically equal is that they have the same coefficients.*

This theorem shows, as was said above, that the terms, coefficients, and degree of a polynomial depend merely on the polynomial itself not on the special way in which it is expressed.

2. Polynomials in More than One Variable. A function of (x, y) is called a polynomial if it is given by an expression of the form

$$c_1 x^{a_1} y^{\beta_1} + c_2 x^{a_2} y^{\beta_2} + \cdots + c_k x^{a_k} y^{\beta_k},$$

where the a's and β's are integers positive or zero.

More generally, a function of $(x_1, x_2, \cdots x_n)$ is called a polynomial if it is determined by an expression of the form

$$(1) \qquad c_1 x_1^{a_1} x_2^{\beta_1} \cdots x_n^{\nu_1} + c_2 x_1^{a_2} x_2^{\beta_2} \cdots x_n^{\nu_2} + \cdots + c_k x_1^{a_k} x_2^{\beta_k} \cdots x_n^{\nu_k},$$

where the a's, β's, $\cdots \nu$'s are integers positive or zero.

Here we may assume without loss of generality that in no two terms are the exponents of the various x's the same ; that is, that if

$$a_i = a_j, \; \beta_i = \beta_j, \; \cdots \mu_i = \mu_j,$$

then

$$\nu_i \neq \nu_j.$$

This assumption being made, $c_i x_1^{a_i} x_2^{\beta_i} \cdots x_n^{\nu_i}$ is called a *term* of the polynomial, c_i its *coefficient*, a_i the *degree of the term in* x_1, β_i in x_2, etc., and $a_i + \beta_i + \cdots + \nu_i$ the *total degree*, or simply the degree, of the term. The highest degree in x_i of any term in the polynomial whose coefficient is not zero is called the *degree of the polynomial in* x_i, and the highest total degree of any term whose coefficient is not zero is called the *degree of the polynomial*.

Here, as in § 1, the conceptions just defined apply for the present not to the function itself but to the special method of representing it by an expression of the form (1). We shall see presently, however, that this method is unique.

Before going farther, we note explicitly that *according to the definition we have given, a polynomial all of whose coefficients are zero has no degree.*

When we speak of a polynomial in n variables, we do not necessarily mean that all n variables are actually present. One or more of them may have the exponent zero in every term, and hence not appear at all. Thus a polynomial in one variable, or even a constant, may be regarded as a special case of a polynomial in any larger number of variables.

A polynomial all of whose terms are of the same degree is said to be *homogeneous.* Such polynomials we will speak of as *forms,**

* There is diversity of usage here. Some writers, following Kronecker, apply the term *form* to all polynomials. On the other hand, homogeneous polynomials are often spoken of as *quantics* by English writers.

distinguishing between *binary, ternary, quaternary,* and in general, *n-ary* forms according to the number of variables involved, binary forms involving two, ternary three, etc.

Another method of classifying forms is according to their degree. We speak here of linear forms, quadratic forms, cubic forms, etc., according as the degree is 1, 2, 3, etc. We will, however, agree that a polynomial all of whose coefficients are zero may also be spoken of indifferently as a linear form, quadratic form, cubic form, etc., in spite of the fact that it has no degree.

If all the coefficients of a polynomial are real, it is called a *real polynomial* even though, in the course of our work, we attribute imaginary values to the variables.

It is frequently convenient to have a polynomial in more than one variable arranged according to the descending powers of some one of the variables. Thus a normal form in which we may write a polynomial in n variables is

$$\phi_0(x_2, \cdots x_n)x_1^{m} + \phi_1(x_2, \cdots x_n)x_1^{m-1} + \cdots + \phi_m(x_2, \cdots x_n),$$

the ϕ's being polynomials in the $n - 1$ variables $(x_2, \cdots x_n)$.

We learn in elementary algebra how to add, subtract, and multiply polynomials, getting as the result new polynomials.

DEFINITION. *Two polynomials in any number of variables are said to be identically equal if they are equal for all values of the variables. A polynomial is said to vanish identically if it vanishes for all values of the variables.*

THEOREM 1. *A necessary and sufficient condition that a polynomial in any number of variables vanish identically is that all its coefficients be zero.*

That this is a sufficient condition is at once obvious. To prove that it is a necessary condition we use the method of mathematical induction. Since we know that the theorem is true in the case of one variable (Theorem 4, § 1), the theorem will be completely proved if we can show that if it is true for a certain number $n - 1$ of variables, it is true for n variables.

Suppose, then, that

$$f(x_1, \cdots x_n) \equiv \phi_0(x_2, \cdots x_n)x_1^{m} + \phi_1(x_2, \cdots x_n)x_1^{m-1} + \cdots + \phi_m(x_2, \cdots x_n)$$

vanishes identically. If we assign to $(x_2, \cdots x_n)$ any fixed values $(x_2', \cdots x_n')$, f becomes a polynomial in x_1 alone, which, by hypothesis, vanishes for all values of x_1. Hence its coefficients must, by Theorem 4, § 1, all be zero:

$$\phi_i(x_2', \cdots x_n') = 0 \qquad (i = 0, 1, \cdots m).$$

That is, the polynomials $\phi_0, \phi_1, \cdots \phi_m$ vanish for all values of the variables, since $(x_2', \cdots x_n')$ was *any* set of values. Accordingly, by the assumption we have made that our theorem is true for polynomials in $n - 1$ variables, all the coefficients of all the polynomials $\phi_0, \phi_1, \cdots \phi_m$ are zero. These, however, are simply the coefficients of f. Thus our theorem is proved.

Since two polynomials are identically equal when and only when their difference is identically zero, we infer now at once the further theorem:

THEOREM 2. *A necessary and sufficient condition that two polynomials be identically equal is that the coefficients of their corresponding terms be equal.*

We come next to

THEOREM 3. *If f_1 and f_2 are polynomials in any number of variables of degrees m_1 and m_2 respectively, the product $f_1 f_2$ will be of degree $m_1 + m_2$.*

This theorem is obviously true in the case of polynomials in one variable. If, then, assuming it to be true for polynomials in $n - 1$ variables we can prove it to be true for polynomials in n variables, the proof of our theorem by the method of mathematical induction will be complete.

Let us look first at the special case in which both polynomials are homogeneous. Here every term we get by multiplying them together by the method of elementary algebra is of degree $m_1 + m_2$. Our theorem will therefore be proved if we can show that there is at least one term in the product whose coefficient is not zero. For this purpose, let us arrange the two polynomials f_1 and f_2 according to descending powers of x_1,

$$f_1(x_1, \cdots x_n) \equiv \phi_0'(x_2, \cdots x_n) x_1^{k_1} + \phi_1'(x_2, \cdots x_n) x_1^{k_1-1} + \cdots,$$

$$f_2(x_1, \cdots x_n) \equiv \phi_0''(x_2, \cdots x_n) x_1^{k_2} + \phi_1''(x_2, \cdots x_n) x_1^{k_2-1} + \cdots.$$

Here we may assume that neither ϕ_0' nor ϕ_0'' vanishes identically. Since f_1 and f_2 are homogeneous, ϕ_0' and ϕ_0'' will also be homogeneous of degrees $m_1 - k_1$ and $m_2 - k_2$ respectively.   In the product $f_1 f_2$ the terms of highest degree in x_1 will be those in the product

$$\phi_0'(x_2, \cdots x_n)\phi_0''(x_2, \cdots x_n)x_1^{k_1+k_2},$$

and since we assume our theorem to hold for polynomials in $n-1$ variables, $\phi_0'\phi_0''$ will be a polynomial of degree $m_1 + m_2 - k_1 - k_2$. Any term in this product whose coefficient is not zero gives us when multiplied by $x_1^{k_1+k_2}$ a term of the product $f_1 f_2$ of degree $m_1 + m_2$ whose coefficient is not zero.   Thus our theorem is proved for the case of homogeneous polynomials.

Let us now, in the general case, write f_1 and f_2 in the forms

$$f_1(x_1, \cdots x_n) \equiv \phi_{m_1}'(x_1, \cdots x_n) + \phi_{m_1-1}'(x_1, \cdots x_n) + \cdots,$$

$$f_2(x_1, \cdots x_n) \equiv \phi_{m_2}''(x_1, \cdots x_n) + \phi_{m_2-1}''(x_1, \cdots x_n) + \cdots,$$

where ϕ_i' and ϕ_j'' are homogeneous polynomials which are either of degrees i and j respectively, or which vanish identically.   Since, by hypothesis, f_1 and f_2 are of degrees m_1 and m_2 respectively, ϕ_{m_1}' and ϕ_{m_2}'' will not vanish identically, but will be of degrees m_1 and m_2.

The terms of highest degree in the product $f_1 f_2$ will therefore be the terms of the product $\phi_{m_1}' \phi_{m_2}''$, and this being a product of homogeneous polynomials comes under the case just treated and is therefore of degree $m_1 + m_2$.   The same is therefore true of the product $f_1 f_2$, and our theorem is proved.

By a successive application of this theorem we infer

COROLLARY.   *If k polynomials are of degrees $m_1, m_2, \cdots m_k$ respectively, their product is of degree $m_1 + m_2 + \cdots + m_k$.*

We mention further, on account of their great importance, the two rather obvious results :

THEOREM 4.   *If the product of two or more polynomials is identically zero, at least one of the factors must be identically zero.*

For if none of them were identically zero, they would all have definite degrees, and therefore their product would, by Theorem 3, have a definite degree, and would therefore not vanish identically.

It is from this theorem that we draw our justification for cancelling out from an identity a factor which we know to be not identically zero.

THEOREM 5.   *If $f(x_1, \cdots x_n)$ is a polynomial which is not identically zero, and if $\phi(x_1, \cdots x_n)$ vanishes at all points where f does not vanish, then ϕ vanishes identically.*

This follows from Theorem 4 when we notice that $f\phi \equiv 0$.

EXERCISES

1. If f and ϕ are polynomials in any number of variables, what can be inferred from the identity $f^2 \equiv \phi^2$ concerning the relation between the polynomials f and ϕ?

2. If f_1 and f_2 are polynomials in $(x_1, \cdots x_n)$ which are of degrees m_1 and m_2 respectively in x_1, prove that their product is of degree $m_1 + m_2$ in x_1.

3. Geometric Interpretations. In dealing with functions of a single real variable, the different values which the variable may take on may be represented geometrically by the points of a line ; it being understood that when we speak of a point x we mean the point which is situated on the line at a distance of x units (to the right or left according as x is positive or negative) from a certain fixed origin O, on the line.   Similarly, in the case of functions of two real variables, the sets of values of the variables may be pictured geometrically by the points of a plane, and in the case of three real variables, by the points of space ; the set of values represented by a point being, in each case, the *rectangular* coördinates of that point. When we come to functions of four or more variables, however, this geometric representation is impossible.

The complex variable $x = \xi + \eta i$ depends on the two independent real variables ξ and η in such a way that to every pair of real values (ξ, η) there corresponds one and only one value of x.   The different values which a single complex variable may take on may, therefore, be represented by the points of a plane in which (ξ, η) are used as cartesian coördinates.   In dealing with functions of more than one complex variable, however, this geometric representation is impossible, since even two complex variables $x = \xi + \eta i$, $y = \xi_1 + \eta_1 i$ are equivalent to four real variables $(\xi, \eta, \xi_1, \eta_1)$.

By the *neighborhood of a point* $x = a$ we mean that part of the line between the points $x = a - \alpha$ and $x = a + \alpha$ (α being an arbitrary positive constant, large or small), or what is the same thing, all points whose coördinates x satisfy the inequality $|x - a| < \alpha$.*

---

*We use the symbol $|Z|$ to denote the absolute value of Z, *i.e.* the numerical value of Z if Z is real, the modulus of Z if Z is imaginary.

Similarly, by the neighborhood of a point (a, b) in a plane, we shall mean all points whose coördinates (x, y) satisfy the inequalities

$$|x - a| < \alpha, \quad |y - b| < \beta,$$

where α and β are positive constants.   This neighborhood thus consists of the interior of a rectangle of which (a, b) is the center and whose sides are parallel to the coördinate axes.

By the neighborhood of a point (a, b, c) in space we mean all points whose coördinates (x, y, z) satisfy the inequalities

$$|x - a| < \alpha, \quad |y - b| < \beta, \quad |z - c| < \gamma.$$

In all these cases it will be noticed that the neighborhood may be large or small according to the choice of the constants α, β, γ.

If we are dealing with a single complex variable $x = \xi + \eta i$, we understand by the neighborhood of a point a all points in the plane of complex quantities whose complex coördinate x satisfies the inequality $|x - a| < \alpha$, α being as before a real positive constant.   Since $|x - a|$ is equal to the distance between x and a, the neighborhood of a now consists of the interior of a circle of radius α described about a as center.

It is found convenient to extend the geometric terminology we have here introduced to the case of any number of real or complex variables.   Thus if we are dealing with n independent variables $(x_1, x_2, \cdots x_n)$, we speak of any particular set of values of these variables as a *point in space of n dimensions*.   Here we have to distinguish between *real points*, that is sets of values of the x's which are all real, and *imaginary points* in which this is not the case.   In using these terms we do not propose even to raise the question whether in any geometric sense there is such a thing as space of more than three dimensions.   We merely use these terms in a wholly conventional algebraic sense because on the one hand they have the advantage of conciseness over the ordinary algebraic terms, and on the other hand, by calling up in our minds the geometric pictures of three dimensions or less, this terminology is often suggestive of new relations which might otherwise not present themselves to us so readily.

By the neighborhood of the point $(a_1, a_2, \cdots a_n)$ we understand all points which satisfy the inequalities

$$|x_1 - a_1| < \alpha_1, \quad |x_2 - a_2| < \alpha_2, \cdots |x_n - a_n| < \alpha_n,$$

where $\alpha_1, \alpha_2, \cdots \alpha_n$ are real positive constants.

If, in particular, $(a_1, a_2, \cdots a_n)$ is a real point, we may speak of the real neighborhood of this point, meaning thereby all real points $(x_1, x_2, \cdots x_n)$ which satisfy the above inequalities.

As an illustration of the use to which the conception of the neighborhood of a point can be put in algebra, we will prove the following important theorem:

THEOREM 1. *A necessary and sufficient condition that a polynomial $f(x_1, \cdots x_n)$ vanish identically is that it vanish throughout the neighborhood of a point $(a_1, \cdots a_n)$.*

That this is a necessary condition is obvious. To prove that it is sufficient we begin with the case $n = 1$.

Suppose then that $f(x)$ vanishes throughout a certain neighborhood of the point $x = a$. If $f(x)$ did not vanish identically, it would be of some definite degree, say k, and therefore could not vanish at more than k points (cf. Theorem 3, § 1). This, however, is not the case, since it vanishes at an infinite number of points, namely all points in the neighborhood of $x = a$. Thus our theorem is proved in the case $n = 1$.

Turning now to the case $n = 2$, let

$$f(x, y) \equiv \phi_0(y)x^k + \phi_1(y)x^{k-1} + \cdots + \phi_k(y)$$

be a polynomial which vanishes throughout a certain neighborhood of the point (a, b), say when

$$|x - a| < \alpha, \quad |y - b| < \beta.$$

Let y_0 be any constant satisfying the inequality

$$|y_0 - b| < \beta.$$

Then $f(x, y_0)$ is a polynominal in x alone which vanishes whenever $|x - a| < \alpha$. Hence, by the case $n = 1$ of our theorem, $f(x, y_0) \equiv 0$. That is,

$$\phi_0(y_0) = \phi_1(y_0) = \cdots = \phi_k(y_0) = 0.$$

Thus all these polynomials ϕ vanish at every point y_0 in the neighborhood of $y = b$, and therefore, by the case $n = 1$ of our theorem, they are all identically zero. From this it follows that for every value of x, $f(x, y)$ vanishes for all values of y, that is $f \equiv 0$, and our theorem is proved.

We leave to the reader the obvious extension of this method of proof to the case of n variables by the use of mathematical induction.

From the theorem just proved we can infer at once the following:

THEOREM 2.   *A necessary and sufficient condition that two polynomials in the variables $(x_1, \cdots x_n)$ be identically equal is that they be equal throughout the neighborhood of a point $(a_1, \cdots a_n)$.*

EXERCISES

1. Theorem 3, § 1 may be stated as follows: If f is a polynomial in one variable which is known not to be of degree higher than n, then if f vanishes at $n + 1$ distinct points, it vanishes identically.

Establish the following generalization of this theorem:

If f is a polynomial in (x, y) which is known not to be of higher degree than n in x, and not of higher degree than m in y, then, if f vanishes at the $(n + 1)(m + 1)$ distinct points:

$$(x_i, y_j) \qquad \left(\begin{matrix} i = 1, 2, \cdots n + 1 \\ j = 1, 2, \cdots m + 1 \end{matrix} \right),$$

it vanishes identically.

2. Generalize the theorem of Exercise 1 to polynomials in any number of variables.

3. Prove Theorem 4, § 2 by means of Theorem 1 of the present section; and from this result deduce Theorem 3, § 2.

4. Do Theorems 1 and 2 of this section hold if we consider only real polynomials and the real neighborhoods of real points?

4. Homogeneous Coördinates. Though only two quantities are necessary in order to locate the position of a point in a plane, it is frequently more convenient to use three, the precise values of the quantities being of no consequence, but only their ratios. We will represent these three quantities by x, y, t, and define their ratios by the equations

$$\frac{x}{t} = X, \qquad \frac{y}{t} = Y,$$

where X and Y are the cartesian coördinates of a point in a plane. Thus $(2, 3, 5)$ will represent the point whose abscissa is $\frac{2}{5}$ and whose ordinate is $\frac{3}{5}$. Any set of three numbers which are proportional to

(2, 3, 5) will represent the same point. So that, while to every set of three numbers (with certain exceptions to be noted below) there corresponds one and only one point, to each point there correspond an infinite number of different sets of three numbers, all of which, however, are proportional.

When $t = 0$ our definition is meaningless; but if we consider the points (2, 3, 1), (2, 3, 0.1), (2, 3, 0.01), (2, 3, 0.001), $\cdots$, which are, in cartesian coördinates, the points (2, 3), (20, 30), (200, 300), (2000, 3000), $\cdots$, we see that they all lie on the straight line through the origin whose slope is $\frac{3}{2}$. Thus as t approaches zero, x and y remaining fixed but not both zero, the point (x, y, t) moves away along a straight line through the origin whose slope is y/x. Hence it is natural to speak of $(x, y, 0)$ as *the point at infinity* on the line whose slope is y/x. If t approaches zero through negative values, the point will move off along *the same* line, but in the opposite direction. We will not distinguish between these two cases, but will speak of *only one* point at infinity on any particular line. It can be easily verified that if a point moves to infinity along any line parallel to the one just considered, its homogeneous coördinates may be made to approach the same values $(x, y, 0)$ as those just obtained. It is therefore natural to speak of the point at infinity in a certain direction rather than on a definite line. Finally we will agree that two points at infinity whose coördinates are proportional shall be regarded as coinciding, since these coördinates may be regarded as the limits of the coördinates of one and the same point which moves further and further off.*

If $x = y = t = 0$, we will not say that we have a point at all, since the coördinates of *any point whatever* may be taken as small as we please, and so (0, 0, 0) might be regarded as the limits of the coördinates of any fixed or variable point.

---

* It should be noticed that in speaking of points at infinity we are, considering the matter from a purely logical point of view, doing exactly the same thing that we did in § 3 in speaking of imaginary points, or points in space of n dimensions ; that is, we are speaking of a set of quantities as a "point" which are not the coördinates of any point. The only difference between the two cases is that the coördinates of our "point at infinity" are the limits of the coördinates of a true point.

Thus, in particular, it is a pure convention, though a desirable and convenient one, when we say that two points at infinity shall be regarded as coincident when and only when their coördinates are proportional. We might, if we chose, regard *all* points at infinity as coincident. There is no logical compulsion in the matter.

The equation

$$AX^2 + BXY + CY^2 + DX + EY + F = 0$$

becomes, in homogeneous coördinates,

$$A\frac{x^2}{t^2} + B\frac{xy}{t^2} + C\frac{y^2}{t^2} + D\frac{x}{t} + E\frac{y}{t} + F = 0,$$

or
$$Ax^2 + Bxy + Cy^2 + Dxt + Eyt + Ft^2 = 0,$$

a homogeneous equation of the second degree; and it is evident that if the coördinates X, Y in any algebraic equation be replaced by the coördinates x, y, t, the resulting equation will be homogeneous, and of the same degree as the original equation. It is to this fact that the system owes its name, as well as one of its chief advantages.

The equation
$$Ax + By + Ct = 0$$

represents, in general, a line, but if $A = B = 0$, $C \neq 0$, it has no true geometric locus. It is, in this case, satisfied by the coördinates of all points at infinity, and by the coördinates of no other point. We shall therefore speak of it as the equation of *the line at infinity*. The reader may easily verify, by using the equation of a line in terms of its intercepts, that if a straight line move further and further away, its homogeneous equation will approach more and more nearly the form $t = 0$.

In space of three dimensions we will represent the point whose cartesian coördinates are X, Y, Z by the four homogeneous coördinates x, y, z, t, whose ratios are defined by the equations

$$\frac{x}{t} = X, \quad \frac{y}{t} = Y, \quad \frac{z}{t} = Z.$$

We will speak of $(x, y, z, 0)$ as "the point at infinity" on a line whose direction cosines are

$$\frac{x}{\sqrt{x^2 + y^2 + z^2}}, \quad \frac{y}{\sqrt{x^2 + y^2 + z^2}}, \quad \frac{z}{\sqrt{x^2 + y^2 + z^2}}.$$

$(0, 0, 0, 0)$ will be excluded, and $t = 0$ will be spoken of as the equation of *the plane at infinity*.

Extending the same terminology to the general case, we shall sometimes find it convenient to speak of $(x_1, x_2, \cdots x_n)$ not as a point in space of n dimensions, but as a point represented by its homogeneous coördinates in space of $n - 1$ dimensions. Two points whose coördinates are proportional will be spoken of as identical, a point whose last coördinate is zero will be spoken of as a point at infinity, and the case $x_1 = \cdots = x_n = 0$ will not be spoken of as a point at all. This terminology will be adopted only in connection with homogeneous polynomials, and even then it must be clearly understood that we are perfectly free to adopt whichever terminology we find most convenient. Thus, for instance, if $f(x_1, x_2, x_3)$ is a homogeneous polynomial of the second degree, the equation $f = 0$ may be regarded either as determining a conic in a plane (x_1, x_2, x_3 being homogeneous coördinates) or a quadric cone in space (x_1, x_2, x_3 being ordinary cartesian coördinates).

Homogeneous coördinates may also be used in space of one dimension. We should then determine the points on a line by two coördinates x, t whose ratio x/t is the non-homogeneous coördinate X, *i.e.* the distance of the point from the origin. It is this representation that is commonly made use of in connection with the theory of binary forms.

5. The Continuity of Polynomials.

DEFINITION. *A function $f(x_1, \cdots x_n)$ is said to be continuous at the point $(c_1, \cdots c_n)$ if, no matter how small a positive quantity ϵ be chosen, a neighborhood of the point $(c_1, \cdots c_n)$ can be found so small that the difference between the value of the function at any point of this neighborhood and its value at the point $(c_1, \cdots c_n)$ is in absolute value less than ϵ.*

That is, f is continuous at $(c_1, \cdots c_n)$ if, having chosen a positive quantity ϵ, it is possible to determine a positive δ such that

$$|f(x_1, \cdots x_n) - f(c_1, \cdots c_n)| < \epsilon$$

for all values of $(x_1 \cdots x_n)$ which satisfy the inequalities,

$$|x_1 - c_1| < \delta, \ |x_2 - c_2| < \delta, \ \cdots |x_n - c_n| < \delta.$$

THEOREM 1. *If two functions are continuous at a point, their sum is continuous at this point.*

Let f_1 and f_2 be two functions continuous at the point $(c_1, \cdots c_n)$ and let k_1 and k_2 be their respective values at this point. Then, no matter how small the positive quantity ϵ may be chosen, we may take δ_1 and δ_2 so small that

$$|f_1 - k_1| < \tfrac{1}{2}\epsilon \qquad \text{when } |x_i - c_i| < \delta_1,$$
$$|f_2 - k_2| < \tfrac{1}{2}\epsilon \qquad \text{when } |x_i - c_i| < \delta_2.$$

Accordingly

$$|f_1 - k_1| + |f_2 - k_2| < \epsilon \qquad \text{when } |x_i - c_i| < \delta,$$

where δ is the smaller of the two quantities δ_1 and δ_2; and, since

$$|A| + |B| \geqq |A + B|, \text{ we have}$$

$$|f_1 - k_1 + f_2 - k_2| = |(f_1 + f_2) - (k_1 + k_2)| < \epsilon \quad \text{when } |x_i - c_i| < \delta.$$

Hence $f_1 + f_2$ is continuous at the point $(c_1, \cdots c_n)$.

COROLLARY. *If a finite number of functions are continuous at a point, their sum is continuous at this point.*

THEOREM 2. *If two functions are continuous at a point, their product is continuous at this point.*

Let f_1 and f_2 be the two functions, and k_1 and k_2 their values at the point $(c_1, \cdots c_n)$ where they are assumed to be continuous. We have to prove that however small ϵ may be, δ can be chosen so small that

$$(1) \qquad |f_1 f_2 - k_1 k_2| < \epsilon \qquad \text{when } |x_i - c_i| < \delta.$$

Let η be a positive constant, which we shall ultimately restrict to a certain degree of smallness, and let us choose two positive constants δ_1 and δ_2 such that

$$|f_1 - k_1| < \eta \qquad \text{when } |x_i - c_i| < \delta_1,$$
$$|f_2 - k_2| < \eta \qquad \text{when } |x_i - c_i| < \delta_2.$$

Now take δ as the smaller of the two quantities δ_1 and δ_2. Then, when $|x_i - c_i| < \delta$,

$$|f_1 f_2 - k_1 k_2| = |f_2(f_1 - k_1) + k_1(f_2 - k_2)|$$
$$\leqq |f_2| |f_1 - k_1| + |k_1| |f_2 - k_2| \leqq \{|f_2| + |k_1|\} \eta.$$

Accordingly since, when $|x_i - c_i| < \delta$,

$$|f_2| = |k_2 + (f_2 - k_2)| \leqq |k_2| + |f_2 - k_2| < |k_2| + \eta,$$

we may write

$$(2) \qquad |f_1 f_2 - k_1 k_2| < \{|k_1| + |k_2|\} \eta + \eta^2.$$

If k_1 and k_2 are not both zero, let us take η small enough to satisfy the two inequalities

$$\eta < \frac{\epsilon}{2\{|k_1| + |k_2|\}}, \qquad \eta < \sqrt{\frac{\epsilon}{2}}.$$

If $k_1 = k_2 = 0$, we will restrict η merely by the inequality

$$\eta < \sqrt{\epsilon}.$$

In either case, inequality (2) then reduces to the form (1) and our theorem is proved.

COROLLARY. *If a finite number of functions are continuous at a point, their product is continuous at this point.*

Referring now to our definition of continuity, we see that any constant may be regarded as a continuous function of $(x_1, \cdots x_n)$ for all values of these variables, and that the same is true of any one of these variables themselves. Hence by the last corollary any function of the form $Cx_1^{k_1} \cdots x_n^{k_n}$, where the k's are integers positive or zero, is continuous at every point. If we now refer to the corollary to Theorem 1, we arrive at the theorem:

THEOREM 3. *Any polynomial is a continuous function for all values of the variables.*

Finally, we give a simple application of this theorem.

THEOREM 4. *If $f(x_1, \cdots x_n)$ is a polynomial and $f(c_1, \cdots c_n) \neq 0$, it is possible to take a neighborhood of the point $(c_1, \cdots c_n)$ so small that f does not vanish at any point in this neighborhood.*

Let $k = f(c_1, \cdots c_n)$. Then, on account of the continuity of f at $(c_1, \cdots c_n)$, a positive quantity δ can be chosen so small that throughout the neighborhood $|x_i - c_i| < \delta$, the inequality

$$|f - k| < \tfrac{1}{2}|k|$$

is satisfied. In this neighborhood f cannot vanish; for at any point where it vanished we should have

$$|f - k| = |k| < \tfrac{1}{2}|k|,$$

which is impossible since by hypothesis $k \neq 0$.

6. The Fundamental Theorem of Algebra. Up to this point no use has been made of what is often known as the fundamental theorem of algebra, namely the proposition that every algebraic equation has a root. This fact we may state in more precise form as follows:

THEOREM 1.   *If $f(x)$ is a polynomial of the nth degree where $n \geq 1$, there exists at least one value of x for which $f(x) = 0$.*

This theorem, fundamental though it is, is not necessary for most of the developments in this book.   Moreover, the methods of proving the theorem are essentially not algebraic, or only in part algebraic.   Accordingly, we will give no proof of the theorem here, but merely refer the reader who desires a formal proof to any of the text-books on the theory of functions of a complex variable.   We shall, however, when we find it convenient to do so, assume the truth of this theorem.   In this section we will deduce a few of its more immediate consequences.

THEOREM 2.   *If $f(x)$ is a polynomial of the nth degree,*

$$f(x) \equiv a_0 x^n + a_1 x^{n-1} + \cdots + a_{n-1} x + a_n \qquad (a_0 \neq 0),$$

there exists one and only one set of constants, α_1, α_2, $\cdots$ α_n, such that

$$f(x) \equiv a_0 (x - \alpha_1)(x - \alpha_2) \cdots (x - \alpha_n).$$

This theorem is seen at once to be true for polynomials of the first degree.   Let us then use the method of mathematical induction and assume the proposition true for all polynomials of degree less than n.   If we can infer that the theorem is true for polynomials of the nth degree, it follows that being true for those of the first degree it is true for those of the second, hence for those of the third, etc.

By Theorem 1 we see that there is at least one value of x for which $f(x) = 0$.   Call such a value α_1.   By Theorem 1, § 1 we may write

$$f(x) \equiv (x - \alpha_1)\phi(x),$$

where

$$\phi(x) \equiv a_0 x^{n-1} + b_1 x^{n-2} + \cdots + b_{n-1}.$$

Since $\phi(x)$ is a polynomial of degree $n - 1$, and since we are assuming our theorem to be true for all such polynomials, there exist $n - 1$ constants α_2, $\cdots$ α_n such that

$$\phi(x) \equiv a_0 (x - \alpha_2) \cdots (x - \alpha_n).$$

Hence

$$f(x) \equiv a_0 (x - \alpha_1)(x - \alpha_2) \cdots (x - \alpha_n).$$

Thus half of our theorem is proved.

Suppose now there were two such sets of constants, $\alpha_1, \cdots \alpha_n$ and $\beta_1, \cdots \beta_n$. We should then have

$$(1) \qquad f(x) \equiv a_0(x - \alpha_1) \cdots (x - \alpha_n) \equiv a_0(x - \beta_1) \cdots (x - \beta_n).$$

Let $x = \alpha_1$ in this identity. This gives

$$a_0(\alpha_1 - \beta_1)(\alpha_1 - \beta_2) \cdots (\alpha_1 - \beta_n) = 0.$$

Accordingly, since $a_0 \neq 0$, α_1 must be equal to one of the quantities $\beta_1, \beta_2, \cdots \beta_n$. Let us suppose the β's to have been taken in such an order that $\alpha_1 = \beta_1$. Now in the identity (1) cancel out the factor $a_0(x - \alpha_1)$ (see Theorem 4, § 2). This gives

$$(x - \alpha_2) \cdots (x - \alpha_n) \equiv (x - \beta_2) \cdots (x - \beta_n).$$

Accordingly, since we have assumed the theorem we are proving to be true for polynomials of degree $n - 1$, the constants $\beta_2, \cdots \beta_n$ are the same, except perhaps for the order, as the constants $\alpha_2, \cdots \alpha_n$, and our theorem is proved.

DEFINITION. *The constants $\alpha_1, \cdots \alpha_n$ determined in the last theorem are called the roots of the polynomial $f(x)$, or of the equation $f(x) = 0$. If k of these roots are equal to one another, but different from all the other roots, this root is called a k-fold root.*

It is at once seen by reference to Theorem 1, § 1 that these roots are the only points at which $f(x)$ vanishes.

THEOREM 3. *If $f(x_1, \cdots x_n)$ is a polynomial which is not identically equal to a constant, there are an infinite number of points $(x_1, \cdots x_n)$ at which $f \neq 0$, and also an infinite number at which $f = 0$, provided $n > 1$.*

The truth of the first part of this theorem is at once obvious, for, since f is not identically zero, a point can be found at which it is not zero, and then a neighborhood of this point can be taken so small that f does not vanish in this neighborhood (Theorem 4, § 5). This neighborhood, of course, consists of an infinite number of points.

To prove that f vanishes at an infinite number of points, let us select one of the variables which enters into f to at least the first degree. Without loss of generality we may suppose this variable to be x_1. We may then write

$$f(x_1, \cdots x_n) \equiv F_0(x_2, \cdots x_n)x_1^k + F_1(x_2, \cdots x_n)x_1^{k-1} + \cdots + F_k(x_2, \cdots x_n),$$

where $k \geq 1$ and F_0 is not identically zero. Let $(c_2, \cdots c_n)$ be any point at which F_0 is not zero. Then $f(x_1, c_2, \cdots c_n)$ is a polynomial of the kth degree in x_1 alone. Accordingly, by Theorem 1, there is at least one value of x_1 for which it vanishes. If c_1 is such a value, $f(c_1, c_2, \cdots c_n) = 0$. Moreover, by the part of our theorem already proved, there are an infinite number of points where $F_0 \neq 0$, that is an infinite number of choices possible for the quantities $c_2, \cdots c_n$. Thus our theorem is completely proved.

Finally, we will state, without proof, for future reference, a theorem which says, in brief, that the roots of an algebraic equation are continuous functions of the coefficients :

THEOREM 4. *If α is a root of the polynomial*

$$a_0 x^n + a_1 x^{n-1} + \cdots + a_{n-1} x + a_n \qquad \begin{pmatrix} a_0 \neq 0 \\ n > 0 \end{pmatrix}^{*},$$

then no matter how small a neighborhood $|x - \alpha| < \epsilon$ of the point α we may consider, it is possible to take in space of $n + 1$ dimensions a neighborhood of the point $(a_0, a_1, \cdots a_n)$ so small that, if $(b_0, b_1, \cdots b_n)$ is any point in this neighborhood, the polynomial

$$b_0 x^n + b_1 x^{n-1} + \cdots + b_{n-1} x + b_n$$

has at least one root β in the neighborhood $|x - \alpha| < \epsilon$ of the point α.

For a proof of this theorem we refer to Weber's *Algebra*, Vol. 1, § 44.

---

* The theorem remains true if we merely assume that the polynomial is of at least the first degree. That is, some of the first coefficients a_0, a_1, $\cdots$ may be zero.

CHAPTER II

A FEW PROPERTIES OF DETERMINANTS

7. Some Definitions. We assume that the reader is familiar with the determinant notation, and will merely recall to him that by a determinant of the nth order

$$\begin{vmatrix} a_{11} & a_{12} & \cdots & a_{1n} \\ a_{21} & a_{22} & \cdots & a_{2n} \\ \cdot & \cdot & \cdot & \cdot \\ \cdot & \cdot & \cdot & \cdot \\ a_{n1} & a_{n2} & \cdots & a_{nn} \end{vmatrix}$$

we understand a certain homogeneous polynomial of the nth degree in the n^2 elements a_{ij}. By the side of these determinants it is often desirable to consider the system of the n^2 elements arranged in the order in which they stand in the determinant, but not combined into a polynomial. Such a square array of n^2 elements we speak of as a *matrix*. In fact, we will lay down the following somewhat more general definition of this term:

DEFINITION 1. *A system of mn quantities arranged in a rectangular array of m rows and n columns is called a matrix. If $m = n$, we say that we have a square matrix of order n.*

It is customary to place double bars on each side of this array. thus:

$$\begin{Vmatrix} a_{11} & a_{12} & \cdots & a_{1n} \\ a_{21} & a_{22} & \cdots & a_{2n} \\ \cdot & \cdot & \cdot & \cdot \\ \cdot & \cdot & \cdot & \cdot \\ a_{m1} & a_{m2} & \cdots & a_{mn} \end{Vmatrix}.$$

Sometimes parentheses are used, thus:

$$
\begin{pmatrix}
a_{11} & a_{12} & \cdots & a_{1n} \\
a_{21} & a_{22} & \cdots & a_{2n} \\
\cdot & \cdot & \cdot & \cdot \\
\cdot & \cdot & \cdot & \cdot \\
a_{m1} & a_{m2} & \cdots & a_{mn}
\end{pmatrix}.
$$

Even when a matrix is square, it must be carefully noticed that it is not a determinant. In fact, a matrix is not a quantity at all,[*] but a system of quantities. This difference between a square matrix and a determinant is clearly brought out if we consider the effect of interchanging columns and rows. This interchange has no effect on a determinant, but gives us a wholly new matrix. In fact. we will lay down the definition:

DEFINITION 2. *Two square matrices*

$$
\begin{Vmatrix}
a_{11} & \cdots & a_{1n} \\
\cdot & \cdot & \cdot \\
\cdot & \cdot & \cdot \\
a_{n1} & \cdots & a_{nn}
\end{Vmatrix},
\qquad
\begin{Vmatrix}
a_{11} & \cdots & a_{n1} \\
\cdot & \cdot & \cdot \\
\cdot & \cdot & \cdot \\
a_{1n} & \cdots & a_{nn}
\end{Vmatrix},
$$

of which either is obtained from the other by interchanging rows and columns are called conjugate [†] *to each other.*

Although, as we have pointed out, square matrices and determinants are wholly different things, every determinant determines a square matrix, the *matrix of the determinant,* and conversely every square matrix determines a determinant, the *determinant of the matrix.*

Every matrix contains other matrices obtained from it by striking out certain rows or columns or both. In particular it contains certain square matrices; and the determinants of these square matrices we will call the determinants of the matrix. If the matrix contains m rows and n columns, it will contain determinants of all orders from 1 (the elements themselves) to the smaller of the two integers and m and n inclusive.[‡] In many important problems all

---

[*] Cf., however, § 21.   [†] Sometimes also *transposed.*

[‡] If $m = n$, there is only one of these determinants of highest order, and it was this which we called above *the* determinant of the square matrix.

of these determinants above a certain order are zero, and it is often of great importance to specify the order of the highest non-vanishing determinant of a given matrix. For this purpose we lay down the following definition :

DEFINITION 3. *A matrix is said to be of rank r if it contains at least one r-rowed determinant which is not zero, while all determinants of order higher than r which the matrix may contain are zero.*

A matrix is said to be of rank 0 if all its elements are zero.

For brevity, we shall speak also of the rank of a determinant, meaning thereby the rank of the matrix of the determinant.

We turn now to certain definitions concerning the minors of determinants; that is, the determinants obtained from the given determinant by striking out certain rows and columns.

It is a familiar fact that to every element of a determinant corresponds a certain first minor; namely, the one obtained by striking out the row and column of the determinant in which the given element lies. Now the elements of a determinant of the nth order may be regarded as its $(n-1)$th minors. Accordingly we have here a method of pairing off each one-rowed minor of a given determinant with one of its $(n-1)$-rowed minors.

Similarly, if M is a two-rowed minor of a determinant of the nth order D, we may pair it off against the $(n-2)$-rowed minor N obtained by striking out from D the two rows and columns which are represented in M. The two minors M and N we will speak of as *complementary*. Thus, in the determinant

$$\begin{vmatrix} a_{11} & a_{12} & a_{13} & a_{14} & a_{15} \\ a_{21} & a_{22} & a_{23} & a_{24} & a_{25} \\ a_{31} & a_{32} & a_{33} & a_{34} & a_{35} \\ a_{41} & a_{42} & a_{43} & a_{44} & a_{45} \\ a_{51} & a_{52} & a_{53} & a_{54} & a_{55} \end{vmatrix},$$

the two minors

$$\begin{vmatrix} a_{21} & a_{23} \\ a_{31} & a_{33} \end{vmatrix}, \qquad \begin{vmatrix} a_{12} & a_{14} & a_{15} \\ a_{42} & a_{44} & a_{45} \\ a_{52} & a_{54} & a_{55} \end{vmatrix},$$

are complementary.

In the same way we pair off with every three-rowed minor an $(n-3)$-rowed minor; etc. In general we lay down

DEFINITION 4. *If D is a determinant of the nth order and M one of its k-rowed minors, then the $(n-k)$-rowed minor N obtained by striking out from D all the rows and columns represented in M is called the complement of M.*

Conversely, M is clearly the complement of N.

Let us go back now for a moment to the case of the one-rowed minors; that is to the elements themselves. Let a_{ij} be the element of the determinant D which stands in the ith row and the jth column. Let D_{ij} represent the corresponding first minor. It will be recalled that we frequently have occasion to consider not this minor D_{ij} but the *cofactor* A_{ij} of a_{ij} defined by the equation $A_{ij} = (-1)^{i+j} D_{ij}$.

Similarly, it is often convenient to consider not the complement of a given minor but its *algebraic complement*, which in the case just mentioned reduces to the cofactor, and which, in general, we define as follows:

DEFINITION 5. *If M is the m-rowed minor of D in which the rows $k_1, \cdots k_m$ and the columns $l_1, \cdots l_m$ are represented, then the algebraic complement of M is defined by the equation*

$$\text{alg. compl. of } M = (-1)^{k_1 + \cdots + k_m + l_1 + \cdots + l_m}[\text{compl. of } M].$$

The following special case is important:

DEFINITION 6. *By a principal minor of a determinant D is understood a minor obtained by striking out from D the same rows as columns.*

Since in this case, using the notation of Definition 5, we have

$$k_1 + \cdots + k_m = l_1 + \cdots + l_m,$$

it follows that *the algebraic complement of any principal minor is equal to its plain complement.*

We have so far assumed tacitly that the orders of the minors we were dealing with were less than the order n of the determinant itself. By the n-rowed minor of a determinant D of the nth order we of course understand this determinant itself. The complement of this minor has, however, by our previous definition no meaning. *We will define the complement in this case to be 1, and, by Definition 5, this will also be the algebraic complement.*

EXERCISE

Prove that, if M and N are complementary minors, either M and N are the algebraic complements of each other, or $- N$ is the algebraic complement of M and $- M$ is the algebraic complement of N.

8. Laplace's Development. Just as the elements of any row or column and their corresponding cofactors may be used to develop a determinant in terms of determinants of lower orders, so the k-rowed minors formed from any k rows or columns may be used, along with their algebraic complements, to obtain a more general development of the determinant, due to Laplace, and which includes as a special case the one just referred to. In order to establish this development, we begin with the following preliminary theorem :

THEOREM 1. *If the rows and columns of a determinant D be shifted in such a way as to bring a certain minor M into the upper left-hand corner without changing the order of the rows and columns either of M or of its complement N, then this shifting will change the sign of D or leave it unchanged according as $- N$ or N is the algebraic complement of M.*

To prove this let us, as usual, number the rows and columns of D, beginning at the upper left-hand corner, and let the numbers of the rows and columns represented in M, arranged in order of increasing magnitude, be $k_1, \cdots k_m$, and $l_1, \cdots l_m$ respectively. In order to effect the rearrangement mentioned in the theorem, we may first shift the row numbered k_1 upward into the first position, thus carrying it over $k_1 - 1$ other rows and therefore changing the sign of the determinant $k_1 - 1$ times. Then shift the row numbered k_2 into the second position. This carries it over $k_2 - 2$ rows and hence changes the sign $k_2 - 2$ times. Proceed in this way until the row numbered k_m has been shifted into the mth position. Then shift the columns in a similar manner. The final result is to multiply D by

$$(-1)^{k_1 + \cdots + k_m + l_1 + \cdots + l_m - 2(1 + 2 + \cdots + m)} = (-1)^{k_1 + \cdots + k_m + l_1 + \cdots + l_m}.$$

Comparing this with Definition 5, § 7, the truth of our theorem is obvious.

LEMMA. *If M is a minor of a determinant D, the product of M by its algebraic complement is identical, when expanded, with some of the terms of the expansion of D.*

Let
$$D = \begin{vmatrix} a_{11} & \cdots & a_{1n} \\ \cdot & \cdot & \cdot \\ \cdot & \cdot & \cdot \\ a_{n1} & \cdots & a_{nn} \end{vmatrix},$$

and call the order of M, m, and its complement N. We will first prove our lemma in the special case in which M stands in the upper left-hand corner of D, so that N, which in this case is the algebraic complement, is in the lower right-hand corner. What we have to show here is that the product of any term of M by a term of N is a term of D, and that this term does not come in twice to the product MN. Any term of M may be written

$$(-1)^{\mu} a_{1l_1} a_{2l_2} \cdots a_{ml_m},$$

where the integers l_1, l_2, $\cdots l_m$ are merely some arrangement of the integers $1, 2, \cdots m$, and μ is the number of inversions of order in this arrangement. Similarly, any term of N may be written

$$(-1)^{\nu} a_{m+1,\, l_{m+1}} a_{m+2,\, l_{m+2}} \cdots a_{n,\, l_n},$$

where $l_{m+1}, \cdots l_n$ is merely some arrangement of the integers $m + 1$, $\cdots n$, and ν is the number of inversions of order in this arrangement. The product of these two terms

$$(-1)^{\mu+\nu} a_{1l_1} a_{2l_2} \cdots a_{nl_n},$$

is a term of D, for the factors a are chosen in succession from the first, second, $\cdots$ nth rows of D, and no two are from the same column, and $\mu + \nu$ is clearly precisely the number of inversions of order in the arrangement l_1, l_2, $\cdots l_n$, as compared to the natural arrangement, $1, 2, \cdots n$, of these integers.

Having thus proved our lemma in the special case in which M lies in the upper left-hand corner of D, we now pass to the general case. Here we may, by shifting rows and columns, bring M into the upper left-hand corner and N into the lower right-hand corner. This has, by Theorem 1, the effect of leaving each term in the expansion of D unchanged, or of reversing the sign of all of them according as N or $-N$ is the algebraic complement of M. Accordingly, since the product MN gives, as we have just seen, terms in the expansion of this rearranged determinant, the product of M by its algebraic complement gives terms in the expansion of D itself, as was to be proved.

Laplace's Development, which may be stated in the form of the following rule, now follows at once:

THEOREM 2. *Pick out any m rows (or columns) from a determinant D, and form all the m-rowed determinants from this matrix. The sum of the products of each of these minors by its algebraic complement is the value of D.*

Since, by our lemma, each of these products when developed consists of terms of D, it remains merely to show that every term of D occurs in one and only one of these products. This is obviously the case; for every term of D contains one element from each of the m rows of D from which our theorem directs us to pick out m-rowed determinants, and, since these elements all lie in different columns, they lie in one and only one of these m-rowed determinants, say M. Since the other elements in this term of D obviously all lie in the complement N of M, this term will be found in the product MN and in none of the other products mentioned in our theorem.

EXERCISES

1. From a square matrix of order n and rank r, s rows (or columns) are selected. Prove that the rank of the matrix thus obtained cannot be less than $r + s - n$.

2. Generalize the theorem of Exercise 1.

9. The Multiplication Theorem. Laplace's Development enables us to write out at once the product of any two determinants as a single determinant whose order is the sum of the orders of the two given determinants

$$
\begin{vmatrix} a_{11} & \cdots & a_{1n} \\ \cdot & \cdot & \cdot \\ a_{n1} & \cdots & a_{nn} \end{vmatrix}
\cdot
\begin{vmatrix} b_{11} & \cdots & b_{1m} \\ \cdot & \cdot & \cdot \\ b_{m1} & \cdots & b_{mm} \end{vmatrix}
=
\begin{vmatrix}
a_{11} & \cdots & a_{1n} & 0 & \cdots & 0 \\
\cdot & \cdot & \cdot & \cdot & \cdot & \cdot \\
a_{n1} & \cdots & a_{nn} & 0 & \cdots & 0 \\
p_{11} & \cdots & p_{1n} & b_{11} & \cdots & b_{1m} \\
\cdot & \cdot & \cdot & \cdot & \cdot & \cdot \\
\cdot & \cdot & \cdot & \cdot & \cdot & \cdot \\
p_{m1} & \cdots & p_{mn} & b_{m1} & \cdots & b_{mm}
\end{vmatrix},
$$

whatever the values of the p's may be. For, expanding the large determinant in terms of the n-rowed minors of the first n rows, all the terms of the expansion are zero except the one written in the first member of the equation.

From this formula we will now deduce a far more important one for expressing the product of two determinants of the same order as a determinant of that order. For this purpose let us choose the p's in the last formula as follows:

$$p_{ij} = 0 \quad \text{when } i \neq j, \quad p_{ii} = -1,$$

and let us consider for simplicity the product of two determinants of the third order. We have

$$\begin{vmatrix} \alpha_1 & \alpha_2 & \alpha_3 \\ \beta_1 & \beta_2 & \beta_3 \\ \gamma_1 & \gamma_2 & \gamma_3 \end{vmatrix} \begin{vmatrix} a_1 & a_2 & a_3 \\ b_1 & b_2 & b_3 \\ c_1 & c_2 & c_3 \end{vmatrix} = \begin{vmatrix} \alpha_1 & \alpha_2 & \alpha_3 & 0 & 0 & 0 \\ \beta_1 & \beta_2 & \beta_3 & 0 & 0 & 0 \\ \gamma_1 & \gamma_2 & \gamma_3 & 0 & 0 & 0 \\ -1 & 0 & 0 & a_1 & a_2 & a_3 \\ 0 & -1 & 0 & b_1 & b_2 & b_3 \\ 0 & 0 & -1 & c_1 & c_2 & c_3 \end{vmatrix}.$$

Let us now reduce this six-rowed determinant by multiplying its first column by a_1 and adding it to the fourth column; then multiply the first column by a_2 and add it to the fifth; then multiply the first column by a_3 and add it to the sixth. In this way we bring zeros into the last three places in the fourth row. Next multiply the second column successively by b_1, b_2, b_3 and add it to the fourth, fifth, and sixth columns respectively. Finally multiply the third column successively by c_1, c_2, c_3 and add it to the fourth, fifth, and sixth columns. The determinant thus takes the form

$$\begin{vmatrix} \alpha_1 & \alpha_2 & \alpha_3 & \alpha_1 a_1 + \alpha_2 b_1 + \alpha_3 c_1 & \alpha_1 a_2 + \alpha_2 b_2 + \alpha_3 c_2 & \alpha_1 a_3 + \alpha_2 b_3 + \alpha_3 c_3 \\ \beta_1 & \beta_2 & \beta_3 & \beta_1 a_1 + \beta_2 b_1 + \beta_3 c_1 & \beta_1 a_2 + \beta_2 b_2 + \beta_3 c_2 & \beta_1 a_3 + \beta_2 b_3 + \beta_3 c_3 \\ \gamma_1 & \gamma_2 & \gamma_3 & \gamma_1 a_1 + \gamma_2 b_1 + \gamma_3 c_1 & \gamma_1 a_2 + \gamma_2 b_2 + \gamma_3 c_2 & \gamma_1 a_3 + \gamma_2 b_3 + \gamma_3 c_3 \\ -1 & 0 & 0 & 0 & 0 & 0 \\ 0 & -1 & 0 & 0 & 0 & 0 \\ 0 & 0 & -1 & 0 & 0 & 0 \end{vmatrix}.$$

and this reduces at once to the three-rowed determinant

$$\begin{vmatrix} \alpha_1 a_1 + \alpha_2 b_1 + \alpha_3 c_1 & \alpha_1 a_2 + \alpha_2 b_2 + \alpha_3 c_2 & \alpha_1 a_3 + \alpha_2 b_3 + \alpha_3 c_3 \\ \beta_1 a_1 + \beta_2 b_1 + \beta_3 c_1 & \beta_1 a_2 + \beta_2 b_2 + \beta_3 c_2 & \beta_1 a_3 + \beta_2 b_3 + \beta_3 c_3 \\ \gamma_1 a_1 + \gamma_2 b_1 + \gamma_3 c_1 & \gamma_1 a_2 + \gamma_2 b_2 + \gamma_3 c_2 & \gamma_1 a_3 + \gamma_2 b_3 + \gamma_3 c_3 \end{vmatrix}.$$

We have thus expressed the product of two determinants of the third order as a single determinant of the third order. The method we have used is readily seen to be entirely general, and we thus get the following rule for multiplying together two determinants of the nth order:

THEOREM. *The product of two determinants of the nth order may be expressed as a determinant of the nth order in which the element which lies in the ith row and jth column is obtained by multiplying each element of the ith row of the first factor by the corresponding element of the jth column of the second factor and adding the results.*

It should be noted that changing rows into columns in either or both of the given determinants, while not affecting the value of the product, will alter its form materially. For example,

$$\begin{vmatrix} 2 & 3 \\ 4 & 5 \end{vmatrix} \cdot \begin{vmatrix} 1 & 7 \\ 6 & 9 \end{vmatrix} = \begin{vmatrix} 20 & 41 \\ 34 & 73 \end{vmatrix} = 66,$$

$$\begin{vmatrix} 2 & 3 \\ 4 & 5 \end{vmatrix} \cdot \begin{vmatrix} 1 & 6 \\ 7 & 9 \end{vmatrix} = \begin{vmatrix} 23 & 39 \\ 39 & 69 \end{vmatrix} = 66,$$

$$\begin{vmatrix} 2 & 4 \\ 3 & 5 \end{vmatrix} \cdot \begin{vmatrix} 1 & 7 \\ 6 & 9 \end{vmatrix} = \begin{vmatrix} 26 & 50 \\ 33 & 66 \end{vmatrix} = 66,$$

$$\begin{vmatrix} 2 & 4 \\ 3 & 5 \end{vmatrix} \cdot \begin{vmatrix} 1 & 6 \\ 7 & 9 \end{vmatrix} = \begin{vmatrix} 30 & 48 \\ 38 & 63 \end{vmatrix} = 66;$$

and similarly the product of any two determinants of the same order may be written in four different forms.

10. Bordered Determinants. If to a determinant of the nth order we add one or more rows and the same number of columns of n quantities each and fill in the vacant corner with zeros, the resulting determinant is called a *bordered determinant*. Thus starting from the two-rowed determinant

$$\begin{vmatrix} \alpha & \beta \\ \gamma & \delta \end{vmatrix}$$

we may form the bordered determinants

$$\begin{vmatrix} a & \beta & u_1 \\ \gamma & \delta & u_2 \\ v_1 & v_2 & 0 \end{vmatrix}, \qquad \begin{vmatrix} \alpha & \beta & u_1 & u_1' \\ \gamma & \delta & u_2 & u_2' \\ v_1 & v_2 & 0 & 0 \\ v_1' & v_2' & 0 & 0 \end{vmatrix}, \qquad \begin{vmatrix} \alpha & \beta & u_1 & u_1' & u_1'' \\ \gamma & \delta & u_2 & u_2' & u_2'' \\ v_1 & v_2 & 0 & 0 & 0 \\ v_1' & v_2' & 0 & 0 & 0 \\ v_1'' & v_2'' & 0 & 0 & 0 \end{vmatrix}, \ \ldots$$

If in the second of these examples we use Laplace's Development to expand the bordered determinant according to the two-rowed determinants of the last two rows, we see that its value is

$$\begin{vmatrix} u_1 & u_1' \\ u_2 & u_2' \end{vmatrix} \cdot \begin{vmatrix} v_1 & v_2 \\ v_1' & v_2' \end{vmatrix},$$

a quantity into which the elements α, β, γ, δ of the original determinant do not enter. Similarly expanding the third of the above bordered determinants according to the three-rowed determinants of its last three rows, we see that its value is zero.

The reasoning we have here used is of general application and leads to the following results:

THEOREM 1. *If a determinant of the nth order is bordered with n rows and n columns, the resulting determinant has a value which depends only on the bordering quantities.*

THEOREM 2. *If a determinant of the nth order is bordered with more than n rows and columns, the resulting determinant always has the value zero.*

The cases of interest are therefore those in which the determinant is bordered with less than n rows and columns. Concerning these we will establish the following fact:

THEOREM 3. *If a determinant of the nth order be bordered by p rows and p columns $(p < n)$ of independent variables, the resulting determinant is a polynomial of degree 2p in the bordering quantities, whose coefficients are the pth minors of the original determinant; and conversely, every pth minor of the original determinant is the coefficient of at least one term of this polynomial.*

Let us consider the special case where $n = 4$ and $p = 2$.

$$D \equiv \begin{vmatrix} a_{11} & a_{12} & a_{13} & a_{14} & u_1 & u_1' \\ a_{21} & a_{22} & a_{23} & a_{24} & u_2 & u_2' \\ u_{31} & a_{32} & a_{33} & a_{34} & u_3 & u_3' \\ a_{41} & a_{42} & a_{43} & a_{44} & u_4 & u_4' \\ v_1 & v_2 & v_3 & v_4 & 0 & 0 \\ v_1' & v_2' & v_3' & v_4' & 0 & 0 \end{vmatrix}.$$

Developing this determinant, by Laplace's method (§ 8), in terms of the two-rowed determinants of the last two rows, we have

$$D \equiv \begin{vmatrix} v_1 & v_2 \\ v_1' & v_2' \end{vmatrix} \begin{vmatrix} a_{13} & a_{14} & u_1 & u_1' \\ a_{23} & a_{24} & u_2 & u_2' \\ a_{33} & a_{34} & u_3 & u_3' \\ a_{43} & a_{44} & u_4 & u_4' \end{vmatrix} + \cdots \text{ to 6 terms.}$$

If now we expand each of these four-rowed determinants, by Laplace's method, in terms of the two-rowed determinants of their last two columns, and then arrange the result as a polynomial in the u's and v's, the truth of the theorem is apparent. We leave it to the reader to fill in the details of the proof here sketched.

11. Adjoint Determinants and their Minors.

DEFINITION.   *If, in the determinant*

$$D = \begin{vmatrix} a_{11} & \cdots & a_{1n} \\ \cdot & \cdot & \cdot & \cdot \\ \cdot & \cdot & \cdot & \cdot \\ a_{n1} & \cdots & a_{nn} \end{vmatrix},$$

A_{ij} is the cofactor of the element a_{ij}, then the determinant

$$D' = \begin{vmatrix} A_{11} & \cdots & A_{1n} \\ \cdot & \cdot & \cdot & \cdot & \cdot \\ \cdot & \cdot & \circ & \cdot & \cdot \\ A_{n1} & \cdots & A_{nn} \end{vmatrix}$$

is called the adjoint of D.

By corresponding minors of D and D', or indeed of any two determinants of the same order, we shall naturally understand minors obtained by striking out the same rows and columns from D as from D'. These definitions being premised, the fundamental theorem here is the following:

THEOREM. *If D' is the adjoint of any determinant D, and M and M' are corresponding m-rowed minors of D and D' respectively, then M' is equal to the product of D^{m-1} by the algebraic complement of M.*

We will prove this theorem first for the special case in which the minors M and M' lie at the upper left-hand corners of D and D' respectively. We may then write

$$M' = \begin{vmatrix} A_{11} & \cdots & A_{1m} & \cdots & \cdots & A_{1n} \\ \cdot & \cdot & \cdot & \cdot & \cdot & \cdot \\ A_{m1} & \cdots & A_{mm} & \cdots & \cdots & A_{mn} \\ 0 & \cdots & 0 & 1 & 0 & \cdots & 0 \\ 0 & \cdots & 0 & 0 & 1 & \cdots & 0 \\ \cdot & \cdot & \cdot & \cdot & \cdot & \cdot & \cdot \\ 0 & \cdots & 0 & 0 & 0 & \cdots & 1 \end{vmatrix}.$$

Let us now interchange the columns and rows of D,

$$D = \begin{vmatrix} a_{11} & \cdots & a_{n1} \\ \cdot & \cdot & \cdot \\ \cdot & \cdot & \cdot \\ a_{1n} & \cdots & a_{nn} \end{vmatrix},$$

and then form the product $M'D$ by the theorem of § 9. This gives

$$M'D = \begin{vmatrix} D & 0 & \cdots & 0 & 0 & \cdots & 0 \\ 0 & D & \cdots & 0 & 0 & \cdots & 0 \\ \cdot & \cdot & \cdot & \cdot & \cdot & \cdot & \cdot \\ 0 & 0 & \cdots & D & 0 & \cdots & 0 \\ a_{1,m+1} & a_{2,m+1} & \cdots & a_{m,m+1} & a_{m+1,m+1} & \cdots & a_{n,m+1} \\ \cdot & \cdot & \cdot & \cdot & \cdot & \cdot & \cdot \\ a_{1n} & a_{2n} & \cdots & a_{mn} & a_{m+1,n} & \cdots & a_{nn} \end{vmatrix}$$

$$= D^m \begin{vmatrix} a_{m+1,m+1} & \cdots & a_{n,m+1} \\ \cdot & \cdot & \cdot \\ \cdot & \cdot & \cdot \\ a_{m+1,n} & \cdots & a_{nn} \end{vmatrix}.$$

Let us here regard $a_{11}, \; \cdots \; a_{nn}$ as n^2 independent variables Then the equation just written becomes an identity, from which D since it is not identically zero, may be cancelled out, and we get

$$(1) \qquad M' \equiv D^{m-1} \begin{vmatrix} a_{m+1,\,m+1} & \cdots & a_{n,\,m+1} \\ \cdot & \cdot \cdot \cdot \cdot & \cdot \\ \cdot & \cdot \cdot \cdot \cdot & \cdot \\ a_{m+1,\,n} & \cdots & a_{n,\,n} \end{vmatrix}.$$

Since the determinant which is written out in (1) is precisely the algebraic complement of M, our theorem is proved in the special case we have been considering. It should be noticed that this proof holds even in the case $m = n$; cf. Corollary 2 below.

Turning now to the case in which the minors M and M' do not lie at the upper left-hand corners of D and D', let us denote by a the sum of the numbers which specify the location of the rows and columns in M or M', the numbering running, as usual, from the upper left-hand corner. Then by Definition 5, § 7,

$$(2) \qquad \text{alg. compl. of } M = (-1)^a \; [\text{compl. of } M].$$

Let us now, by shifting rows and columns, bring the determinant M into the upper left-hand corner of D. Calling the determinant D, as thus rearranged, D_1, we have (cf. Theorem 1, § 8)

$$(3) \qquad D_1 = (-1)^a D.$$

The cofactors in D_1 are equal to $(-1)^a A_{ij}$, since the interchange of two adjacent rows or columns of a determinant changes the sign of every one of its cofactors. Accordingly the adjoint of D_1, which we will call D_1', may be obtained from D' by rearranging its rows and columns in the same way as the rows and columns of D were rearranged to give D_1, and then prefixing the factor $(-1)^a$ to each element.

Let us now apply the special case already established of our theorem to the determinant D_1 and its adjoint D_1', the m-rowed minors M_1 and M_1' being those which are situated in the upper left-hand corner of D_1 and D_1' respectively. We thus get

$$(4) \qquad M_1' = D_1^{m-1} \; [\text{alg. compl. of } M_1].$$

Now, since M_1 is a principal minor, its algebraic complement is the same as its ordinary complement, and this in turn is the same as the ordinary complement of the minor M in D. Accordingly, using (2), we may write

(5)
$$\text{alg. compl. of } M_1 = (-1)^a \, [\text{alg. compl. of } M].$$

Since the elements of M_1' differ from those of M' only in having the factor $(-1)^a$ prefixed to each, it follows that

(6)
$$M_1' = (-1)^{ma} M'.$$

We may now reduce (4) by means of (3), (5), and (6). We thus get
$$(-1)^{ma} M' = (-1)^{a(m-1)} D^{m-1}(-1)^a \, [\text{alg. compl. of } M].$$

Cancelling out the factor $(-1)^{ma}$ from both sides of this equation, we see that our theorem is proved.

We proceed now to point out a number of special cases of this theorem which are worth noting on account of their frequent occurrence.

COROLLARY 1. *If a_{ij} is any element of a determinant D of the nth order, and if α_{ij} is the cofactor of the corresponding element A_{ij} in the adjoint of D, then*
$$\alpha_{ij} = D^{n-2} a_{ij}.$$

This is merely the special case of our general theorem in which $m = n - 1$, modified, however, slightly in statement by the use of the cofactor α_{ij} in place of the $(n-1)$-rowed minor $(-1)^{i+j} \alpha_{ij}$.

COROLLARY 2. *If D is any determinant of the nth order and D' its adjoint, then*
$$D' = D^{n-1}.$$

This is the special case $m = n$.

COROLLARY 3. *If D is any determinant, and S is the second minor obtained from it by striking out its ith and kth rows and its jth and lth columns, and if we denote by A_{ij} the cofactor of the element which stands in the ith row and the jth column of D, then*
$$\begin{vmatrix} A_{ij} & A_{il} \\ A_{kj} & A_{kl} \end{vmatrix} = (-1)^{i+j+k+l} DS.$$

This is the special case $m = 2$.

CHAPTER III

THE THEORY OF LINEAR DEPENDENCE

12. **Definitions and Preliminary Theorems**. Two sets of constants $(a_1, b_1, c_1, d_1,)$ and (a_2, b_2, c_2, d_2) are usually said to be proportional to one another if every element of one set may be obtained from the corresponding element of the other by multiplying by the same constant factor. For example, $(1, 2, 3, 4)$ and $(2, 4, 6, 8)$ are proportional. It is ordinarily assumed that *either* set may be thus obtained from the other, and in most cases this is true; but in the case of the two sets $(1, 2, 3, 4)$ and $(0, 0, 0, 0)$ we can pass from the first to the second by multiplying by 0, but we cannot pass from the second to the first.

A more convenient definition, for many purposes, and one which is easily seen to be equivalent to the above-mentioned one, is the following :

DEFINITION 1. *The two sets of constants*

$$x_1', x_2', \cdots x_n',$$
$$x_1'', x_2'', \cdots x_n'',$$

are said to be proportional to each other if two constants c_1 and c_2, not both zero, exist such that

$$c_1 x_i' + c_2 x_i'' = 0 \qquad\qquad (i = 1, 2, \cdots n).$$

If $c_1 \neq 0$, we have

$$x_1' = -\frac{c_2}{c_1} x_1'', \; x_2' = -\frac{c_2}{c_1} x_2'', \cdots x_n' = -\frac{c_2}{c_1} x_n'',$$

and if $c_2 \neq 0$, we have

$$x_1'' = -\frac{c_1}{c_2} x_1', \; x_2'' = -\frac{c_1}{c_2} x_2', \cdots x_n'' = -\frac{c_1}{c_2} x_n'.$$

The two sets of constants $\quad x_1', x_2', \cdots x_n',$
$$0, \; 0, \cdots \; 0,$$

are evidently proportional, since if we take $c_1 = 0$ and $c_2 =$ any constant not zero, we have a pair of c's which fulfill the requirements of our definition.

Linear dependence may be regarded as a generalization of the conception of proportionality. Instead of two sets of constants we now consider m sets, and give the following :

DEFINITION 2. *The m sets of n constants each,*

$$x_1^{[i]},\ x_2^{[i]},\ \cdots\ x_n^{[i]} \qquad\qquad (i = 1,\ 2,\ \cdots\ m),$$

are said to be linearly dependent if m constants $c_1,\ c_2,\ \cdots\ c_m$, not all zero, exist such that

$$c_1 x_j' + c_2 x_j'' + \cdots + c_m x_j^{[m]} = 0 \qquad\qquad (j = 1,\ 2,\ \cdots\ n).$$

If this is not the case, the sets of quantities are said to be linearly independent.

In the same way we generalize the familiar conception of the proportionality of two polynomials as follows:

DEFINITION 3. *The m polynomials (in any number of independent variables) $f_1, f_2, \cdots f_m$ are said to be linearly dependent if m constants $c_1,\ c_2,\ \cdots\ c_m$, not all zero, exist such that*

$$c_1 f_1 + c_2 f_2 + \cdots + c_m f_m \equiv 0.$$

*If this is not the case, the polynomials are said to be linearly independent.**

The following theorems about linear dependence, while almost self-evident, are of sufficient importance to deserve explicit statement :

THEOREM 1. *If m sets of constants (or if m polynomials) are linearly dependent, it is always possible to express one — but not necessarily any one — of them linearly in terms of the others. This set of constants (or this polynomial) is then said to be linearly dependent on the others.*

This is seen at once if we remember that at least one of the c's is not zero. The relations (or relation) in which the c's occur can, then, be divided through by this c.

---

* We might clearly go farther and consider the linear dependence of m sets of n polynomials each. The two cases of the text would be merely special cases from this general point of view.

THEOREM 2. *If there exist among the sets of constants (or among the polynomials) a smaller number of sets (or of polynomials) which are linearly dependent, then the m sets (or the m polynomials) are linearly dependent.*

For suppose there are l sets of constants (or l polynomials) which are linearly dependent ($l < m$), then we may take for our set of m c's, the l c's which must exist for the l sets (or polynomials) and $(m - l)$ zeros

THEOREM 3. *If any one of the m sets of constants consists exclusively of zeros (or if any one of the polynomials is identically zero), the m sets (or the m polynomials) are linearly dependent.*

For we may take for the c corresponding to this particular set (or polynomial) any constant whatever, except zero, and for the other $(m - 1)$ c's, $(m - 1)$ zeros.

13. The Condition for Linear Dependence of Sets of Constants. In considering m sets of n constants each,

$$(1) \qquad x_1^{[i]}, x_2^{[i]}, \cdots x_n^{[i]} \qquad\qquad (i = 1, 2, \cdots m),$$

it will be convenient to distinguish between the two cases $m \leqq n$ and $m > n$.

(*a*) $m \leqq n$. We wish here to prove the following fundamental theorem :

THEOREM 1. *A necessary and sufficient condition for the linear dependence of the m sets (1) of n constants each, when $m \leqq n$, is that all the m-rowed determinants of the matrix*

$$\left\|\begin{array}{cccc} x_1' & x_2' & \cdots & x_n' \\ x_1'' & x_2'' & \cdots & x_n'' \\ \cdot & \cdot & \cdot & \cdot \\ \cdot & \cdot & \cdot & \cdot \\ x_1^{[m]} & x_2^{[m]} & \cdots & x_n^{[m]} \end{array}\right\|$$

should vanish.

That this is a necessary condition is at once obvious; for if the m sets of constants are linearly dependent, one of the rows can be expressed as a linear combination of the others. Accordingly if in any of the m-rowed determinants we subtract from the elements of this row the corresponding elements of the other rows after each row has been multiplied by a suitable constant, the elements of this row will reduce to zero. The determinant therefore vanishes.

We come now to the proof that the vanishing of these determinants is also a sufficient condition. We assume, therefore, that all the m-rowed determinants of the above matrix vanish. Let us also assume that the rank of the matrix is $r > 0$* (cf. Definition 3, § 7). Without any real loss of generality we may (and will) assume that *the r-rowed determinant which stands in the upper left-hand corner of the matrix does not vanish;* for by changing the order of the sets of constants and the order of the constants in each set (and these orders are clearly quite immaterial) we can bring one of the non-vanishing r-rowed determinants into this position.

We will now prove that the first $(r + 1)$ sets of constants are linearly dependent. From this the linear dependence of the m sets follows by Theorem 2, § 12.

Let us denote by c_1, c_2, $\cdots$ c_{r+1} the cofactors in the $(r + 1)$-rowed determinant which stands in the upper left-hand corner of the matrix, and which correspond to the elements of its last column. If we remember that all the $(r+1)$-rowed determinants vanish, we get the relations

$$c_1 x_j' + c_2 x_j'' + \cdots + c_{r+1} x_j^{[r+1]} = 0 \qquad (j = r + 1, r + 2, \cdots n).$$

Since the sum of the products of the elements of any column of a determinant by the cofactors of the corresponding elements of *another* column is zero, this equation is also true when $j = 1, 2, \cdots r$.

This establishes the linear dependence of the first $(r + 1)$ sets of constants, since c_{r+1}, being the r-rowed determinant which stands in the upper left-hand corner of the matrix, is not zero.

(b) $m > n$. This case can be reduced to the one already considered by the following simple device. Add to each set of n constants $m - n$ zeros. We then have m sets of m constants each. Their matrix contains only one m-rowed determinant, and this vanishes since one, at least, of its columns is composed of zeros. Therefore these m sets of m constants each are linearly dependent; and hence the original m sets of n constants each were linearly dependent. Thus we get the theorem:

THEOREM 2. *m sets of n constants each are always linearly dependent if $m > n$.*

---

* In general we shall have $r = m - 1$, but r may have any value less than m. The only case which we here exclude is that in which all the elements of the matrix are zero, a case in which the linear dependence is at once obvious.

EXERCISES

Determine whether the following sets of constants are linearly dependent or not :

1. $\begin{cases} 3\,a, & -2\,b, & -3\,c, & 6\,d, \\ a, & 0, & -\,c, & 4\,d, \\ 0, & -\,b, & 0, & -3\,d. \end{cases}$

2. $\begin{cases} 1, & 0, & 0, & 5, \\ 1, & 2, & 6, & 7, \\ 3, & 1, & 3, & 16. \end{cases}$

3. $\begin{cases} 5, & 2, & 1, & 3, & 4, \\ 0, & 3, & 0, & 0, & 8, \\ 15, & 7, & 3, & 9, & 7. \end{cases}$

4. $\begin{cases} 5, & -7, & 0, & 1, & -1, \\ 1, & -3, & -2, & 3, & -1, \\ 4, & 0, & 7, & -9, & 2. \end{cases}$

14. The Linear Dependence of Polynomials. Suppose we have m polynomials,

$$f_1, f_2, \cdots f_m,$$

in any number of independent variables. A necessary and sufficient condition for the linear dependence of these polynomials is evidently the linear dependence of their m sets of coefficients. Thus the conditions deduced in the last section can be applied at once to the case of polynomials.

EXERCISES

Determine whether the following polynomials are linearly dependent or not:

1. $\begin{cases} 16\,x & +\,30\,z, \\ 6\,x + 2\,y + 5\,z - 4, \\ 15\,x + 9\,y & -\,18. \end{cases}$

2. $\begin{cases} 3\,x_1 + 4\,x_2 - 4\,x_3 + 6\,x_4, \\ 7\,x_1 & +\,3\,x_3 + 7\,x_4, \\ 2\,x_1 - \ x_2 & -\,3, \\ -\,5\,x_1 + 9\,x_2 - \ x_3 + 4\,x_4 + 8. \end{cases}$

3. $\begin{cases} 2\,x^2 + 8\,xy + 6\,y^2 + 14\,x + 12\,y - 4, \\ 7\,x^2 & + \ y^2 + 6\,x - 4\,y, \\ 3\,x^2 - 6\,xy + 3\,y^2 - 5\,x & + 7, \\ 5\,x^2 + 20\,xy + 15\,y^2 + 35\,x + 30\,y - 10. \end{cases}$

15. Geometric Illustrations. The sets of n constants with which we had to deal in §§ 12, 13 may, provided that not all the constants in any one set are zero, advantageously be regarded as the homogeneous coördinates of points in space of $n-1$ dimensions. It will then be convenient to speak of the linear dependence or independence of these points. The geometric meaning of linear dependence will be at once evident from the following theorems for the case $n = 4$.

Two points will here be represented by two sets of four constants each,

$$x_1,\ y_1,\ z_1,\ t_1,$$
$$x_2,\ y_2,\ z_2,\ t_2,$$

which will be linearly dependent when, and only when, they are proportional, that is, when the points coincide. Hence:

THEOREM 1. *Two points are linearly dependent when, and only when, they coincide.*

If we have three points in space, P_1, P_2, P_3, whose coördinates are $(x_1,\ y_1,\ z_1,\ t_1)$, $(x_2,\ y_2,\ z_2,\ t_2)$, $(x_3,\ y_3,\ z_3,\ t_3)$, respectively, and which are linearly dependent, there must exist three constants c_1, c_2, c_3, not all zero, such that

$$c_1 x_1 + c_2 x_2 + c_3 x_3 = 0,$$
$$c_1 y_1 + c_2 y_2 + c_3 y_3 = 0,$$
$$c_1 z_1 + c_2 z_2 + c_3 z_3 = 0,$$
$$c_1 t_1 + c_2 t_2 + c_3 t_3 = 0.$$

Let us suppose the order of the points to be so taken that $c_3 \neq 0$, and solve for x_3, y_3, z_3, t_3:

$$(1)\qquad \begin{cases} x_3 = k_1 x_1 + k_2 x_2, \\ y_3 = k_1 y_1 + k_2 y_2, \\ z_3 = k_1 z_1 + k_2 z_2, \\ t_3 = k_1 t_1 + k_2 t_2, \end{cases}$$

where $k_1 = -c_1/c_3$, $k_2 = -c_2/c_3$. Now if

$$Ax + By + Cz + Dt = 0$$

is the equation of any plane through the points P_1 and P_2, we have

$$Ax_1 + By_1 + Cz_1 + Dt_1 = 0,$$
$$Ax_2 + By_2 + Cz_3 + Dt_2 = 0.$$

Multiplying the first of these equations by k_1, the second by k_2, and adding, we have, by means of the equations (1),

$$Ax_3 + By_3 + Cz_3 + Dt_3 = 0.$$

Hence every plane through P_1 and P_2 passes through P_3 also, and the three points are collinear.

Now, in order to prove conversely that any three collinear points are linearly dependent, let us suppose the three points P_1, P_2, P_3 collinear. We may assume that these three points are distinct, as otherwise their linear dependence would follow from Theorem 1. We have seen that when three points are linearly dependent, the line through two of them contains the third. Hence if we let

$$x' = k_1 x_1 + k_2 x_2,$$
$$y' = k_1 y_1 + k_2 y_2,$$
$$z' = k_1 z_1 + k_2 z_2,$$
$$t' = k_1 t_1 + k_2 t_2,$$

where k_1 and k_2 are two constants, not both zero, the point (x', y', z', t') or P' lies on the line $P_1 P_2$, and our theorem will be established if we can show that the constants k_1 and k_2 can be so chosen that the points P' and P_3 coincide. Now let $ax + by + cz + dt = 0$ be the equation of any plane through the point P_3 but *not* through P_1 or P_2. Thus P_3 is determined as the intersection of this plane with the line $P_1 P_2$, so that if P', which we know lies on $P_1 P_2$, can be made to lie in this plane, it must coincide with P_3 and the proof is complete. The condition for P' to lie in this plane is $ax' + by' + cz' + dt' = 0$. Substituting for x', y', z', t' their values given above, we have

$$k_1(ax_1 + by_1 + cz_1 + dt_1) + k_2(ax_2 + by_2 + cz_2 + dt_2) = 0.$$

But neither of these parentheses is zero, since the plane does not pass through P_1 or P_2, hence we may give to k_1 and k_2 values different from zero for which this equation is satisfied. We have thus proved

THEOREM 2. *Three points are linearly dependent when, and only when, they are collinear.*

The proofs of the following theorems are left to be supplied by the reader. It will be found that some of them are readily proved from the definition of linear dependence, as above, while for others it is more convenient to use the condition for linear dependence obtained in § 13.

THEOREM 3. *Four points are linearly dependent when, and only when, they are complanar.*

THEOREM 4. *Five or more points are always linearly dependent.*

Another geometric application is suggested by the following considerations:

A set of n ordinary * quantities is nothing more nor less than a complex quantity with n components (cf. § 21). Our first definition of linear dependence is therefore precisely equivalent to the following:

The m complex quantities $$a_1,\ a_2,\ \cdots\ a_m$$

are said to be linearly dependent if m ordinary quantities $c_1, c_2, \cdots c_m$, not all zero, exist such that :

$$c_1 a_1 + c_2 a_2 + \cdots + c_m a_m = 0.$$

Now the simplest geometric interpretation for a complex quantity with n components is as a vector in space of n dimensions,† and we are thus led to the conception of linear dependence of vectors. The geometric meaning of this linear dependence will be seen from the following theorems for the case $n = 3$:

THEOREM 5. *Two vectors are linearly dependent when, and only when, they are collinear.*

THEOREM 6. *Three vectors are linearly dependent when, and only when, they are complanar.*

THEOREM 7. *Four or more vectors are always linearly dependent.*

In order to get a geometric interpretation of the linear dependence of polynomials, we must consider, not the polynomials themselves, but the equations obtained by equating them to zero. We speak of these equations as being linearly dependent if the polynomials are

---

* Two different standpoints are here possible according as we understand the term *ordinary quantity* to mean *real quantity*, or *ordinary complex quantity*.

† There are of course other possible geometric interpretations. Thus in the case $n = 4$ we may regard our complex quantities as quaternions, and consider the meaning of linear dependence of two, three, or four quaternions.

linearly dependent. If then we regard the independent variables as rectangular coördinates, these equations give us geometric loci in space of as many dimensions as there are independent variables. Thus, in the cases of two and three variables, we have plane curves and surfaces respectively. The case of two loci is of no interest, as they must coincide in order to be linearly dependent. In the case of three linearly dependent loci it is easily shown that any one must meet the other two in all their common points and in no others. The following theorems will serve to illustrate the geometric meaning of linear dependence :

(1) In the plane :

THEOREM 8. *Three circles are linearly dependent when, and only when, they belong to the same coaxial family.*

THEOREM 9. *Four circles are linearly dependent when, and only when, they have a (real or imaginary) common orthogonal circle.*

THEOREM 10. *Four circles are linearly dependent when, and only when, the points of intersection of the first and second, and the points of intersection of the third and fourth, lie on a common circle.*

THEOREM 11. *Five or more circles are always linearly dependent.*

(2) In space (using homogeneous coördinates):

THEOREM 12. *Three planes are linearly dependent when, and only when, they intersect in a line.*

THEOREM 13. *Four planes are linearly dependent when, and only when, they intersect in a point.*

THEOREM 14. *Five or more planes are always linearly dependent.*

CHAPTER IV

LINEAR EQUATIONS

16. Non-homogeneous Linear Equations. In every elementary treatment of determinants, however brief, it is explained how to solve by determinants a system of n equations of the first degree in n unknowns, provided that the determinant of the coefficients of the unknowns is not zero. Cramer's Rule, by which this is done, is this:

CRAMER'S RULE. *If in the equations*

$$a_{11}x_1 + \cdots + a_{1n}x_n = k_1,$$
$$\cdot \quad \cdot \quad \cdot \quad \cdot \quad \cdot \quad \cdot \quad \cdot \quad \cdot$$
$$\cdot \quad \cdot \quad \cdot \quad \cdot \quad \cdot \quad \cdot \quad \cdot \quad \cdot$$
$$a_{n1}x_1 + \cdots + a_{nn}x_n = k_n,$$

the determinant

$$a = \begin{vmatrix} a_{11} & \cdots & a_{1n} \\ \cdot & \cdot & \cdot & \cdot \\ a_{n1} & \cdots & a_{nn} \end{vmatrix}$$

is not zero, the equations have one and only one solution, namely:

$$x_1 = \frac{a_1}{a}, \; x_2 = \frac{a_2}{a}, \; \cdots \; x_n = \frac{a_n}{a},$$

where a_i is the n-rowed determinant obtained from a by replacing the elements of the ith column by the elements k_1, k_2, $\cdots k_n$.

This rule, whose proof we assume to be known,* is of fundamental importance in the general theory of linear equations to which we now proceed.

---

* The proof as given in most English and American text-books merely establishes the fact that *if the equations have a solution* it is given by Cramer's formulæ. That these formulæ really satisfy the equations in all cases is not commonly proved, but may be easily estabished by direct substitution. We leave it for the reader to do this.

Consider the system of m linear equations in n variables:

$$a_{11}x_1 + \cdots + a_{1n}x_n + b_1 = 0,$$
$$\cdots \cdots \cdots \cdots \cdots$$
$$\cdots \cdots \cdots \cdots \cdots$$
$$a_{m1}x_1 + \cdots + a_{mn}x_n + b_m = 0,$$

where m and n may be any positive integers.   Three cases arise:

(1) The equations may have no solution, in which case they are said to be *inconsistent*.

(2) They may have just one solution.

(3) They may have more than one solution, in which case it will presently appear that they necessarily have an infinite number of solutions.

Let us consider the two matrices:

$$\mathbf{a} = \begin{Vmatrix} a_{11} & \cdots & a_{1n} \\ \cdot & \cdot & \cdot \\ \cdot & \cdot & \cdot \\ a_{m1} & \cdots & a_{mn} \end{Vmatrix}, \qquad \mathbf{b} = \begin{Vmatrix} a_{11} & \cdots & a_{1n} & b_1 \\ \cdot & \cdot & \cdot & \cdot \\ \cdot & \cdot & \cdot & \cdot \\ a_{m1} & \cdots & a_{mn} & b_m \end{Vmatrix}.$$

We will call $\mathbf{a}$ the matrix of the system of equations, $\mathbf{b}$ the *augmented matrix*.

It is evident that the rank of the matrix $\mathbf{a}$ cannot be greater than that of the matrix $\mathbf{b}$, since every determinant contained in $\mathbf{a}$ is also contained in $\mathbf{b}$.   We have, then, two cases:

I.   Rank of $\mathbf{a}$ = Rank of $\mathbf{b}$.

II.   Rank of $\mathbf{a}$ < Rank of $\mathbf{b}$.

We will consider Case II first.

Let r be the rank of $\mathbf{b}$.   Then $\mathbf{b}$ must contain at least one r-rowed determinant which is not zero.   Moreover, this determinant must contain a column of $\mathbf{b}$'s, since otherwise it would be contained in $\mathbf{a}$ also, which is contrary to our hypothesis.   Suppose for definiteness that this non-vanishing r-rowed determinant is the one situated in the upper right-hand corner of $\mathbf{b}$.   There is no loss of generality in assuming this, since by writing the equations in a different order and changing the order of the variables $x_1, \cdots x_n$ we can always bring the determinant into this position.   Now for brevity let us represent the polynomials forming the first members of our given equations by $F_1, F_2, \cdots F_m$ respectively, and the homogeneous polynomials obtained by omitting the constant terms in each of these equations by $f_1, f_2, \cdots f_m$.   Then we have the identities:

$$F_i \equiv f_i + b_i, \qquad\qquad (i = 1, 2, \ldots m).$$

Consider the first r of these identities. Since the rank of **a** is less than r, the polynomials $f_1, f_2, \cdots f_r$ are linearly dependent,

$$c_1 f_1 + c_2 f_2 + \cdots + c_r f_r \equiv 0,$$

hence
$$c_1 F_1 + \cdots + c_r F_r \equiv c_1 b_1 + \cdots + c_r b_r = C.$$

But since the rank of **b** is r, the polynomials $F_1, \cdots F_r$ are linearly independent and therefore $C \neq 0$. Hence *the given equations are inconsistent*, for if they were consistent all the F's would be zero for some suitably chosen values of $x_1, \cdots x_n$, and if we substitute these values in the last written identity we should have

$$0 = C \neq 0.$$

Let us now consider Case I. Let r be the common rank of **a** and **b**, then there is at least one r-rowed determinant in **a** which is not zero. This same determinant also occurs in **b**. Suppose it to be situated in the upper left-hand corner of each matrix. Since all $(r+1)$-rowed determinants of either matrix are zero, the first $(r+1)$ of the F's are linearly dependent, and we have

$$c_1 F_1 + c_2 F_2 + \cdots + c_r F_r + c_{r+1} F_{r+1} \equiv 0 \; ;$$

and, since $F_1, \cdots F_r$ are linearly independent, c_{r+1} cannot be zero; hence we may divide through by it and express F_{r+1} linearly in terms of $F_1, \cdots F_r$. The same argument holds if instead of F_{r+1} we take F_{r+2} or any other one of the remaining F's. Hence

$$F_{r+l} \equiv k_1^{[l]} F_1 + \cdots + k_r^{[l]} F_r \qquad (l = 1, 2, \cdots m - r).$$

From these identities it is obvious that at any point $(x_1, \cdots x_n)$ where $F_1, \cdots F_r$ all vanish, the remaining F's also vanish. In other words, any solution which the first r equations of the given system may have is necessarily a solution of the whole system.

Now consider the first r of the given equations. Assign to $x_{r+1}, \cdots x_n$ any fixed values $x'_{r+1}, \cdots x'_n$, and transpose all the terms after the rth in each equation to the second member,

$$a_{11} x_1 + \cdots + a_{1r} x_r = - a_{1, r+1} x'_{r+1} - \cdots - a_{1n} x'_n - b_1,$$
$$\cdot \quad \cdot \quad \cdot \quad \cdot \quad \cdot \quad \cdot \quad \cdot \quad \cdot \quad \cdot \quad \cdot \quad \cdot \quad \cdot \quad \cdot \quad \cdot$$
$$a_{r1} x_1 + \cdots + a_{rr} x_r = - a_{r, r+1} x'_{r+1} - \cdots - a_{rn} x'_n - b_r.$$

Remembering that the right-hand sides of these equations are known constants, and that the determinant of the coefficients on the left is not zero, we see that we have the case to which Cramer's Rule applies, and that this system of equations has therefore just one solution. Hence the given system of equations is consistent, and we have the theorem:

THEOREM 1. *A necessary and sufficient condition for a system of linear equations to be consistent is that the matrix of the system have the same rank as the augmented matrix.*

From the foregoing considerations we have also

THEOREM 2. *If in a system of linear equations the matrix of the system and the augmented matrix have the same rank r, the values of $n - r$ of the unknowns may be assigned at pleasure and the others will then be uniquely determined.*

The $n - r$ unknowns whose values may be assigned at pleasure may be chosen in any way provided that the matrix of the coefficients of the remaining unknowns is of rank r.

EXERCISES

Solve completely the following systems of equations:

1.
$$\begin{cases} 2x - y + 3z - 1 = 0, \\ 4x - 2y - z + 3 = 0, \\ 2x - y - 4z + 4 = 0, \\ 10x - 5y - 6z + 10 = 0. \end{cases}$$

2.
$$\begin{cases} 4x - y + z + 5 = 0, \\ 2x - 3y + 5z + 1 = 0, \\ x + y - 2z + 2 = 0, \\ 5x - z + 2 = 0. \end{cases}$$

3.
$$\begin{cases} 2x - 3y + 4z - w = 3, \\ x + 2y - z + 2w = 1, \\ 3x - y + 2z - 3w = 4, \\ 3x - y + z - 7w = 4. \end{cases}$$

17. **Homogeneous Linear Equations.** We will now consider the special case where the equations of the last section are *homogeneous*, *i.e.* where all the b's are zero,

$$a_{11}x_1 + \cdots + a_{1n}x_n = 0,$$
$$\cdot \quad \cdot \quad \cdot \quad \cdot \quad \cdot \quad \cdot \quad \cdot$$
$$\cdot \quad \cdot \quad \cdot \quad \cdot \quad \cdot \quad \cdot \quad \cdot$$
$$a_{m1}x_1 + \cdots + a_{mn}x_n = 0.$$

The matrices **a** and **b** of the last section differ here only by a column of zeros ; hence they always have the same rank and this is called *the rank of the system of equations.* Theorems 1 and 2 of the last section become

THEOREM 1. *A system of homogeneous linear equations always has one or more solutions.*

THEOREM 2. *If the rank of a system of homogeneous linear equations in n variables is r, the values of $n - r$ of the unknowns may be assigned at pleasure and the others will then be uniquely determined.**

If the rank of the equations is n, there will therefore be only one solution, and this solution is obviously $x_1 = x_2 = \cdots = x_n = 0$. Since the rank can never be greater than n, we have

THEOREM 3. *A necessary and sufficient condition for a system of homogeneous linear equations in the n variables $(x_1, \cdots x_n)$ to have a solution other than $x_1 = x_2 = \cdots = x_n = 0$ is that their rank be less than n.*

COROLLARY 1. *If there are fewer equations than unknowns, the equations always have solutions other than $x_1 = x_2 = \cdots = x_n = 0$.*

COROLLARY 2. *If the number of equations is equal to the number of unknowns, a necessary and sufficient condition for solutions other than $x_1 = x_2 = \cdots = x_n = 0$ is that the determinant of the coefficients be zero.*

In the special case where the number of equations is just one less than the number of unknowns and the equations are linearly independent, we will prove the following :

THEOREM 4. *Every set of values of $x_1, \cdots x_n$ which satisfies a system of $n - 1$ linearly independent,† homogeneous linear equations in*

---

* Cf. also the closing lines of Theorem 2, § 16.

† The theorem is still true if the equations are linearly dependent, but it is then trivial, since the determinants in question are all zero.

n unknowns is proportional to the set of $(n-1)$-rowed determinants taken alternately with plus and minus signs, and obtained by striking out from the matrix of the coefficients first the first column, then the second, etc.

Let us denote by a_i the $(n-1)$-rowed determinant obtained by striking out the ith column from the matrix of the equations. Since the equations are linearly independent, there must be at least one of the determinants $a_1,\ a_2,\ \cdots a_n$ which is not zero. Let it be a_i. Now assign to x_i any fixed value, c, and transpose the ith term of each equation to the second member and we have

$$a_{11}x_1 + \cdots + a_{1,\,i-1}x_{i-1}\ + \ a_{1,\,i+1}\,x_{i+1}\ + \cdots + a_{1n}\,x_n\ = -\,a_{1i}c,$$

$$\cdot\ \ \cdot\ \ \cdot\ \ \cdot\ \ \cdot\ \ \cdot\ \ \cdot\ \ \cdot\ \ \cdot\ \ \cdot\ \ \cdot\ \ \cdot\ \ \cdot\ \ \cdot\ \ \cdot\ \ \cdot\ \ \cdot\ \ \cdot\ \ \cdot$$

$$a_{n-1,\,1}x_1 + \cdots + a_{n-1,\,i-1}x_{i-1} + a_{n-1,\,i+1}x_{i+1} + \cdots + a_{n-1,\,n}x_n = -\,a_{n-1,\,i}c.$$

Hence : $$x_k = \frac{(-1)^{i-k}\cdot c\cdot a_k}{a_i} \qquad k = 1,\ 2,\ \cdots n),$$

from which it is clear that $(x_1,\ \cdots x_n)$ are proportional to the determinants $(a_1,\ -a_2,\ a_3,\ \cdots (-1)^{n-1}a_n)$, as was to be proved.

The theory of homogeneous linear equations has here been deduced from the theory of linear dependence. It can, however, in turn be used to obtain further results in this last-mentioned theory. As an example of this we will deduce the following theorem, which we shall find useful later :

THEOREM 5. *If a set of points $(x_1,\ \cdots x_n)$, finite or infinite in number, have the property that k points can be found among them upon which every other point of the set is linearly dependent, then any $k+1$ points of the set will be linearly dependent.*

Let $(x_1',\ \cdots x_n')$, $(x_1'',\ \cdots x_n'')$, $\cdots (x_1^{[k]},\ \cdots x_n^{[k]})$ be the k points upon which every other point of the set is linearly dependent, and let

$$(X_1',\ \cdots X_n'),\ (X_1'',\ \cdots X_n''),\ \cdots (X_1^{[k+1]},\ \cdots X_n^{[k+1]})$$

be any $k+1$ points of the set. Then we may write

$$(1)\qquad
\begin{cases}
X_1^{[i]} = c_1^{[i]}\,x_1' + c_2^{[i]}\,x_1'' + \cdots + c_k^{[i]}x_1^{[k]}, \\
\cdot\ \ \cdot\ \ \cdot\ \ \cdot\ \ \cdot\ \ \cdot\ \ \cdot\ \ \cdot\ \ \cdot\ \ \cdot \\
\cdot\ \ \cdot\ \ \cdot\ \ \cdot\ \ \cdot\ \ \cdot\ \ \cdot\ \ \cdot\ \ \cdot\ \ \cdot \\
X_n^{[i]} = c_1^{[i]}x_n' + c_2^{[i]}x_n'' + \cdots + c_k^{[i]}x_n^{[k]}.
\end{cases}
\qquad (i = 1,\ 2,\ \cdots k+1)$$

This is true by hypothesis if $(X_1^{[i]}, \cdots X_n^{[i]})$ is not one of the first k points, and if it is one of these points, it is obviously true. We have then to prove that $k + 1$ constants, C_1, C_2, $\cdots C_{k+1}$, not all zero, can be found such that

$$C_1 X_j' + C_2 X_j'' + \cdots + C_{k+1} X_j^{[k+1]} = 0 \qquad (j = 1, 2, \cdots n).$$

By substituting here the values of the X's from (1), we see that these equations will be fulfilled if

$$C_1 c_1' + C_2 c_1'' + \cdots + C_{k+1} c_1^{[k+1]} = 0,$$
$$\cdots \cdots \cdots \cdots \cdots$$
$$\cdots \cdots \cdots \cdots \cdots$$
$$C_1 c_k' + C_2 c_k'' + \cdots + C_{k+1} c_k^{[k+1]} = 0,$$

and this is a system of fewer equations than unknowns, which is therefore satisfied by a set of C's not all zero. (Cf. Theorem 3. Cor. 1.)

EXERCISES

Solve completely the following systems of equations:

$$
\mathbf{1.} \quad
\begin{cases}
11\,x + 8\,y - 2\,z + 3\,w = 0, \\
2\,x + 3\,y - z + 2\,w = 0, \\
7\,x - y + z - 3\,w = 0, \\
4\,x - 11\,y + 5\,z - 12\,w = 0.
\end{cases}
$$

$$
\mathbf{2.} \quad
\begin{cases}
2\,x - 3\,y + 5\,z + 3\,w = 0, \\
4\,x - y + z + w = 0, \\
3\,x - 2\,y + 3\,z + 4\,w = 0.
\end{cases}
$$

18. Fundamental Systems of Solutions of Homogeneous Linear Equations. If $(x_1', \cdots x_n')$ is a solution of the system of equations

$$
(1) \qquad
\begin{cases}
a_{11}x_1 + \cdots + a_{1n}x_n = 0, \\
\cdots \cdots \cdots \cdots \cdots \\
\cdots \cdots \cdots \cdots \cdots \\
a_{m1}x_1 + \cdots + a_{mn}x_n = 0,
\end{cases}
$$

then $(cx_1', \cdots cx_n')$ is also a solution, and by giving to c different values we get thus (except in the special case in which the x''s are all zero) an infinite number of solutions. These may include all the solutions of (1) (cf. Theorem 4 of the last section), but in general this will not be the case.

Suppose, again, that $(x'_1, \cdots x'_n)$ and $(x''_1, \cdots x''_n)$ are two solutions of (1), then $(c_1 x'_1 + c_2 x''_1, \cdots c_1 x'_n + c_2 x''_n)$ is also a solution. If the two given solutions are proportional to each other, this clearly gives us nothing more than what we had above by starting from a single solution; but if these two solutions are linearly independent, we build up from them, by allowing c_1 and c_2 to take on all values, a doubly infinite system of solutions; but even this system will usually not include all the solutions of (1). Similarly we see that, if we can find three linearly independent solutions, we can build up from them a triply infinite system of solutions, etc. If, proceeding in this way, we succeed in finding a finite number of linearly independent solutions in terms of which all solutions can be expressed, this finite number of solutions is said to form a *fundamental system*.

DEFINITION. *If* $(x_1^{[i]}, \cdots x_n^{[i]})$ $(i = 1, 2, \cdots k)$ *are a system of k solutions of* (1) *which satisfy the following two conditions, they are said to form a fundamental system:*

(a) *They shall be linearly independent.*

(b) *Every solution of* (1) *shall be expressible in the form*

$$(c_1 x'_1 + c_2 x''_1 + \cdots + c_k x_1^{[k]}, \;\cdots\cdots\; c_1 x'_n + c_2 x''_n + \cdots + c_k x_n^{[k]}).$$

THEOREM 1. *If the equations* (1) *are of rank $r < n$, they possess an infinite number of fundamental systems each of which consists of $n - r$ solutions.*

Suppose the r-rowed determinant which stands in the upper left-hand corner of the matrix of the equations (1) does not vanish, and let us consider the first r of these equations. Any solution of these will be a solution of all the others. Transpose all terms after the rth to the second members, and let $(x_{r+1}, \cdots x_n)$ have any fixed set of values $(x'_{r+1}, \cdots x'_n)$, not all zero; then these r equations will have just one solution given by Cramer's Rule. Call it $(x'_1, \cdots x'_n)$. Now let $(x_{r+1}, \cdots x_n)$ have any *other* fixed set of values $(x''_{r+1}, \cdots x''_n)$, not all zero, and we get another solution, $(x''_1, \cdots x''_n)$. Continue in this way until we have $n - r$ solutions

$$\begin{aligned}
&x'_1, \;\cdots\; x'_r, \; x'_{r+1}, \;\cdots\; x'_n, \\
&\;\cdot \quad\cdot \quad\cdot \quad\cdot \quad\cdot \quad\cdot \quad\cdot \\
&\;\cdot \quad\cdot \quad\cdot \quad\cdot \quad\cdot \quad\cdot \quad\cdot \\
&x_1^{[n-r]}, \;\cdots\; x_r^{[n-r]}, x_{r+1}^{[n-r]}, \;\cdots\; x_n^{[n-r]}.
\end{aligned}$$

If we have chosen these $n-r$ sets of values for $(x_{r+1}, \cdots x_n)$ so that the determinant

$$(2) \qquad \begin{vmatrix} x'_{r+1} & \cdots & x'_n \\ \cdot & \cdot & \cdot & \cdot \\ \cdot & \cdot & \cdot & \cdot \\ x^{[n-r]}_{r+1} & \cdots & x^{[n-r]}_n \end{vmatrix}$$

is not zero, — and this may clearly be done in an infinite variety of ways, — these $n-r$ solutions will be linearly independent. That is to say, we may thus obtain an infinite number of sets of $n-r$ solutions each, each of which satisfies condition (a) of our definition for a fundamental system.

To prove that these sets of solutions also satisfy condition (b), let us suppose that $(X_1, \cdots X_n)$ is any solution of the r equations we are considering. The last $n-r$ of these X's are linearly dependent on the $n-r$ sets of values we have chosen for $(x_{r+1}, \cdots x_n)$ since we have here more sets of constants than there are elements in each set (cf. Theorem 2, §13), and the determinant (2) is not zero. Thus

$$(3) \quad X_i = c_1 x'_i + c_2 x''_i + \cdots + c_{n-r} x^{[n-r]}_i \qquad (i = r+1,\ r+2,\ \cdots n).$$

Let us now solve the first r equations (1) by Cramer's Rule, regarding $x_{r+1}, \cdots x_n$ as known. We thus get results of the form

$$x_j = A'_j x_{r+1} + A''_j x_{r+2} + \cdots + A^{[n-r]}_j x_n \qquad (j = 1, 2, \cdots r).$$

By assigning special values here to $x_{r+1}, \cdots x_n$, we get

$$(4) \quad \begin{cases} x'_j = A'_j x'_{r+1} + A''_j x'_{r+2} + \cdots + A^{[n-r]}_j x'_n, \\ \cdot \ \cdot \ \cdot \ \cdot \ \cdot \ \cdot \ \cdot \ \cdot \ \cdot \ \cdot \ \cdot \ \cdot \ \cdot \ \cdot \\ \cdot \ \cdot \ \cdot \ \cdot \ \cdot \ \cdot \ \cdot \ \cdot \ \cdot \ \cdot \ \cdot \ \cdot \ \cdot \\ x^{[n-r]}_j = A'_j x^{[n-r]}_{r+1} + A''_j x^{[n-r]}_{r+2} + \cdots + A^{[n-r]}_j x^{[n-r]}_n, \\ X_j = A'_j X_{r+1} + A''_j X_{r+2} + \cdots + A^{[n-r]}_j X_n. \end{cases} \qquad (j = 1,\ 2,\ \cdots r)$$

If we multiply the first $n-r$ of these equations by $c_1, \cdots c_{n-r}$ respectively and add, we get, by (3),

$$c_1 x'_j + \cdots + c_{n-r} x^{[n-r]}_j = A'_j X_{r+1} + \cdots + A^{[n-r]}_j X_n.$$

Consequently, by the last equation (4),

$$(5) \qquad X_j = c_1 x'_j + \cdots c_{n-r} x^{[n-r]}_j \qquad (j = 1,\ 2,\ \cdots r)$$

Equations (3) and (5) together prove our theorem.

We thus see that the totality of all solutions of the system (1)
forms a set of points satisfying the conditions of Theorem 5, §17.
Consequently,

THEOREM 2. *If the rank of a system of homogeneous linear equations in n variables is r, then any $n - r + 1$ solutions are linearly dependent.*

Finally we will prove the theorem.

THEOREM 3. *A necessary and sufficient condition that a set of solutions of a system of homogeneous linear equations of rank r in n variables form a fundamental system is that they be*

(a) *linearly independent,*
(b) $n - r$ *in number.*

By definition, (a) is a necessary condition. To see that (b) also
is necessary, notice that by Theorem 2 there cannot be more
than $n - r$ linearly independent solutions. We have, then,
merely to show that l linearly independent solutions never form
a fundamental system when $l < n - r$. If they did, then by
Theorem 5, § 17, any set of $l + 1$ solutions would be linearly
dependent, and therefore the same would be true of any set of
$n - r$ solutions (since $n - r \geqq l + 1$). But by Theorem 1, this
is not true.

In order now to prove that conditions (a) and (b) are also
sufficient, let

$$(x_1^{[i]},\ x_2^{[i]},\ \dots\ x_n^{[i]}) \qquad (i = 1,\ 2,\ \dots\ n - r)$$

be any system of $n - r$ linearly independent solutions of our system
of equations, and let $(x_1,\ \dots\ x_n)$ be any solution of the system.
Then, by Theorem 2, we have $n - r + 1$ constants $(c_1,\ \dots\ c_{n-r+1})$,
not all zero, and such that

$$c_1 x_j' + c_2 x_j'' + \dots + c_{n-r} x_j^{[n-r]} + c_{n-r+1} x_j = 0 \qquad (j = 1,\ 2,\ \dots\ n)$$

But since the $n - r$ given points are linearly independent, $c_{n-r+1} \neq 0$;
accordingly these last equations enable us to express the solution $(x_1,\ \dots\ x_n)$ linearly in terms of the $n - r$ given solutions,
and this shows that these $n - r$ solutions form a fundamental
system.

EXERCISES

1. Prove that all the fundamental systems of solutions of a system of homogeneous linear equations are included in the infinite number obtained in the proof of Theorem 1.

2. Given three planes in space by their equations in homogeneous coördinates. What are their relative positions when the rank of the system of equations is 3? when it is 2? when it is 1?

3. Given three planes in space by their equations in non-homogeneous coördinates. What are their relative positions for the different possible pairs of values of the ranks of the matrices and augmented matrices?

CHAPTER V

SOME THEOREMS CONCERNING THE RANK OF A MATRIX

19. General Matrices. In order to show that a given matrix is of rank r, we have first to show that at least one r-rowed determinant of the matrix is not zero, and secondly that all $(r+1)$-rowed determinants are zero. This latter work may be considerably shortened by the following theorem :

THEOREM 1. *If in a given matrix a certain r-rowed determinant is not zero, and all the $(r+1)$-rowed determinants of which this r-rowed determinant is a first minor are zero, then all the $(r+1)$-rowed determinants of the matrix are zero.*

We will assume, as we may do without loss of generality, that the non-vanishing r-rowed determinant stands in the upper left-hand corner of the matrix. Let the matrix be

$$\left\| \begin{array}{ccc} a_{11} & \cdots & a_{1n} \\ \cdot & & \cdot \\ \cdot & & \cdot \\ a_{m1} & \cdots & a_{mn} \end{array} \right\| ,$$

and consider the $r+1$ sets of n quantities each which lie in the first $r+1$ rows of this matrix. These $r+1$ sets of quantities are linearly dependent, as will be seen by reference to the proof of Theorem 1, §13, for although we knew there that *all* the $(r+1)$-rowed determinants were zero, we made use of this fact only for those $(r+1)$-rowed determinants which we now assume to be zero. Moreover, since the r sets of constants which stand in the first r rows of our matrix are linearly independent, it follows that the $(r+1)$th row is linearly dependent on the first r. Precisely the same reasoning shows that each of the subsequent rows is linearly dependent on the first r rows. Accordingly, by Theorem 5, §17, any $r+1$ rows are linearly dependent ; and therefore, by Theorem 1, §13, all the $(r+1)$-rowed determinants of our matrix are zero, as was to be proved.

54

Still another method of facilitating the determination of the rank of a matrix is by changing the form of the matrix in certain ways which do not change its rank. In order to explain this method, we begin by laying down the following definition :

DEFINITION 1. *By an elementary transformation of a matrix we understand a transformation of any one of the following forms :*
(a) the interchange of two rows or of two columns ;
(b) the multiplication of each element of a row (or column) by the same constant not zero ;
(c) the addition to the elements of one row (or column) of the products of the corresponding elements of another row (or column) by one and the same constant.

It is clear that if we can pass from a matrix **a** to a matrix **b** by one of these transformations, we can pass back from **b** to **a** by an elementary transformation.

DEFINITION 2. *Two matrices are said to be equivalent if it is possible to pass from one to the other by a finite number of elementary transformations.*

THEOREM 2. *If two matrices are equivalent, they have the same rank.*

It is evident that the transformations (a) and (b) of Definition 1 do not change the rank of a matrix, since they do not affect the vanishing or non-vanishing of any determinant of the matrix. In order to prove our theorem, it is therefore sufficient to prove that the rank of a matrix is not changed by a transformation (c).

Suppose this transformation consists in adding to the elements of the pth row of a matrix **a** k times the elements of the qth row, thus giving the matrix **b**. Let r be the rank of the matrix **a**. We will first show that this rank cannot be increased by the transformation, that is, that all $(r+1)$-rowed determinants of the matrix **b** are zero. By hypothesis all the $(r+1)$-rowed determinants of the matrix **a** are zero, and some of these determinants are clearly not changed by the transformation, namely, those which do not contain the pth row, or which contain both the pth and the qth row. The other determinants, which contain the pth row but not the qth, take on after the transformation the form $A \pm kB$ where A and B are $(r+1)$-rowed determinants of **a**, and are therefore zero. Thus we see that the transformation (c) never increases the rank of a matrix.

Moreover, the rank of **b** cannot be less than that of **a**, for then the transformation (c) which carries **b** into **a** would increase the rank of **b**, and this we have just seen is impossible.

This theorem can often be used to advantage in determining the rank of a matrix, for by means of elementary transformations it is often easy to simplify the matrix very materially.

EXERCISES

Determine the ranks of the following matrices:

1.

$$\begin{Vmatrix} 14 & 12 & 6 & 8 & 2 \\ 6 & 104 & 21 & 9 & 17 \\ 7 & 6 & 3 & 4 & 1 \\ 35 & 30 & 15 & 20 & 5 \end{Vmatrix}$$

2.

$$\begin{Vmatrix} 75 & 0 & 116 & -39 & 0 \\ 171 & -69 & 402 & 123 & 45 \\ 301 & 0 & 87 & -417 & -169 \\ 114 & -46 & 268 & 82 & 30 \end{Vmatrix}$$

3. Prove that any matrix of rank r can be reduced by means of elementary transformations to a form where the element in the ith row and ith column is 1 when $i \leqq r$, while all the other elements of the matrix are zero.

4. Hence prove that two matrices with m rows and n columns each are always equivalent when they have the same rank.

5. Prove that a necessary and sufficient condition that the matrix

$$\begin{Vmatrix} a_{11} & \cdots & a_{1n} \\ \cdot & \cdot & \cdot \\ \cdot & \cdot & \cdot \\ a_{m1} & \cdots & a_{mn} \end{Vmatrix}$$

be of rank 0 or 1 is that there exist $m + n$ constants $a_1, \cdots a_m, \beta_1, \cdots \beta_n$ such that $a_{ij} = a_i \beta_j$.

20. Symmetrical Matrices.

DEFINITION. *The square matrix*

$$\mathbf{a} = \begin{Vmatrix} a_{11} & a_{12} & \cdots & a_{1n} \\ a_{21} & a_{22} & \cdots & a_{2n} \\ \cdot & \cdot & \cdot & \cdot \\ \cdot & \cdot & \cdot & \cdot \\ a_{n1} & a_{n2} & & a_{nn} \end{Vmatrix}$$

(and also its determinant) is said to be symmetrical if the pairs of terms which are situated symmetrically with respect to the principal diagonal are equal. That is, if $a_{ij} = a_{ji}$.

We will denote by M_i an i-rowed principal minor of **a**. It is our main object in this section to show how the rank of the symmetrical matrix may be determined by an examination of the principal minors only. This may be done by means of the following three theorems.

THEOREM 1. *If an r-rowed principal minor M_r of the symmetrical matrix **a** is not zero, while all the principal minors obtained by adding one row and the same column, and also all those obtained by adding two rows and the same two columns, to M_r are zero, then the rank of **a** is r.*

Let the non-vanishing minor be the one which stands in the upper left-hand corner of **a**, and let $B_{\alpha\beta}$ denote the determinant obtained by adding the αth row and the βth column to M_r. If we can show that $B_{\alpha\beta} = 0$ for all unequal values of α and β our theorem will be proved. Cf. Theorem 1, § 19. Give to the integers α and β any two unequal values, and let C denote the determinant obtained by adding to M_r the αth and βth rows and the αth and βth columns of **a**. Then we have, by hypothesis, $M_r \neq 0$, $B_{\alpha\alpha} = 0$, $B_{\beta\beta} = 0$, $C = 0$. Let M_2' be the two-rowed principal minor of the adjoint of C which corresponds to the complement of M_r in C. Then by Corollary 3, § 11, we have

$$M_2' = CM_r = 0.$$

But
$$M_2' = B_{\alpha\alpha}B_{\beta\beta} - B_{\alpha\beta}^2.$$

Therefore
$$B_{\alpha\beta} = 0.$$

THEOREM 2. *If all the $(r+1)$-rowed principal minors of the symmetrical matrix **a** are zero, and also all the $(r+2)$-rowed principal minors, then the rank of **a** is r or less.*

If $r = 0$, all the elements in the principal diagonal are zero and all the two-rowed principal minors are zero.

That is,
$$a_{ii} \cdot a_{jj} - a_{ij}^2 = 0,$$

and therefore, since $a_{ii} = a_{jj} = 0$, $a_{ij} = 0$. That is, every element is zero and hence the rank is zero, and the theorem is true in this special case.

Now, assume it true when $r = k$; that is, we assume that when all $(k+1)$-rowed principal minors are zero and all $(k+2)$-rowed principal minors are zero, the rank of **a** is less than $k+1$. Then it follows that when all $(k+2)$-rowed, and all $(k+3)$-rowed principal minors are zero.

the rank of $\mathbf{a}$ is less than $k + 2$.   For in this case, if all $(k + 1)$-rowed principal minors are zero, the rank is less than $k + 1$, by hypothesis, and if some $(k+1)$-rowed principal minor is *not* zero, the rank is exactly $k + 1$, by the last theorem.   We see then that if the theorem is true for $r = k$ it is true for $r = k + 1$.   But we have proved it true for $r = 0,$ hence it is true for all values of r.

THEOREM 3.   *If the rank of the symmetrical matrix $\mathbf{a}$ is $r > 0$, there is at least one r-rowed principal minor of $\mathbf{a}$ which is not zero.*

For all $(r+1)$-rowed principal minors are zero, and, if all r-rowed principal minors were zero also, the rank of $\mathbf{a}$ would be $r-1$ or less, by the last theorem.

We close with a theorem of a somewhat special character which will be found useful later (cf. Exercises 4–6, § 50).

THEOREM 4.   *If the rank of the symmetrical matrix $\mathbf{a}$ is $r > 0$, we may shift the rows (at the same time shifting the columns in the same way, thus keeping $\mathbf{a}$ symmetrical) in such a way that no consecutive two of the set of quantities* $M_0,\ M_1,\ M_2,\ \dots M_r$

shall be zero and $M_r \neq 0$; M_0 being unity, and the other M's being the principal minors of $\mathbf{a}$ of orders indicated by their subscripts, which stand in the upper left-hand corner of $\mathbf{a}$ after the shifting.

By definition we have $M_0 \neq 0$.   Leaving aside for the moment the special case in which all the elements of the principal diagonal are zero, let us suppose the element a_{ii} is not zero.   Then by shifting the ith row and column to the first place, we have $M_1 \neq 0$.   We have tnus fixed the first row and column, but we are still at liberty to shift all the others.   Now consider the two-rowed principal minor obtained by adding to M_1 one row and the same column.   Leaving aside still the special case in which these are all zero, let us suppose that the two-rowed determinant obtained by striking out all the rows and columns except those numbered 1 and i_1 is not zero.   Then, by shifting the i_1th row and column into the second place, we have $M_2 \neq 0$.   We next have to consider the three-rowed principal minors of which M_2 is a first minor.   We can evidently proceed in this way until we have so shifted our rows and columns that none of the quantities $M_0,\ M_1,\ \dots M_r$ are zero, unless at a certain stage we find that all the principal minors of a certain order which we have to consider are zero.   In this case we should have so shifted our first k rows and columns that none of the quantities $M_0,\ M_1,\ \dots M_k$ are zero, but we should then find that all $(k+1)$-rowed principal minors of which M_k is a first minor vanish, so that, however we may shift the last $n-k$ rows and columns, we have $M_{k+1}=0$. Let us then examine the $(k+2)$-rowed principal minors of which M_k is a second minor.* These can (by Theorem 1) not all be zero as otherwise the rank of **a** would be $k<r$. That is, if $M_{k+1}=0$, we can so arrange the rows and columns that $M_{k+2}\neq 0$. Thus we see that the rows and columns of **a** may be so shifted that no consecutive two of the M's are zero. Now, if $M_{r-1}=0$, the above proof shows that we can make $M_r\neq 0$. But even though $M_{r-1}\neq 0$ we can still make $M_r\neq 0$, for by hypothesis † all the determinants obtained by adding to M_{r-1} two rows and the same two columns vanish, and if all those obtained by adding one row and the same column were zero also, the rank of **a** would be $r-1$, by Theorem 1.

A symmetrical matrix is said to be arranged in *normal form* when no consecutive two of the M's of Theorem 4 are zero and $M_r\neq 0$.

EXERCISES

1. Determine the ranks of the following matrices:

$$\left\|\begin{array}{cccc} 2 & 1 & 11 & 2 \\ 1 & 0 & 4 & -1 \\ 11 & 4 & 56 & 5 \\ 2 & -1 & 5 & -6 \end{array}\right\|, \quad \left\|\begin{array}{cccc} 0 & 4 & 10 & 1 \\ 4 & 8 & 18 & 7 \\ 10 & 18 & 40 & 17 \\ 1 & 7 & 17 & 3 \end{array}\right\|,$$

$$\left\|\begin{array}{ccccc} 1 & 0 & 0 & 1 & 4 \\ 0 & 1 & 0 & 2 & 5 \\ 0 & 0 & 1 & 3 & 6 \\ 1 & 2 & 3 & 14 & 32 \\ 4 & 5 & 6 & 32 & 77 \end{array}\right\|, \quad \left\|\begin{array}{cccc} 0 & 1 & b & d \\ 1 & 0 & c & e \\ b & c & 2bc & cd+be \\ d & e & cd+be & 2de \end{array}\right\|.$$

2. By a skew-symmetric determinant, or matrix, is meant one in which $a_{ij}=-a_{ji}$ (and therefore $a_{ii}=0$).

Establish for such matrices theorems similar to Theorems 1, 2, 3 of this section.

3. By considering the effect of changing rows into columns, prove that a skew-symmetric determinant of odd order is always zero.

4. Prove that the rank of a skew-symmetric matrix is always even.

---

*The tacit assumption is here made that when $k=r-1$, $r<n$, as otherwise M_{k+2} would have no meaning. The case $r=n$ can, however, obviously not occur here, for then we should have $M_{k+1}=a\neq 0$.

† Here again we assume that $r<n$, for if $r=n$, $M_r=a\neq 0$.

CHAPTER VI

LINEAR TRANSFORMATIONS AND THE COMBINATION OF MATRICES

21. Matrices as Complex Quantities. We have said in § 7 that a matrix of m rows and n columns is not a quantity, but a set of mn quantities. This statement is true only if we restrict the term *quantity* to the real and complex quantities of ordinary algebra. A moment's reflection, however, will show that the conception of quantity as used in arithmetic and algebra has been gradually enlarged from the primitive conception of the positive integer by using the word *quantity* to denote entities which, at an earlier stage, would not have been regarded as quantities at all, as, for instance, negative quantities. We will consider here only one of these extensions, namely the introduction of complex quantities, as this will lead us to look at our matrices from a broader point of view.

If we have objects of two or more different kinds which can be counted or measured, and if we consider aggregates of such objects, we get concrete examples of complex quantities, as, for instance, 5 horses, 3 cows, and 7 sheep. A convenient way to write such a complex quantity is (5, 3, 7), it being agreed that, in the illustration we are considering, the first place shall always indicate horses, the second cows, and the third sheep. In the abstract theory of complex quantities we do not specify any concrete objects such as horses, cows, etc., but merely consider sets of quantities (couples, triplets, etc.), distinguishing these quantities by the position they occupy in our symbol. Such a complex quantity we often find it convenient to designate by a single letter,

$$\alpha = (a,\ b,\ c)$$

just as in ordinary algebra we denote a fraction ($\frac{2}{3}$ for instance), which really involves two numbers, by a single letter. We speak here of the simple quantities $a,\ b,\ c$ of which α is composed as its first, second, third components ; and we call two complex quantities equal when and only when the components of one are equal respectively to the corresponding components of the other. Similarly a complex quantity is said to vanish when and only when all of its components are zero.

What makes it worth while to speak of such sets of quantities as complex quantities is that it is found useful to perform certain algebraic operations on them.   By the sum and difference of two complex quantities

$$\alpha_1 = (a_1,\, b_1,\, c_1), \quad \alpha_2 = (a_2,\, b_2,\, c_2)$$

we mean the two new complex quantities

$$\alpha_1 + \alpha_2 = (a_1 + a_2,\, b_1 + b_2,\, c_1 + c_2),\ \alpha_1 - \alpha_2 = (a_1 - a_2,\, b_1 - b_2,\, c_1 - c_2).^{*}$$

When it comes to the question of defining what we shall understand by the product of two complex quantities, things are by no means so simple.   It is necessary here to lay down some rule according to which, when two complex quantities are given, a third, which we call their product, is determined.   Such rules may be laid down in an infinite variety of ways, and each such rule gives us a different system of complex quantities.†

We come now to the subject of matrices. A matrix of m rows and n columns being merely a set of mn quantities (which we assume to be either real quantities or the ordinary complex quantities of elementary algebra) arranged in a definite order, is, according to the point of view we have explained, a complex quantity with mn components; and it is only a special application of the theory of complex quantities which we have sketched, when we lay down the following definitions:

DEFINITION 1. *A matrix is said to be zero when and only when all of its elements are zero.*

DEFINITION 2. *Two matrices are said to be equal when and only when they have the same number of rows and of columns, and every element of one is equal to the corresponding element of the other.*

---

* That this is the natural meaning to be attached to the terms *sum* and *difference* will be seen by reference to the concrete illustration given above.

† If, in particular, we wish to introduce the ordinary system of complex quantities of elementary algebra, we use a system of couples, and define the product of two couples, $\quad\alpha_1 = (a_1,\, b_1), \quad \alpha_2 = (a_2,\, b_2),$ by the formula $\quad\alpha_1\alpha_2 = (a_1 a_2 - b_1 b_2,\ a_1 b_2 + a_2 b_1).$

For further details cf. Burkhardt's *Funktionentheorie*, §§ 2, 3.

DEFINITION 3. *By the sum (or difference) of two matrices of m rows and n columns each, we understand a matrix of m rows and n columns, each of whose elements is the sum (or difference) of the corresponding elements of the given matrices.*

In order to distinguish them from matrices, we will call the ordinary quantities of algebra (real quantities and ordinary complex quantities) *scalars.*

Before proceeding, as we shall do in the next section, to the definition of the product of two matrices, we will define the product of a matrix and a scalar.

DEFINITION 4. *If* a *is a matrix * and k a scalar, then by the product* ka *or* ak *we understand the matrix each of whose elements is k times the corresponding element of* a.

As an obvious consequence of our definitions we state the theorem:

THEOREM. *All the laws of ordinary algebra hold for the addition or subtraction of matrices and their multiplication by scalars.*

For instance, if a, b, c are matrices, and k, l scalars,

$$a + b = b + a,$$
$$a + (b + c) = (a + b) + c,$$
$$ka + kb = k(a + b),$$
$$ka + la = (k + l)a.\dagger$$

EXERCISE

If r_1 and r_2 are the ranks of two matrices and R the rank of their sum, prove that

$$R \leqq r_1 + r_2.$$

22. The Multiplication of Matrices. Up to this point we have considered matrices with m rows and n columns. For the sake of simplicity of statement, we shall confine our attention from now on to square matrices, that is to the case $m = n$. This involves no real loss

---

* The notation here used, matrices being denoted by heavy-faced type, will be systematically followed in this book.

† We add that, as a matter of notation, we shall write

$$(-1)\, a = -\, a.$$

of generality provided we agree to consider a matrix of m rows and n columns, where $m \neq n$, as equivalent to a square matrix of order equal to the larger of the two integers m, n and obtained from the given matrix by filling in the lacking rows or columns with zeros.

The question now presents itself: How shall we define the product of two square matrices of the same order? It must be clearly understood that we are logically free to lay down here such definition as we please, and that the definition we select is preferable to others not on any *a priori* grounds, but only because it turns out to be more useful. We select the following definition, which is suggested * by the multiplication theorem for determinants:

DEFINITION 1. *The product* **ab** *of two square matrices of the nth order is a square matrix of the nth order in which the element which lies in the ith row and jth column is obtained by multiplying each element of the ith row of* **a** *by the corresponding element of the jth column of* **b** *and adding the results.*

Let us denote by a_{ij} and b_{ij} the elements in the ith row and jth column of **a** and **b** respectively, or, as we will say for brevity, the element (i, j) of these matrices. Then, according to our definition, the element (i, j) of the product **ab** is

$$(1) \qquad a_{i1}b_{1j} + a_{i2}b_{2j} + \cdots + a_{in}b_{nj},$$

while the element (i, j) in the matrix **ba** is

$$(2) \qquad a_{1j}b_{i1} + a_{2j}b_{i2} + \cdots + a_{nj}b_{in}.$$

Since the two quantities (1) and (2) are not in general equal, we obtain

THEOREM 1. *The multiplication of matrices is not in general commutative, that is, in general*
$$\mathbf{ab} \neq \mathbf{ba}.$$

Let us now consider a third matrix **c** whose element (i, j) is c_{ij}, and form the product (**ab**)**c**. The element (i, j) of this matrix is

$$
(3) \qquad
\begin{aligned}
&(a_{i1}b_{11} + a_{i2}b_{21} + \cdots + a_{in}b_{n1})c_{1j} \\
&+ (a_{i1}b_{12} + a_{i2}b_{22} + \cdots + a_{in}b_{n2})c_{2j} \\
&+ \quad \cdot \quad \cdot \quad \cdot \quad \cdot \quad \cdot \quad \cdot \quad \cdot \\
&+ (a_{i1}b_{1n} + a_{i2}b_{2n} + \cdots + a_{in}b_{nn})c_{nj}.
\end{aligned}
$$

* Historically this definition was suggested to Cayley by the consideration of the composition of linear transformations; cf. § 23.

On the other hand, the element (i, j) of the matrix $\mathbf{a}(\mathbf{bc})$ is

$$
(4) \quad
\begin{aligned}
&a_{i1}(b_{11}c_{1j} + b_{12}c_{2j} + \cdots + b_{1n}c_{nj}) \\
&+ a_{i2}(b_{21}c_{1j} + b_{22}c_{2j} + \cdots + b_{2n}c_{nj}) \\
&+ \cdots \cdots \cdots \cdots \\
&+ a_{in}(b_{n1}c_{1j} + b_{n2}c_{2j} + \cdots + b_{nn}c_{nj}).
\end{aligned}
$$

Since the two quantities (3) and (4) are equal, we have established

THEOREM 2. *The multiplication of matrices is associative, that is,*

$$(\mathbf{ab})\mathbf{c} = \mathbf{a}(\mathbf{bc}).$$

Finally, since the element (i, j) of the matrix $\mathbf{a}(\mathbf{b} + \mathbf{c})$ is clearly equal to the sum of the elements (i, j) of the matrices $\mathbf{ab}$ and $\mathbf{ac}$, we have the result

THEOREM 3. *The multiplication of matrices is distributive, that is,*

$$\mathbf{a}(\mathbf{b} + \mathbf{c}) = \mathbf{ab} + \mathbf{ac}.$$

Besides the commutative, associative, and distributive laws, there is one other principle of elementary algebra which is of constant use, namely, the principle that a product cannot vanish unless at least one of the factors is zero. Simple examples show that this is not true in the algebra of matrices. We have, for instance,

$$
(5) \quad
\begin{Vmatrix} a_{11} & a_{12} & 0 \\ a_{21} & a_{22} & 0 \\ a_{31} & a_{32} & 0 \end{Vmatrix}
\cdot
\begin{Vmatrix} 0 & 0 & 0 \\ 0 & 0 & 0 \\ b_{31} & b_{32} & b_{33} \end{Vmatrix}
=
\begin{Vmatrix} 0 & 0 & 0 \\ 0 & 0 & 0 \\ 0 & 0 & 0 \end{Vmatrix}
= 0,
$$

whatever the values of the a's and b's may be. Hence

THEOREM 4. *From the vanishing of the product of two or more matrices, we cannot infer that one of the factors is zero.*

The process of cancelling out non-vanishing factors which enter throughout an equation will, therefore, be inadmissible in the algebra of matrices.

We next state a result which follows at once from the similarity between the theorem for the multiplication of determinants and our definition of the product of two matrices:

THEOREM 5. *The determinant of a matrix which is obtained by multiplying together two or more matrices is equal to the product of the determinants of these matrices.*

The conception of the conjugate of a matrix, as defined in § 7, Definition 2, is an important one, and the following theorem concerning it is often useful:

THEOREM 6. *The conjugate of the product of any number of matrices is the product of their conjugates taken in the reverse order.*

In order to prove this theorem we first notice that its truth in the case of two matrices follows at once from the definition of the product of two matrices. Its truth will therefore follow in all cases if, assuming the theorem to be true for the product of $n-1$ matrices, we can prove that it is true for the product of n matrices. Let us write

$$b = a_2 a_3 \cdots a_n.$$

Then, from what we have assumed,

$$b' = a_n' \cdots a_3' \, a_2',$$

where we use accents to denote conjugates. Accordingly,

$$(a_1 a_2 \cdots a_n)' = (a_1 b)' = b' a_1' = a_n' \cdots a_2' a_1',$$

and our theorem is proved.

In conclusion we lay down the following:

DEFINITION 2. *A square matrix is said to be singular if its determinant is zero.*

According to the convention made at the beginning of this section, it will be seen that *all* matrices which are not square are singular.

EXERCISES

1. DEFINITION. *A matrix* a *is called a divisor of zero if a matrix* b *different from zero exists such that either* $ab = 0$ *or* $ba = 0$.

Prove that every matrix one of whose rows or columns is composed wholly of zeros is a divisor of zero.

2. If it is possible to pass from a to b by means of an elementary transformation (cf. § 19, Definition 1), prove that there either exists a non-singular matrix c such that

$$ac = b,$$

or a non-singular matrix d such that

$$da = b.$$

3. If all the elements of a matrix are real, and if the product of this matrix and its conjugate is zero, prove that the matrix itself is zero.

4. If the corresponding elements of two matrices **a** and **b** are conjugate imaginaries, and, **b**′ being the matrix conjugate to **b**, if

$$\mathbf{ab}' = 0, \text{ then } \mathbf{a} = \mathbf{b} = 0.$$

23. Linear Transformation. Before going farther with the theory of matrices we will take up, in this section and the next, the closely allied subject of linear transformation, which may be regarded as one of the most important applications of the theory of matrices.

In algebra and analysis we frequently have occasion to introduce, in place of the unknowns, or variables, we had originally to deal with, certain functions of these quantities which we regard as new unknowns or variables. Such a transformation, or change of variables, is particularly simple, and for many purposes particularly important, if the functions in question are homogeneous linear polynomials. It is then called a homogeneous linear transformation, or, as we shall say for brevity, simply a *linear transformation*. If $x_1, \cdots x_n$ are the original variables, and $x_1', \cdots x_n'$ the new ones, we have, as the formulæ for the transformation,

$$\begin{cases} x_1' = a_{11}x_1 + \cdots + a_{1n}x_n, \\ \cdot \quad \cdot \quad \cdot \quad \cdot \quad \cdot \quad \cdot \quad \cdot \\ x_n' = a_{n1}x_1 + \cdots + a_{nn}x_n. \end{cases}$$

The square matrix
$$\mathbf{a} = \left\| \begin{matrix} a_{11} \cdots a_{1n} \\ \cdot \quad \cdot \quad \cdot \\ a_{n1} \cdots a_{nn} \end{matrix} \right\|$$

is called the matrix of the transformation, and the determinant of this matrix, which we will represent by a, is called the determinant of the transformation. Inasmuch as the transformation is completely determined by its matrix, no confusion will arise if we speak of the transformation **a**.

In most cases where we have occasion to use a transformation it is important for us to be able, in the course of our work, to pass back to the original variables, and for this purpose it must be possible, not merely to express $x_1', \cdots x_n'$ as functions of $x_1, \cdots x_n$, but also to express $x_1, \cdots x_n$ as functions of $x_1', \cdots x_n'$. In the case of linear transformations this can in general be done. For the equations of the transformation may be regarded as non-homogeneous linear equations in $x_1, \cdots x_n$, and if the determinant a of the transformation is not zero, they can be solved and give

$$
\mathbf{A} \begin{cases} x_1 = \dfrac{A_{11}}{a}\, x_1' + \cdots + \dfrac{A_{n1}}{a}\, x_n', \\[2mm] \;\cdot\quad\cdot\quad\cdot\quad\cdot\quad\cdot\quad\cdot\quad\cdot\quad\cdot \\[2mm] x_n = \dfrac{A_{1n}}{a}\, x_1' + \cdots + \dfrac{A_{nn}}{a}\, x_n', \end{cases}
$$

where $A_{11}, \cdots A_{nn}$ are the cofactors of $a_{11}, \cdots a_{nn}$ in a.

This transformation $\mathbf{A}$ is called the *inverse* of the transformation $\mathbf{a}$, but it must be remembered that it exists only if $a \neq 0$. A linear transformation for which $a = 0$ is called a *singular transformation*. If $\mathbf{a}$ is non-singular, its inverse $\mathbf{A}$ is also non-singular, since the determinant of $\mathbf{A}$ is a^{-1} (cf. Corollary 2, § 11).

DEFINITION. *The special linear transformation*

$$
x_1' = x_1, \;\; x_2' = x_2, \;\; \cdots \;\; x_n' = x_n,
$$

whose matrix is

$$
\mathbf{I} = \begin{Vmatrix} 1 & 0 & \cdots & 0 \\ 0 & 1 & \cdots & 0 \\ \cdot & \cdot & \cdot\ \cdot\ \cdot & \cdot \\ \cdot & \cdot & \cdot\ \cdot\ \cdot & \cdot \\ 0 & 0 & \cdots & 1 \end{Vmatrix},
$$

is called the identical transformation.

The determinant of this transformation is 1.

We turn now to the subject of the composition of linear transformations. If we introduce a new set of variables x' as functions of the original variables x, and then make a second transformation by introducing a third set of variables x'' as functions of the variables x', these two transformations can obviously be combined and the variables x'' expressed directly in terms of the x's. If the two transformations which we combine are linear transformations, it is readily seen that the resulting transformation will also be linear. The precise formulæ are important here, and for the sake of simplicity we will write them in the case of three variables, a case which will be seen to be perfectly typical of the general case.

Let

$$\mathbf{a}\begin{cases} x_1' = a_{11}x_1 + a_{12}x_2 + a_{13}x_3, \\ x_2' = a_{21}x_1 + a_{22}x_2 + a_{23}x_3, \\ x_3' = a_{31}x_1 + a_{32}x_2 + a_{33}x_3, \end{cases} \qquad \mathbf{b}\begin{cases} x_1'' = b_{11}x_1' + b_{12}x_2' + b_{13}x_3', \\ x_2'' = b_{21}x_1' + b_{22}x_2' + b_{23}x_3', \\ x_3'' = b_{31}x_1' + b_{32}x_2' + b_{33}x_3', \end{cases}$$

be two linear transformations.   Replacing the x''s in $\mathbf{b}$ by their values from $\mathbf{a}$, we get

$$\begin{cases} x_1'' = (a_{11}b_{11} + a_{21}b_{12} + a_{31}b_{13})x_1 \\ \quad + (a_{12}b_{11} + a_{22}b_{12} + a_{32}b_{13})x_2 \\ \quad + (a_{13}b_{11} + a_{23}b_{12} + a_{33}b_{13})x_3, \\[4pt] x_2'' = (a_{11}b_{21} + a_{21}b_{22} + a_{31}b_{23})x_1 \\ \quad + (a_{12}b_{21} + a_{22}b_{22} + a_{32}b_{23})x_2 \\ \quad + (a_{13}b_{21} + a_{23}b_{22} + a_{33}b_{23})x_3, \\[4pt] x_3'' = (a_{11}b_{31} + a_{21}b_{32} + a_{31}b_{33})x_1 \\ \quad + (a_{12}b_{31} + a_{22}b_{32} + a_{32}b_{33})x_2 \\ \quad + (a_{13}b_{31} + a_{23}b_{32} + a_{33}b_{33})x_3. \end{cases}$$

It will be seen that the matrix of this transformation is $\mathbf{ba}$. Hence,

THEOREM. *If we pass from the variables x to the variables x' by a linear transformation of matrix $\mathbf{a}$, and from the variables x' to the variables x'' by another linear transformation of matrix $\mathbf{b}$, then the linear transformation of matrix $\mathbf{ba}$ will carry us directly from the variables x to the variables x''.* *

24. Collineation.   We come now to an important geometrical application of the subject of linear transformation.   For the sake of simplicity we begin with the case of three variables, which we will regard as the homogeneous coördinates of points in a plane.

The equations

(1)
$$\begin{cases} x' = a_1x + b_1y + c_1t, \\ y' = a_2x + b_2y + c_2t, \\ t' = a_3x + b_3y + c_3t \end{cases}$$

----

* This result may be remembered conveniently by means of the following symbolic notation, which is often convenient.   Let us denote the transformation $\mathbf{a}$ by the symbolic equation $x' = \mathbf{a}(x)$, and the transformation $\mathbf{b}$ by $x'' = \mathbf{b}(x')$.   The result of combining these two transformations is then $x'' = \mathbf{b}(\mathbf{a}(x))$ or simply $x'' = \mathbf{ba}(x)$.

may be regarded as defining a transformation of the points of the plane; that is, if (x, y, t) is an arbitrarily given point, we can compute, by means of (1), the coördinates (x', y', t') of a second point into which we regard the first point as being transformed. The only exception is when the computed values of x', y', t' are all three zero, in which case there is no point into which the given point is transformed. This exceptional case can clearly occur only when the determinant of the transformation (1) is zero. Let us then confine our attention to non-singular linear transformations. In this case, not only does every point (x, y, t) correspond to a definite point (x', y', t'), but conversely, every point (x', y', t') corresponds to a definite point (x, y, t), since the transformation (1) now has an inverse

$$
(2) \qquad
\begin{cases}
x = \dfrac{A_1}{D} x' + \dfrac{A_2}{D} y' + \dfrac{A_3}{D} t', \\[2ex]
y = \dfrac{B_1}{D} x' + \dfrac{B_2}{D} y' + \dfrac{B_3}{D} t', \\[2ex]
t = \dfrac{C_1}{D} x' + \dfrac{C_2}{D} y' + \dfrac{C_3}{D} t',
\end{cases}
$$

where D is the determinant of (1), and A_i, B_i, C_i are the cofactors in D.

The points (x, y, t) of the line

$$(3) \qquad ax + \beta y + \gamma t = 0$$

are transformed by means of the non-singular transformation (1) into points of another line,

$$(4) \qquad \frac{\alpha A_1 + \beta B_1 + \gamma C_1}{D} x' + \frac{\alpha A_2 + \beta B_2 + \gamma C_2}{D} y' + \frac{\alpha A_3 + \beta B_3 + \gamma C_3}{D} t' = 0,$$

as we see by using formulæ (2). Conversely every point of the line (4) corresponds, as we see by using (1), to a point on (3). That is, the transformation establishes a one-to-one correspondence between the points on the two lines (3) and (4), or, as we say, it transforms the line (3) into the line (4). On account of this property of transforming straight lines into straight lines, the transformation is called a *collineation*. The transformation is also known as a *projective transformation*, for it may be shown that it can be effected by projecting one plane on to another by means of straight lines radiating from a point in space.

What we have here said in the case of two dimensions applies with no essential change to three dimensions. The transformation

$$
(5) \quad
\begin{cases}
x' = a_1 x + b_1 y + c_1 z + d_1 t, \\
y' = a_2 x + b_2 y + c_2 z + d_2 t, \\
z' = a_3 x + b_3 y + c_3 z + d_3 t, \\
t' = a_4 x + b_4 y + c_4 z + d_4 t
\end{cases}
$$

gives us, provided its determinant is not zero, a one-to-one transformation of the points of space, which carries over planes into planes, and therefore also straight lines into straight lines, and is called a collineation or projective transformation of space. The same idea can be extended to spaces of higher dimensions.

Quite as important is the case of one dimension. The transformation

$$
(6) \quad
\begin{cases}
x' = a_1 x + b_1 t, \\
t' = a_2 x + b_2 t
\end{cases}
$$

gives us, provided its determinant is not zero, a one-to-one transformation of the points on a line. This we call a projective transformation of the line, the term *collineation* being in this case obviously inadequate.

It is possible, although for most purposes not desirable, to express the projective transformations (6), (1), (5) in one, two, and three dimensions in terms of non-homogeneous, instead of homogeneous coördinates. We thus get the formulæ

$$
(7) \quad X' = \frac{a_1 X + b_1}{a_2 X + b_2},
$$

$$
(8) \quad
\begin{cases}
X' = \dfrac{a_1 X + b_1 Y + c_1}{a_3 X + b_3 Y + c_3}, \\[2ex]
Y' = \dfrac{a_2 X + b_2 Y + c_2}{a_3 X + b_3 Y + c_3},
\end{cases}
$$

$$
(9) \quad
\begin{cases}
X' = \dfrac{a_1 X + b_1 Y + c_1 Z + d_1}{a_4 X + b_4 Y + c_4 Z + d_4}, \\[2ex]
Y' = \dfrac{a_2 X + b_2 Y + c_2 Z + d_2}{a_4 X + b_4 Y + c_4 Z + d_4}, \\[2ex]
Z' = \dfrac{a_3 X + b_3 Y + c_3 Z + d_3}{a_4 X + b_4 Y + c_4 Z + d_4}.
\end{cases}
$$

These fractional forms may, in particular, be used to advantage in case their denominators reduce to mere constants. This special case, which is known as an *affine transformation*, may clearly be characterized by saying that all finite points go into finite points.*

* If we consider the still more special case in which the constant terms in the numerators of (8) and (9) are zero, that is, affine transformations in which the origin is transformed into itself, we see that our formulæ (8) and (9) have the form (6) and

These affine transformations are of much importance in mechanics, where they are known as *homogeneous strains;* cf., for instance, Webster's *Dynamics* (Leipzig, Teubner), pp. 427–444.

Although we propose to leave the detailed discussion of singular transformations to the reader (see Exercise 1 at the end of this section), we will give one theorem concerning them.

THEOREM 1. *If the points P_1, P_2, $\cdots$ are carried over by a singular projective transformation into the points P'_1, P'_2, $\cdots$, then, if our transformation is in one dimension, the points P' will all coincide; if in two dimensions, they will all be collinear; if in three dimensions, they will all be complanar, etc.*

Suppose, for instance, that we have to deal with two dimensions. Since the determinant of the collineation (1) is supposed to be zero, the three polynomials in the second members of (1) are linearly dependent; that is, there exist three constants, k_1, k_2, k_3, not all zero, and such that for all values of x, y, t,

$$(10) \qquad k_1 x' + k_2 y' + k_3 t' = 0.$$

Accordingly all points (x', y', t') obtained by this transformation lie on the line (10).

Similar proofs apply to the cases of one dimension and of three or more dimensions.

THEOREM 2. *Any three distinct points on a line may be carried over respectively into any three distinct points on the line by one, and only one, projective transformation.*

Let the three initial points be P_1, P_2, P_3, with homogeneous coördinates (x_1, t_1), (x_2, t_2), (x_3, t_3) respectively, and let the points into which we wish them transformed be P'_1, P'_2, P'_3 with coördinates (x'_1, t'_1), (x'_2, t'_2), (x'_3, t'_3). The projective transformation

$$x' = \alpha x + \beta t,$$
$$t' = \gamma x + \delta t$$

(1) respectively. Thus (6) may be regarded either as the general projective transformation of a line (if x, t are regarded as homogeneous coördinates) or as a special affine transformation of the plane (if x, t are regarded as non-homogeneous coördinates). Similarly (1) may be regarded either as the general projective transformation of a plane, or as a special affine transformation of space.

carries over any given point (x, t) into a point (x', t') whose position depends on the values of the constants α, β, γ, δ. Our theorem is true if it is possible to find one, and, except for a constant factor which may be introduced throughout, only one, set of seven constants — four, α, β, γ, δ, and three others, ρ_1, ρ_2, ρ_3, none of which is zero — which satisfy the six equations

$$\begin{cases} \rho_1 x_1' = \alpha x_1 + \beta t_1, \\ \rho_1 t_1' = \gamma x_1 + \delta t_1, \end{cases} \quad \begin{cases} \rho_2 x_2' = \alpha x_2 + \beta t_2, \\ \rho_2 t_2' = \gamma x_2 + \delta t_2, \end{cases} \quad \begin{cases} \rho_3 x_3' = \alpha x_3 + \beta t_3, \\ \rho_3 t_3' = \gamma x_3 + \delta t_3. \end{cases}$$

Since the x's and t's are all known, we have here six homogeneous linear equations in seven unknowns. Hence there are always solutions other than zeros, the number of independent ones depending on the rank of the matrix of the coefficients. Transposing and rearranging the equations, we have

$$\begin{aligned}
x_1 \alpha + t_1 \beta \qquad\qquad - x_1' \rho_1 \qquad\qquad &= 0, \\
x_1 \gamma + t_1 \delta - t_1' \rho_1 \qquad\qquad &= 0, \\
x_2 \alpha + t_2 \beta \qquad\qquad - x_2' \rho_2 \qquad\quad &= 0, \\
x_2 \gamma + t_2 \delta \qquad - t_2' \rho_2 \qquad\quad &= 0, \\
x_3 \alpha + t_3 \beta \qquad\qquad\qquad - x_3' \rho_3 &= 0, \\
x_3 \gamma + t_3 \delta \qquad\qquad\qquad - t_3' \rho_3 &= 0.
\end{aligned}$$

The matrix of these equations is of rank six. For consider the determinant of the first six columns with its sign reversed,

$$D = \begin{vmatrix} x_1 & t_1 & 0 & 0 & x_1' & 0 \\ x_2 & t_2 & 0 & 0 & 0 & x_2' \\ 0 & 0 & x_1 & t_1 & t_1' & 0 \\ 0 & 0 & x_2 & t_2 & 0 & t_2' \\ x_3 & t_3 & 0 & 0 & 0 & 0 \\ 0 & 0 & x_3 & t_3 & 0 & 0 \end{vmatrix}.$$

Since P_1, P_2, P_3 are distinct, there exist two constants c_1, c_2, neither of which is zero, such that

$$c_1 x_1 + c_2 x_2 + x_3 = 0,$$
$$c_1 t_1 + c_2 t_2 + t_3 = 0.$$

Hence, adding to the fifth row of D c_1 times the first row and c_2 times the second, and to the sixth row c_1 times the third row and c_2 times the fourth, we have

$$D = \begin{vmatrix} x_1 & t_1 & 0 & 0 & x_1' & 0 \\ x_2 & t_2 & 0 & 0 & 0 & x_2' \\ 0 & 0 & x_1 & t_1 & t_1' & 0 \\ 0 & 0 & x_2 & t_2 & 0 & t_2' \\ 0 & 0 & 0 & 0 & c_1 x_1' & c_2 x_2' \\ 0 & 0 & 0 & 0 & c_1 t_1' & c_2 t_2' \end{vmatrix} = c_1 c_2 \begin{vmatrix} x_1 & t_1 \\ x_2 & t_2 \end{vmatrix}^2 \cdot \begin{vmatrix} x_1' & x_2' \\ t_1' & t_2' \end{vmatrix},$$

and this is not zero, since P_1' and P_2' are distinct as well as P_1 and P_2.

In the same way we see that the determinants obtained by striking out the sixth and the fifth columns respectively of the matrix are not zero. Accordingly, by Theorem 4, § 17, we see that the equations have a solution in which none of the quantities ρ_1, ρ_2, ρ_3 are zero, and that every solution is proportional to this one. All these solutions obviously yield the same projective transformation of the line.

Corollary. *The transformation just determined is non-singular.*

This follows, by a reference to Theorem 1, from the fact that it does not carry P_1, P_2, P_3 into a single point.

EXERCISES

1. Discuss singular projective transformations in one, two, and three dimensions; noting, in particular, the effect of the rank of the matrix of the transformation, first, on the distribution of the points which have no corresponding points after the transformation, and secondly, on the distribution of the points into which no points are carried over by the transformation.

2. Prove that any four complanar points no three of which are collinear may be carried over into any four points in the plane, no three of which are collinear, by one and only one collineation.

3. State and prove the corresponding theorem in n dimensions.

4. Prove that the transformation from a first system of homogeneous coördinates to a second is effected by a non-singular linear transformation. Consider the case of one, two, and three dimensions.

5. Prove that a projective transformation in space effects on every plane a two-dimensional, and on every line a one-dimensional, projective transformation, while at the same time the positions of the plane and line are changed.

[SUGGESTION. If p and p' are any two corresponding planes, assume in any way a pair of perpendicular axes in each of them, and denote by (x_1, y_1, t_1), and (x_1', y_1', t_1') respectively the systems of two-dimensional homogeneous coördinates based on these axes. Then show, by using the result of Exercise 4, that the transformation of one plane on the other will be expressed by writing x_1', y_1', t_1' as homogeneous linear polynomials in x_1, y_1, t_1.]

25. Further Development of the Algebra of Matrices. We proceed to establish certain further properties of matrices, leaving, however, much to the reader in the shape of exercises at the end of the section.

The theory of linear transformations suggests to us at once certain properties of matrices.   The first of these is :

THEOREM 1.   *The matrix*

$$\mathbf{I} = \begin{Vmatrix} 1 & 0 & \cdots\cdots 0 \\ 0 & 1 & \cdots\cdots 0 \\ \cdot & \cdot & \cdot \quad \cdot \\ \cdot & \cdot & \cdot \quad \cdot \\ 0 & 0 & \cdots\cdots 1 \end{Vmatrix},$$

has the property that if $\mathbf{a}$ *is any matrix whatever*

$$\mathbf{Ia} = \mathbf{aI} = \mathbf{a}.$$

For the linear transformation of which $\mathbf{a}$ is the matrix will evidently not be changed by being either followed or preceded by the identical transformation of which $\mathbf{I}$ is the matrix.

If we do not wish to use the idea of linear transformation, we may prove the theorem directly by actually forming the products $\mathbf{Ia}$ and $\mathbf{aI}$.

This theorem tells us that $\mathbf{I}$ plays in the algebra of matrices the same rôle that is played by 1 in ordinary algebra.   For this reason $\mathbf{I}$ is sometimes called the *unit matrix* or *idemfactor*.

Let us now consider any non-singular linear transformation and its inverse.   These two transformations performed in succession in either order obviously lead to the identical transformation.   This gives us the theorem :

THEOREM 2.   *If*
$$\mathbf{a} = \begin{Vmatrix} a_{11} & \cdots & a_{1n} \\ \cdot & \cdot & \cdot \\ \cdot & \cdot & \cdot \\ a_{n1} & \cdots & a_{nn} \end{Vmatrix}$$

is a non-singular matrix of determinant a, and if A_{ij} denote in the ordi-nary way the cofactors of the elements of a, the matrix

$$\left\| \begin{array}{ccc} \dfrac{A_{11}}{a} & \cdots\cdots & \dfrac{A_{n1}}{a} \\[2ex] \cdot & \cdot & \cdot \\[1ex] \cdot & \cdot & \cdot \\[2ex] \dfrac{A_{1n}}{a} & \cdots\cdots & \dfrac{A_{nn}}{a} \end{array} \right\|,$$

called the inverse of a, *and denoted by* a^{-1}, *is a non-singular matrix which has the property that*

$$aa^{-1} = a^{-1}a = I.$$

This suggests that we define positive and negative integral powers of matrices as follows :

DEFINITION 1. *If p is any positive integer and* a *any matrix we understand by* a^p *the product* aa $\cdots$ a *to p factors. If* a *is a non-singular matrix, we define its negative and zero powers by the for-mulæ*

$$a^{-p} = (a^{-1})^p, \quad a^0 = I.$$

From this definition we infer at once

THEOREM 3. *The laws of indices*

$$a^p a^q = a^{p+q}, \quad (a^p)^q = a^{pq}$$

hold for all matrices when the indices p and q are positive integers, and for all non-singular matrices when p and q are any integers.

We turn now to the question of the division of one matrix by another. We naturally define division as the inverse of multiplica-tion, and, since multiplication is not commutative, we thus get two distinct kinds of division ; a divided by b being on the one hand a matrix x such that

$$a = bx,$$

on the other hand a matrix y such that

$$a = yb.$$

On account of this ambiguity, the term *division* is not ordinarily used here. We have, however, as is easily seen, the following theorem :

THEOREM 4. *If* a *is any matrix and* b *any non-singular matrix there exists one, and only one, matrix* x *which satisfies the equation*

$$a = bx,$$

and one, and only one, matrix y *which satisfies the equation*

$$a = yb,$$

and these matrices are given respectively by the formulæ

$$x = b^{-1}a, \qquad y = ab^{-1}.$$

A special class of matrices is of some importance ; namely, those of the type

$$\left\|
\begin{matrix}
k & 0 & \cdots & 0 \\
0 & k & \cdots & 0 \\
\cdot & \cdot & \cdot & \cdot & \cdot \\
\cdot & \cdot & \cdot & \cdot & \cdot \\
0 & 0 & \cdots & k
\end{matrix}
\right\|.$$

Such matrices we will call *scalar matrices* for a reason which will presently appear.

If we denote by k the scalar matrix just written, and by a any matrix of the same order as k, we obtain readily the formula

$$(1) \qquad\qquad ka = ak = k\mathbf{a}.$$

If now, besides the scalar matrix k, we have a second scalar matrix 1 in which each element in the principal diagonal is l, we have the two formulæ

$$(2) \qquad\qquad k + 1 = 1 + k = (k + l)\,I,$$

$$(3) \qquad\qquad kl = lk = kl\,I.$$

Formula (1) shows that scalar matrices may be replaced by ordinary scalars when they are to be multiplied by other matrices ; while formulæ (2) and (3) show that scalar matrices combined with one another not only obey all the laws of ordinary scalars, but that each scalar matrix may in such cases be replaced by the scalar which occurs in each element of its principal diagonal provided that at the end of the work the resulting scalar be replaced by the corresponding scalar matrix.

For these reasons we may, in the algebra of matrices, replace all scalar matrices by the corresponding scalars, and then consider that all scalars which enter into our work stand for the corresponding scalar matrices. If we do this, the unit matrix $\mathbf{I}$ will be represented by the symbol $\mathbf{1}$.

DEFINITION 2. *By the adjoint* $\mathbf{A}$ *of a matrix* $\mathbf{a}$ *is understood another matrix of the same order in which the element in the ith row and jth column is the cofactor of the element in the jth row and ith column of* $\mathbf{a}$.*

It will be seen that when $\mathbf{a}$ is non-singular,

$$(4) \qquad\qquad \mathbf{A} = a\mathbf{a}^{-1},$$

but it should be noticed that while every matrix has an adjoint, only non-singular matrices have inverses.

Equation (4) may be written in the form

$$(5) \qquad\qquad \mathbf{A}\mathbf{a} = \mathbf{a}\mathbf{A} = a\,\mathbf{I},$$

a form in which it is true not merely when $\mathbf{a}$ is non-singular, but also, as is seen by direct multiplication, when the determinant of $\mathbf{a}$ is zero.

Finally we come to a few important theorems concerning the rank of the matrix obtained by multiplying together two given matrices. In the first place, we notice that the rank of the product is not always completely determined by the ranks of the factors. This may be shown by numerous examples, for instance, in formula (5), § 22, the ranks of the factors are in general two and one, and the rank of the product is zero, while in the formula

$$
\begin{Vmatrix} a_{11} & a_{12} & 0 \\ a_{21} & a_{22} & 0 \\ a_{31} & a_{32} & 0 \end{Vmatrix} \cdot
\begin{Vmatrix} 0 & 0 & 0 \\ 0 & 0 & 1 \\ 0 & 0 & 0 \end{Vmatrix} =
\begin{Vmatrix} 0 & 0 & a_{12} \\ 0 & 0 & a_{22} \\ 0 & 0 & a_{32} \end{Vmatrix}
$$

the ranks of the factors are in general the same, namely two and one, while the rank of the product is one.

But though, as this example shows, the ranks of the factors (even together with the order of the matrices) do not suffice to determine the rank of the product, there are, nevertheless, important inequalities between these ranks, one of which we now proceed to deduce.

---

* Notice the interchange of rows and columns here, which in the case of adjoint determinants, being immaterial and sometimes inconvenient, was not made.

For this purpose consider the two matrices

$$\mathbf{a} = \begin{Vmatrix} a_{11} & \cdots & a_{1n} \\ \cdot & \cdots & \cdot \\ \cdot & \cdots & \cdot \\ a_{n1} & \cdots & a_{nn} \end{Vmatrix}, \qquad \mathbf{b} = \begin{Vmatrix} b_{11} & \cdots & b_{1n} \\ \cdot & \cdots & \cdot \\ \cdot & \cdots & \cdot \\ b_{n1} & \cdots & b_{nn} \end{Vmatrix},$$

and their product **ab**.

THEOREM 5. *Any k-rowed determinant of the matrix* **ab** *is equal to an aggregate of k-rowed determinants of* **b** *each multiplied into a polynomial in the a's, and also to an aggregate of k-rowed determinants of* **a** *each multiplied by a polynomial in the b's.*

For any k-rowed determinant of **ab** may be broken up into a sum of determinants of the kth order in such a way that each column of each determinant has one of the b's as a common factor.* After taking out these common factors from each determinant, we have left a determinant in the x's which, if it does not vanish identically, is a k-rowed determinant of **a**. Or, on the other hand, we may break up the k-rowed determinant of **ab** into a sum of determinants of the kth order in such a way that each row of each determinant has one of the a's as a common factor. After taking out these common factors from each determinant, we have left a determinant in the b's which, if it does not vanish identically, is a k-rowed determinant of **b**.

From the theorem just proved it is clear that if all the k-rowed determinants of **a** or of **b** are zero, the same will be true of all the k-rowed determinants of **ab**. Hence

THEOREM 6. *The rank of the product of two matrices cannot exceed the rank of either factor.†*

---

* The truth of this statement and the following will be evident if the reader actually writes out the matrix **ab**.

† Thus if r_1 and r_2 are the ranks of the two factors and R is the rank of the product, we have $R \leq r_1$, $R \leq r_2$. This is one half of Sylvester's " Law of Nullity," of which the other half may be stated in the form $R \geq r_1 + r_2 - n$, where n is the order of the matrices ; cf. Exercise 8 at the end of this section. Sylvester defines the nullity of a matrix as the difference between its order and its rank, so that his statement of the law of nullity is : The nullity of the product of two matrices is at least as great as the nullity of either factor, and at most as great as the sum of the nullities of the factors.

There is one important case in which this theorem enables us to determine completely the rank of the product, namely, the case in which one of the two matrices **a** or **b** is non-singular. Suppose first that **a** is non-singular, and denote the ranks of **b** and **ab** by r and R respectively. By Theorem 6, $R \leqq r$. We may, however, also regard **b** as the product of $\mathbf{a}^{-1}$ and **ab**, and hence, applying Theorem 6 again, we have $r \leqq R$. Combining these two results, we see that $r = R$.

On the other hand, if **b** is non-singular, and we denote the ranks of **a** and **ab** respectively by r and R, we get from Theorem 6, $R \leqq r$; and, applying this theorem again to the equation

$$(\mathbf{ab})\, \mathbf{b}^{-1} = \mathbf{a},$$

we have $r \leqq R$. Thus again we get $r = R$.

We have thus established the result:

THEOREM 7. *If a matrix of rank r is multiplied in either order by a non-singular matrix, the rank of the product is also r.*

EXERCISES

1. Prove that a necessary and sufficient condition that two matrices **a** and **b** of the same order be equivalent is that there exist two non-singular matrices **c** and **d** such that
$$\mathbf{dac} = \mathbf{b}.$$
Cf. § 22, Exercise 2, and § 19, Exercise 4.

2. Prove that a necessary and sufficient condition that two matrices **a** and **b** of the same order be equivalent is that there exist four matrices **c**, **d**, **e**, **f** such that
$$\mathbf{dac} = \mathbf{b}, \qquad \mathbf{a} = \mathbf{fbe}.$$

3. Prove that every matrix of rank r can be written as the sum of r matrices of rank one.*

[SUGGESTION. Notice that the special matrix mentioned in § 19, Exercise 3, can be so written.]

---

* A matrix of rank one has been called by Gibbs a *dyad*, since it may (cf. § 19, Ex. 5) be regarded as a product of two complex quantities $(a_1, a_2, \cdots a_n)$ and $(b_1, b_2, \cdots b_n)$. The sum of any number of dyads is called a *dyadic polynomial*, or simply a *dyadic*. Every matrix is therefore a dyadic, and *vice versa*. Gibbs's theory of dyadics, in the case $n = 3$, is explained in the *Vector Analysis* of Gibbs-Wilson, Chap. V. Geometric language is used here exclusively, the complex quantities (a_1, a_2, a_3) and (b_1, b_2, b_3) from which the dyads are built up being interpreted as vectors in space of three dimensions. This theory is equivalent to Hamilton's theory of the Linear Vector Function in Quaternions.

4. Prove that a necessary and sufficient condition that a matrix be a divisor of zero (cf. § 22, Exercise 1) is that it be singular.

[SUGGESTION. Consider equivalent matrices.]

5. Prove that the inverse of the product of any number of non-singular matrices is the product of the inverses of these matrices taken in the reverse order.

Hence deduce a similar theorem concerning the adjoint of a product of any number of matrices, whether these matrices are singular or not.

What theorem concerning determinants can be inferred?

6. Prove that the conjugate of the inverse of a non-singular matrix is the inverse of the conjugate; and that the conjugate of the adjoint of any matrix is the adjoint of the conjugate.

7. Prove that if a matrix has the property that its product with every matrix of the same order is commutative, it is necessarily a scalar matrix.

8. If r_1 and r_2 are the ranks of two matrices of order n, and R the rank of their product, prove that

$$R \geqq r_1 + r_2 - n.*$$

[SUGGESTION. Prove this theorem first on the supposition that one of the two matrices which are multiplied together is of the form mentioned in Exercise 3, § 19, using also at this point Exercise 1, § 8. Then reduce the general case to this one by means of Exercise 1 of this section.]

26. Sets, Systems, and Groups. These three words are the technical names for conceptions which are to be met with in all branches of mathematics. In fact the first two are of such generality that they may be said to form the logical foundation on which all mathematics rests.† In this section we propose, after having given a brief explanation of these three conceptions, to show how they apply to the special subjects considered in this chapter.

The objects considered in mathematics — we use the word *object* in the broadest possible sense — are of the most varied kinds. We have, on the one hand, to mention a few of the more important ones, the different kinds of *quantities* ranging all the way from the positive integers to complex quantities and matrices. Next we have in geometry not only points, lines, curves, and surfaces but also such

---

* Cf. the footnote to Theorem 6.

† For a popular exposition of the point of view here alluded to, see my address on *The Fundamental Conceptions and Methods of Mathematics*, St. Louis Congress of Arts and Science, 1904. Reprinted in Bull. Amer. Math. Soc., December, 1904.

things as displacements (rotations, translations, etc.), collineations, and, in fact, geometrical transformations in general. Then in various parts of mathematics we have to deal with the Theory of Substitutions, that is, with the various changes which can be made in the order of certain objects, and these substitutions themselves may be regarded as objects of mathematical study. Finally, in mechanics we have to deal with such objects as forces, couples, velocities, etc.

These objects, and all others which are capable of mathematical consideration, are constantly presenting themselves to us, not singly, but in *sets*. Such sets (or, as they are sometimes called, *classes*) of objects may consist of a finite or an infinite number of objects, or *elements*. We mention as examples :

(1) All prime numbers.
(2) All lines which meet two given lines in space.
(3) All planes of symmetry of a given cube.
(4) All substitutions which can be performed on five letters.
(5) All rotations of a plane about a given line perpendicular to it.

Having thus gained a slight idea of the generality of the conception of a set, we next notice that in many cases in which we have to deal with a set in mathematics, there are one or more rules by which pairs of elements of the set may be combined so as to give objects, either belonging to the set or not as the case may be. As examples of such rules of combination, we mention addition and multiplication both in ordinary algebra and in the algebra of matrices ; the process by which two points, in geometry, determine a line ; the process of combining two displacements to give another displacement, etc.

Such a set, with its associated rules of combination, we will call a *mathematical system*, or simply a system.*

We come now to a very important kind of system known as a *group*, which we define as follows :

* This definition is sufficiently general for our immediate purposes. In general. however, it is desirable to admit, not merely rules of combination, but also *relations* between the elements of a system. In fact we may have merely one or more relations and no rules of combination at all. From this point of view the positive integers with the relation of greater and less would form a system, even though we do not introduce any rule of combination such as addition or multiplication. It may be added that rules of combination may be regarded as merely relations between three objects ; cf. the address referred to above.

DEFINITION. *A system consisting of a set of elements and one rule of combination, which we will denote by* ∘, *is called a group if the following conditions are satisfied:*

(1) *If a and b are any elements of the set, whether distinct or not, $a \circ b$ is also an element of the set.**

(2) *The associative law holds; that is, if a, b, c are any elements of the set,*
$$(a \circ b) \circ c = a \circ (b \circ c).$$

(3) *The set contains an element, i, called the identical element, which is such that every element is unchanged when combined with it,*
$$i \circ a = a \circ i = a.$$

(4) *If a is any element, the set also contains an element a′, called the inverse of a, such that*
$$a' \circ a = a \circ a' = i.$$

Thus, for example, the positive and negative integers with zero form a ·group if the rule of combination is addition. In this case zero is the identical element, and the inverse of any element is its negative. These same elements, however, do not form a group if the rule of combination is multiplication, for while conditions (1), (2), and (3) are fulfilled (the identical element being 1 in this case), condition (4) is not, since zero has no reciprocal.

Again, the set of all real numbers forms a group if the rule of combination is addition, but not if it is multiplication, since in this case zero has no inverse. If we exclude zero from the set, we have a group if the rule of combination is multiplication, but not if it is addition.

As an example of a group with a finite number of elements we mention the four numbers
$$+1, \; -1, \; +\sqrt{-1}, \; -\sqrt{-1}$$

with multiplication as the law of combination.

In order to get an example of a group of geometrical operations, let us consider the translations of a plane, regarded as a rigid lamina, in the directions of its own lines. Every such translation may be represented both in magnitude and in direction by the length and

---

* A system satisfying condition (1) is sometimes said to have "the group property." In the older works on the subject this condition was the only one to be explicitly mentioned, the others, however, being tacitly assumed.

direction of an arrow lying in the plane in question. Two such translations performed in succession are obviously equivalent to a translation of the same sort represented by the arrow obtained by combining the two given arrows according to the law of the parallelogram of forces. The set of all translations with the law of combination just explained is readily seen to form a group if we include in it the *null translation, i.e.* the transformation which leaves every point in the plane fixed. This null translation is then the identical element, and two translations are the inverse of each other if they are equal in magnitude and opposite in direction.

All the groups we have so far mentioned satisfy, not only the four conditions stated in the definition, but also a fifth condition, viz. that the law of combination is commutative. Such groups are called *commutative* or *Abelian groups*. In general, however, groups do not have this property. As examples of non-Abelian groups, we may mention first the group of all non-singular matrices of a given order, the rule of combination being multiplication ; and secondly the group of all matrices of a given order whose determinants have the value ± 1, the rule of combination being again multiplication. This second group is called a *subgroup* of the first, since all its elements are also elements of the first group, and the rule of combination is the same in both cases. A subgroup of the group last mentioned is the group of all matrices of a given order whose determinants have the value $+1$,* the rule of combination being multiplication.

We add that non-Abelian groups may readily be built up whose elements are linear transformations, or collineations. On the other hand, Abelian groups may be formed from matrices if we take as our rule of combination addition instead of multiplication.

27. Isomorphism.

DEFINITION. *Two groups are said to be isomorphic†* *if it is possible to establish a one-to-one correspondence between their elements of such a*

---

* These are called *unimodular matrices ;* or, more accurately, *properly unimodular matrices* to distinguish them from the *improperly unimodular matrices* whose determinants have the value -1. It should be noticed that these last matrices taken by themselves do not constitute a group, since they do not even have the group property.

† *Simply isomorphic* would be the more complete term. We shall, however, not be concerned with isomorphism which is not simple.

*sort that if a, b are any elements of the first group and a', b' the corresponding elements of the second, then a' ∘ b' corresponds to a ∘ b.**

We proceed to illustrate this definition by some examples, leaving to the reader the proofs of the statements we make. In each case we omit the statement of the rule of combination in the case of transformations, where no misunderstanding is possible.

FIRST EXAMPLE. (*a*) The group of the four elements

$$1, \sqrt{-1}, -1, -\sqrt{-1},$$

the rule of combination being multiplication.

(*b*) The group of four rotations about a given line through angles of 0°, 90°, 180°, 270°.

These two groups may be proved to be isomorphic by pairing the elements against one another in the order in which they have just been written.

SECOND EXAMPLE. (*a*) The group of the four matrices

$$\begin{pmatrix} 1 & 0 \\ 0 & 1 \end{pmatrix}, \begin{pmatrix} -1 & 0 \\ 0 & -1 \end{pmatrix}, \begin{pmatrix} 1 & 0 \\ 0 & -1 \end{pmatrix}, \begin{pmatrix} -1 & 0 \\ 0 & 1 \end{pmatrix},$$

the rule of combination being multiplication.

(*b*) The group of the following four transformations : the identical transformation ; reflection in a plane ; reflection in a second plane at right angles to the first ; rotation through 180° about the line of intersection of these two planes.

(*c*) The group consisting of the identical transformation and of three rotations through angles of 180° about three straight lines through a point at right angles to one another.

The two groups of Example 1 are not isomorphic with the three of Example 2 in spite of the fact that there are the same number of elements in all the groups. This follows from the presence of two elements in the groups of Example 1 whose squares are not the identical element.

---

* This idea of isomorphism may obviously be extended to the case of any two systems provided merely that there are the same number of rules of combination in both cases. Thus the system of all scalar matrices on the one hand and of all scalars on the other, the rules of combination being in both cases addition and multiplication, are obviously isomorphic. It is for this reason that no confusion arises if no distinction is made between scalar matrices and scalars.

THIRD EXAMPLE. (*a*) The group of all real quantities; the rule of combination being addition.

(*b*) The group of all scalar matrices of order k with real elements; the rule of combination being addition.

(*c*) The group of all translations of space parallel to a given line.

FOURTH EXAMPLE. (*a*) The group of all non-singular matrices of order n, with multiplication as the rule of combination.

(*b*) The group of all non-singular homogeneous linear transformations in n variables.

We might be tempted to mention as a group of geometrical transformations isomorphic with the last two groups, the group of all non-singular collineations in space of $n-1$ dimensions. This, however, would be incorrect, for the correspondence we have established between collineations and linear transformations is not one-to-one; to every linear transformation corresponds one collineation, but to every collineation correspond an infinite number of linear transformations, whose coefficients are proportional to one another.* In order to get a group of geometrical transformations isomorphic with the group of non-singular matrices of the nth order it is sufficient to interpret the variables $x_1, \cdots x_n$ as non-homogeneous coördinates in space of n dimensions, and to consider the geometric transformation effected by non-singular homogeneous linear transformations of these x's. These transformations are those affine transformations of space of n dimensions which leave the origin unchanged; cf. the footnote on p. 70. Thus the group of all non-singular matrices of the nth order is isomorphic with a certain subgroup of the group of collineations in space of n dimensions, not with the group of all non-singular collineations in space of $n-1$ dimensions.

An essential difference between these two groups is that one

* This does not really prove that the groups are not isomorphic, since it is conceivable that some other correspondence might be established between their elements which would be one-to-one and of such a sort as to prove isomorphism. Even the fact, to be pointed out presently, that the groups depend on a different number of parameters does not settle the question. A reference to the result stated in Exercise 7, § 25, shows that the groups are not isomorphic; for, according to it, the only non-singular collineation which is commutative with all collineations is the identical transformation, whereas all linear transformations with scalar matrices have this property.

depends on n^2 parameters (the n^2 coefficients of the linear transformation) while the other depends only on $n^2 - 1$ parameters (the ratios of the coefficients of the collineation).

We can, however, by looking at the subject a little differently, obtain a group of matrices isomorphic with the group of all non-singular collineations in space of $n - 1$ dimensions. For this purpose we need merely to regard two matrices as equal whenever the elements of one can be obtained from those of the other by multiplying all the elements by the same quantity not zero. When we take this point of view with regard to matrices, it is desirable to indicate it by a new terminology and notation. According to a suggestion of E. H. Moore of Chicago, we will call such matrices *fractional* matrices, and write them

$$\left\| \frac{a_{11} \quad a_{12}}{a_{21} \quad a_{22}} \right\| , \qquad \left\| \frac{a_{11} \quad a_{12} \quad a_{13}}{\dfrac{a_{21} \quad a_{22} \quad a_{23}}{a_{31} \quad a_{32} \quad a_{33}}} \right\| , \text{ etc.}$$

Agreeing that fractional matrices are to be added and multiplied according to the same rules as ordinary matrices, we may now say that the group of all non-singular collineations in space of $n - 1$ dimensions is isomorphic with the group of all fractional matrices of the nth order whose determinants are not zero.[*]

To take another example, the groups in the second example above are isomorphic with the group whose elements are the four fractional matrices

$$\left\| \frac{1 \quad 0}{0 \quad 1} \right\| , \quad \left\| \frac{1 \quad 0}{0 \quad -1} \right\| , \quad \left\| \frac{0 \quad 1}{1 \quad 0} \right\| , \quad \left\| \frac{0 \quad 1}{-1 \quad 0} \right\| ,$$

and where the law of combination is multiplication. These four matrices, if regarded as ordinary matrices, would not even satisfy the first condition for a group.

The reader wishing to get a further insight into the theory of groups of linear transformations will find the following three treat-

---

[*] It should be noticed that we cannot speak of the *value* of the determinant of a fractional matrix unless this value is zero, for if we multiply all the elements of the matrix by c we do not change the matrix, but do multiply the determinant by c^n. There is in particular no such thing as a unimodular fractional matrix. We may, however, speak of the rank of a fractional matrix.

ments interesting and instructive. They duplicate each other to only a very slight extent.

Weber, *Algebra*, Vol. II.

Klein, *Vorlesungen über das Ikosaeder.*

Lie-Scheffers, *Vorlesungen über continuirliche Gruppen.*

EXERCISES

1. DEFINITION. *A group is said to be of order n if it contains n, and only n, elements.*

If a group of order n has a subgroup, prove that the order of this subgroup is a factor of n.

[SUGGESTION. Denote the elements of the subgroup by $a_1 \cdots a_k$, and let b be any other element of the group. Show that $ba_1, ba_2, \cdots ba_k$ are all elements of the group distinct from each other and distinct from the a's. If there are still other elements, let c be one and consider the elements $ca_1, \cdots ca_k$, etc.]

2. Prove that if a is any element of a group of finite order, it is possible by multiplying a by itself a sufficient number of times to get the identical element.

DEFINITION. *The lowest power to which a can be raised so as to give the identical element is called the period of a.*

3. Prove that every element of a group of order n has as its period a factor of n (1 and n included).

4. DEFINITION. *A group is called cyclic if all its elements are powers of a single element.*

Prove that all cyclic groups of order n are isomorphic with the group of rotations about an axis through angles $0, \omega, 2\,\omega, \cdots (n-1)\,\omega$, where $\omega = 2\,\pi/n$, and that conversely every such group of rotations is a cyclic group.

5. Prove that every group whose order is a prime number is a cyclic group.

6. Prove that all groups of order 4 are either cyclic or isomorphic with the groups of the second example above. A group of this last kind is called a *fours group* (*Vierergruppe*).

7. Obtain groups with regard to one or the other of which all groups of order 6 are isomorphic.

8. Obtain groups with regard to one or the other of which all groups of order 8 are isomorphic.

CHAPTER VII

INVARIANTS. FIRST PRINCIPLES AND ILLUSTRATIONS

28. Absolute Invariants, Geometric, Algebraic, and Arithmetical.
If we subject a geometric figure to a transformation, we find that,
while many properties of the figure have been altered, others have
not. If we consider, not a single transformation, but a set of transformations, then those properties of figures which are not changed by
any of the transformations of the set are said to be invariant properties with regard to this set of transformations. Thus if our set of
transformations is the group of all displacements, the property of
two lines being parallel or perpendicular to each other and the
property of a curve being a circle are invariant properties, since
after the transformation the lines will still be parallel or perpendicular and the curve will still be a circle. If, however, we consider,
not the group of displacements, but the group of all non-singular
collineations, none of the properties just mentioned will be invariant
properties. Properties invariant with regard to all non-singular
collineations have played such an important part in the development
of geometry that a special name has been given to them, and they
are called *projective* or *descriptive* properties. As examples of such
projective properties we mention the collinearity and complanarity
of points, the complanarity and concurrence of lines, etc.; or, on
the other hand, the contact of a line with a curve or a surface or
the contact of two curves or of two surfaces, or of a curve and a
surface.

DEFINITION 1. *If there is associated with a geometric figure a
quantity which is unchanged by all the transformations of a certain
set, then this quantity is called an invariant with regard to the transformations of the set.*

For instance, if our set of transformations is the group of displacements, the distance between two points and the angle between
two lines would be two examples of invariants.

The geometric invariants so far considered lead up naturally to the subject of algebraic invariants. Thus let us consider the two polynomials

(1)
$$\begin{cases} A_1 x + B_1 y + C_1, \\ A_2 x + B_2 y + C_2, \end{cases}$$

and subject the variables (x, y) to the transformations of the set

(2)
$$\begin{cases} x' = x\cos\theta + y\sin\theta + \alpha, \\ y' = -x\sin\theta + y\cos\theta + \beta, \end{cases}$$

where α, β, θ are parameters which may have any values. The transformation (2) carries over the polynomials (1) into two new polynomials:

(3)
$$\begin{cases} A_1' x' + B_1' y' + C_1', \\ A_2' x' + B_2' y' + C_2'. \end{cases}$$

The coefficients of (3) may be readily expressed in terms of the coefficients of (1) and the parameters α, β, θ. Using these expressions, we easily obtain the formulæ

(4)
$$\begin{cases} A_1' B_2' - A_2' B_1' = A_1 B_2 - A_2 B_1, \\ A_1' A_2' + B_1' B_2' = A_1 A_2 + B_1 B_2. \end{cases}$$

We shall therefore speak of the two expressions

(5)
$$A_1 B_2 - A_2 B_1, \quad A_1 A_2 + B_1 B_2$$

as invariants of the system of polynomials (1) with regard to the set of transformations (2) according to the following general definition:

DEFINITION 2. *If we have a system of polynomials in the variables $(x, y, z, \ldots)$ and a set of transformations of these variables, then any function of the coefficients of the polynomials is called an invariant (or more accurately an absolute invariant) with regard to these transformations if it is unchanged when the polynomials are subjected to all the transformations of the set.*

The relation of the example considered above to the subject of geometric invariants becomes obvious when we notice that the algebraic transformations (2) may be regarded as expressing the displacements of plane figures in their plane when (x, y) are rectangular coördinates of points in the plane. If now we consider, not the polynomials (1), but the two lines determined by setting them equal to zero, we have to deal with the displacements of these two lines. The invariants (5) have themselves no geometric significance, but by equating them to zero, we get the necessary and sufficient conditions that the two lines be respectively parallel and perpendicular, and these, as we noticed above, are invariant properties with regard to displacements. Finally we may notice that the ratio of the two invariants (5) gives the tangent of the angle between the lines,—a geometric invariant.

As a second example, let us consider, not two lines, but a line and a point. Algebraically this means that we start with the system

$$(6) \qquad \begin{cases} Ax + By + C, \\ \quad (x_1, y_1), \end{cases}$$

consisting of a polynomial and a pair of variables. We shall wish to demand here that whenever the variables (x, y) are subjected to a transformation, the variables (x_1, y_1) be subjected to the same transformation, or as we say according to Definition 3 below, that (x, y) and (x_1, y_1) be cogredient variables. If we subject the system (6) to any transformation of the set (2), we get a new system

$$(7) \qquad \begin{cases} A'x' + B'y' + C', \\ \quad (x_1', y_1'), \end{cases}$$

and it is readily seen that

$$A'x_1' + B'y_1' + C' = Ax_1 + By_1 + C.$$

Accordingly we shall call $Ax_1 + By_1 + C$ a covariant of the system (6) according to Definition 4 below. This covariant has also no direct geometric meaning, but its vanishing gives the necessary and sufficient condition for an invariant property, namely, that the point (x_1, y_1) lie on the line $Ax + By + C = 0$.

In the light of this example we may lay down the following general definitions:

DEFINITION 3. *If we have several sets of variables*

$$(x, y, z, \cdots), \quad (x_1, y_1, z_1, \cdots), \quad (x_2, y_2, z_2, \cdots), \quad \cdots$$

and agree that whenever one of these sets is subjected to a transformation every other set shall be subjected to the same transformation, then we say that we have sets of cogredient variables.

DEFINITION 4. *If we have a system consisting of a number of polynomials in $(x, y, z, \cdots)$ and of a number of sets of variables cogredient to $(x, y, z, \cdots)$, then any function of the coefficients of the polynomials and of the cogredient variables which is unchanged when the variables $(x, y, z, \cdots)$ are subjected to all the transformations of a certain set is called a covariant (or more accurately an absolute covariant) of this system with regard to the transformations of this set.*

It will be seen that invariants may be regarded as special cases of covariants.

Among the geometric invariants there are some which from their nature are necessarily integers, and which we will speak of as *arithmetical invariants.* An example would be the number of vertices of a polygon if our set of transformations was either the group of displacements or the group of non-singular collineations. Another example is the largest number of real points in which an algebraic curve can be cut by a line, if our set of transformations is the group of *real* non-singular collineations.

These arithmetical invariants also play, as we shall see, an important part in algebra. We mention here as an example the degree of an *n*-ary form, which is an invariant with regard to all non-singular linear transformations.*

EXERCISES

1. Prove that $(x_2 - x_1)^2 + (y_2 - y_1)^2$, and $\begin{vmatrix} x_1 & y_1 & 1 \\ x_2 & y_2 & 1 \\ x_3 & y_3 & 1 \end{vmatrix}$

are covariants of the system
$$(x_1, y_1), (x_2, y_2), (x_3, y_3)$$
with regard to the transformations (2).

2. Prove that $A + B$ and $B^2 - AC$
are invariants of the polynomial
$$Ax^2 + 2Bxy + Cy^2 + 2Dx + 2Ey + F$$
with regard to the transformations (2).

What geometric meaning can be attached to these invariants?

3. Prove that $A^2 + B^2$ is an invariant of the polynomial
$$Ax + By + C$$
with regard to the transformations (2).

Hence show that
$$\frac{Ax_1 + By_1 + C}{\sqrt{A^2 + B^2}}$$
is a covariant of the system (6). Note its geometric meaning.

---

*It is, in fact, an invariant with regard to all linear transformations except the one in which all the coefficients of the transformation are zero.

29. Equivalence.

DEFINITION 1. *If A and B denote two geometric configurations or two algebraic expressions, or sets of expressions, then A and B shall be said to be equivalent with regard to a certain set of transformations when, and only when, there exists a transformation of the set which carries over A into B and also a transformation of the set which carries over B into A.*

To illustrate this definition we notice that the conception of equivalence of geometric figures with regard to displacements is identical with the Euclidean conception of the equality or congruence of figures.

Again, we see from Theorem 2, § 24, that on a straight line two sets of three points each are always equivalent with regard to nonsingular projective transformations.

In both of the illustrations just mentioned the set of transformations forms a group. In such cases the condition for equivalence can be decidedly simplified, for the transformation which carries A into B has an inverse belonging to the set, and this inverse necessarily carries B into A. Thus we have the

THEOREM. *A necessary and sufficient condition for the equivalence of A and B with regard to a group of transformations is that a transformation of the group carry over A into B.*

This theorem will be of great importance, as the question of equivalence will present itself to us only when the set of transformations we are considering forms a group.

Let us consider, for the sake of greater definiteness, a group of geometric transformations. If two geometric configurations are equivalent with regard to this group, every invariant of the first configuration must be equal to the corresponding invariant of the second. Thus, for instance, if two triangles are equivalent with regard to the group of displacements, all the sides and angles of the first will be equal to the corresponding sides and angles of the second. The same will be true of the altitudes, lengths of the medial lines, radius of the inscribed circle, etc., all of these being invariants. Now one of the first problems in geometry is to pick out from among these invariants of the triangle as small a number as possible whose equality for two triangles insures the equivalence of the triangles. This can be done, for instance, by taking two sides and the included angle, or two angles and the included side, or three sides. Any one of these three elements may be called *a complete system of invariants* for a triangle with regard to the group of displacements, since two triangles having these invariants in common are equivalent and therefore have all other invariants in common. The conception we have here illustrated may be defined in general terms as follows :

DEFINITION 2. *A set of invariants of a geometric configuration or an algebraic expression are said to form a complete system of invariants if two configurations or expressions having these invariants in common are necessarily equivalent.**

It will be seen from this definition that all the invariants of a geometric configuration or of an algebraic expression are uniquely determined by any complete system of invariants.

Finally we will glance at an application to matrices of the ideas of invariants and equivalence. Let us consider matrices of the nth order,† and consider transformations of the following form which transform the matrix $\mathbf{A}$ into the matrix $\mathbf{B}$:

$$(1) \qquad\qquad \mathbf{aAb} = \mathbf{B},$$

where $\mathbf{a}$ and $\mathbf{b}$ are any non-singular matrices of the nth order. This transformation may be denoted by the symbol $(\mathbf{a}, \mathbf{b})$, and these symbols must obviously be combined by the formula

$$(\mathbf{a}_2,\, \mathbf{b}_2)(\mathbf{a}_1,\, \mathbf{b}_1) = (\mathbf{a}_2\mathbf{a}_1,\, \mathbf{b}_1\mathbf{b}_2).$$

By means of this formula it may readily be shown that these transformations form a group.

According to our general definition of equivalence, two matrices $\mathbf{A}$ and $\mathbf{B}$ must therefore be said to be equivalent when, and only when, two non-singular matrices $\mathbf{a}$ and $\mathbf{b}$ exist which satisfy (1). That this definition of equivalence amounts to the same thing as our earlier definition is seen by a reference to Exercise 1, § 25.

---

* In the classical theory of algebraic invariants this term is used in a different **and** much more restricted sense. There we have to deal with integral rational relative invariants (cf. § 31). By a complete system of such invariants of a system of algebraic forms is there understood a set of such invariants in terms of which every invariant of this sort of the system of forms can be expressed integrally and rationally. Cf. for instance Clebsch, *Binäre Formen*, p. 109.

† We may, if we choose, confine our attention throughout to matrices with real elements.

30. The Rank of a System of Points or a System of Linear Forms as an Invariant. Let (x_1, y_1, z_1, t_1), (x_2, y_2, z_2, t_2), (x_3, y_3, z_3, t_3) be any three distinct collinear points, so that the rank of the matrix

$$\left\| \begin{array}{cccc} x_1 & y_1 & z_1 & t_1 \\ x_2 & y_2 & z_2 & t_2 \\ x_3 & y_3 & z_3 & t_3 \end{array} \right\|$$

is two. Now subject space to a non-singular collineation and we get three new points which will also be distinct and collinear, and hence the rank of their matrix will also be two. Thus we see that in this special case the rank of the system of points is unchanged by a non-singular collineation.

Again, let

$$a_1 x + b_1 y + c_1 z + d_1 t = 0,$$
$$a_2 x + b_2 y + c_2 z + d_2 t = 0,$$
$$a_3 x + b_3 y + c_3 z + d_3 t = 0,$$
$$a_4 x + b_4 y + c_4 z + d_4 t = 0$$

be any four planes which have one, and only one, point in common, so that the rank of their matrix is three. After a non-singular collineation we have four new planes which will also have one, and only one, point in common, and hence the rank of the matrix of their coefficients will be three. The rank of this system of planes is therefore unchanged by such a transformation.

We proceed now to generalize these facts.

THEOREM 1. *The rank of the matrix of m points*

$$(x_1^{[i]}, x_2^{[i]}, \cdots x_n^{[i]}), \qquad (i = 1, 2, \cdots m)$$

is an invariant with regard to non-singular linear transformations.

Let

$$(1) \qquad \left\{ \begin{array}{l} X_1 = c_{11} x_1 + \cdots + c_{1n} x_n, \\ \cdot \quad \cdot \quad \cdot \quad \cdot \quad \cdot \quad \cdot \quad \cdot \quad \cdot \\ \cdot \quad \cdot \quad \cdot \quad \cdot \quad \cdot \quad \cdot \quad \cdot \quad \cdot \\ X_n = c_{n1} x_1 + \cdots + c_{nn} x_n \end{array} \right.$$

be a non-singular linear transformation which carries the points $(x_1^{[i]}, \cdots x_n^{[i]})$ over into the points $(X_1^{[i]}, \cdots X_n^{[i]})$. Now suppose any k of the points $(x_1^{[i]}, \cdots x_n^{[i]})$, which for convenience we will take as the first k, are linearly dependent. Then there exist k constants $(c_1, \cdots c_k)$ not all zero, such that

$$(2) \qquad c_1 x_j' + c_2 x_j'' + \cdots + c_k x_j^{[k]} = 0, \qquad (j = 1, 2, \cdots n).$$

By means of the transformation (1) we have

$$X_j^{[i]} = c_{j1}x_1^{[i]} + \cdots + c_{jn}x_n^{[i]},$$

hence $\quad c_1 X_j' + c_2 X_j'' + \cdots + c_k X_j^{[k]} = c_{j1}(c_1 x_1' + c_2 x_1'' + \cdots + c_k x_1^{[k]}) +$

$$\cdots + c_{jn}(c_1 x_n' + c_2 x_n'' + \cdots + c_k x_n^{[k]}) \qquad (j = 1, 2, \cdots n).$$

Since this vanishes on account of (2), the first k of the points $(X_1^{[i]}, \cdots X_n^{[i]})$ are linearly dependent. Since (1) is a non-singular transformation, it is immaterial which set of points we consider as the initial set. Thus we have shown that if any k points of either set are linearly dependent, the corresponding k points of the other set will be, also.

Now if the rank of the matrix of the x's is r, at least one set of r of the x-points is linearly independent, but every set of $(r + 1)$ of them is linearly dependent. Consequently the same is true for the X-points, and therefore their matrix must also be of rank r.

THEOREM 2. *The rank of the matrix of m linear forms*

$$f_i(x_1, \cdots x_n) \equiv a_{i1}x_1 + a_{i2}x_2 + \cdots + a_{in}x_n \qquad (i = 1, 2, \cdots m)$$

is an invariant with regard to non-singular linear transformations.

The proof of this theorem, which is very similar to the proof of Theorem 1, we leave to the reader.

It will be noticed that the invariants we have considered in this section are examples of what we have called arithmetical invariants.

31. Relative Invariants and Covariants. We will begin by considering a system of n linear forms in n variables

$$(1) \quad \begin{cases} a_{11}x_1 + a_{12}x_2 + \cdots + a_{1n}x_n, \\ a_{21}x_1 + a_{22}x_2 + \cdots + a_{2n}x_n, \\ \cdot \quad \cdot \quad \cdot \quad \cdot \quad \cdot \quad \cdot \quad \cdot \quad \cdot \\ \cdot \quad \cdot \quad \cdot \quad \cdot \quad \cdot \quad \cdot \quad \cdot \quad \cdot \\ a_{n1}x_1 + a_{n2}x_2 + \cdots + a_{nn}x_n. \end{cases}$$

DEFINITION 1. *The determinant*

$$\begin{vmatrix} a_{11} & \cdots & a_{1n} \\ \cdot & \cdot & \cdot \\ \cdot & \cdot & \cdot \\ a_{n1} & \cdots & a_{nn} \end{vmatrix}$$

is called the resultant of the system (1).

Let us now subject the system (1) to the linear transformation

$$(2) \qquad \begin{cases} x_1 = c_{11} x_1' + \cdots + c_{1n} x_n', \\ \cdot \quad \cdot \quad \cdot \quad \cdot \quad \cdot \quad \cdot \quad \cdot \quad \cdot \\ \cdot \quad \cdot \quad \cdot \quad \cdot \quad \cdot \quad \cdot \quad \cdot \quad \cdot \\ x_n = c_{n1} x_1' + \cdots + c_{nn} x_n'. \end{cases}$$

This gives the new system of forms

$$(3) \qquad \begin{cases} a_{11}' x_1' + \cdots + a_{1n}' x_n', \\ \cdot \quad \cdot \quad \cdot \quad \cdot \quad \cdot \quad \cdot \quad \cdot \\ \cdot \quad \cdot \quad \cdot \quad \cdot \quad \cdot \quad \cdot \quad \cdot \\ a_{n1}' x_1' + \cdots + a_{nn}' x_n', \end{cases}$$

where $\qquad a_{ij}' = a_{i1} c_{1j} + a_{i2} c_{2j} + \cdots + a_{in} c_{nj}.$

From these formulæ and the law of multiplication of matrices we infer that

$$(4) \qquad \left\| \begin{matrix} a_{11}' \cdots a_{1n}' \\ \cdot \quad \cdot \quad \cdot \\ \cdot \quad \cdot \quad \cdot \\ a_{n1}' \cdots a_{nn}' \end{matrix} \right\| = \left\| \begin{matrix} a_{11} \cdots a_{1n} \\ \cdot \quad \cdot \quad \cdot \\ \cdot \quad \cdot \quad \cdot \\ a_{n1} \cdots a_{nn} \end{matrix} \right\| \cdot \left\| \begin{matrix} c_{11} \cdots c_{1n} \\ \cdot \quad \cdot \quad \cdot \\ \cdot \quad \cdot \quad \cdot \\ c_{n1} \cdots c_{nn} \end{matrix} \right\|.$$

This result we state as follows :

THEOREM 1. *If a system of n linear forms in n variables with matrix* **a** *is subjected to a linear transformation with matrix* **c**, *the resulting system has as its matrix* **ac**.

Taking the determinants of both sides of (4), we see that the resultant of (1) is not an absolute invariant. It is, however, changed in only a very simple manner by a linear transformation, namely, by being multiplied by the determinant of the transformation. This leads us to the following definition:

DEFINITION 2. *A rational function* * *of the coefficients of a form or system of forms which, when these forms are subjected to any non-singular linear transformation, is merely multiplied by the μth power (μ an integer †) of the determinant of the transformation is called a relative invariant of weight μ of the form or system of forms.* ‡ *The forms themselves are called the ground-forms.*

---

* Besides these rational invariants we may also consider irrational ones (cf. § 90), in which case the exponent μ will not necessarily be an integer.

† The condition that μ be an integer need not be included as a part of our hypothesis, since it may be proved. The proof that μ cannot be a fraction is simple. The proof that μ cannot be irrational or imaginary would take us outside of the domain of algebra.

‡ From this definition it is clear that every relative invariant is an absolute invariant with regard to the group of linear transformations of determinant $+1$. Cf. Exercise 7, § 81.

It will be seen that absolute invariants are simply relative invariants of weight zero.

The fact pointed out above concerning the resultant may now be stated in the following form :

THEOREM 2. *The resultant of a set of n linear forms in n variables is a relative invariant of weight* 1.

We pass on now to relative covariants :

DEFINITION 3. *If we have a system consisting of a number of n-ary forms and of a number of points $(y_1, \ldots y_n)$, $(z_1, \ldots z_n)$, $\ldots$, the coördinates of each of which are cogredient with the variables $(x_1 \ldots x_n)$ of the forms, then any rational function of the coefficients of the forms and the coördinates of the points which is merely multiplied by the μth power $(\mu$ an integer) of the determinant of the transformation when the x's are subjected to any non-singular linear transformation is called a relative covariant of weight μ of the system of forms and points.*

We may regard an invariant as the extreme case of a covariant where the number of points is zero. The other extreme case is that in which the number of forms is zero. Here we have the theorem:

THEOREM 3. *The determinant*

$$\begin{vmatrix} x_1' & \cdots & x_n' \\ \cdot & \cdot & \cdot \\ x_1^{[n]} & \cdots & x_n^{[n]} \end{vmatrix}$$

is a relative covariant of weight -1 *of the system of points*

$$(x_1', \cdots x_n'), \ (x_1'', \cdots x_n''), \ \cdots (x_1^{[n]}, \cdots x_n^{[n]}).$$

For applying the transformation

$$x_1 = c_{11} X_1 + \cdots + c_{1n} X_n,$$
$$\cdot \quad \cdot \quad \cdot \quad \cdot \quad \cdot \quad \cdot \quad \cdot \quad \cdot$$
$$\cdot \quad \cdot \quad \cdot \quad \cdot \quad \cdot \quad \cdot \quad \cdot \quad \cdot$$
$$x_n = c_{n1} X_1 + \cdots + c_{nn} X_n,$$

---

* In most books where the subject of covariants is treated, the same letters $(x_1, \cdots x_n)$ are used for one of the points as for the variables of the forms. There is no objection to this, and it is sometimes convenient. We prefer to use a notation which shall make it perfectly clear that the variables of the forms have no connection with the coördinates of the points except that they are cogredient with them.

we have

$$\begin{vmatrix} x_1' & \cdots & x_n' \\ \cdot & \cdot & \cdot \\ \cdot & \cdot & \cdot \\ x_1^{[n]} & \cdots & x_n^{[n]} \end{vmatrix} = \begin{vmatrix} c_{11}X_1' + \cdots + c_{1n}X_n' & \cdots & c_{n1}X_1' + \cdots + c_{nn}X_n' \\ \cdot & & \cdot \\ \cdot & & \cdot \\ c_{11}X_1^{[n]} + \cdots + c_{1n}X_n^{[n]} & \cdots & c_{n1}X_1^{[n]} + \cdots + c_{nn}X_n^{[n]} \end{vmatrix}$$

$$= \begin{vmatrix} c_{11} & \cdots & c_{1n} \\ \cdot & \cdot & \cdot \\ \cdot & \cdot & \cdot \\ c_{n1} & \cdots & c_{nn} \end{vmatrix} \cdot \begin{vmatrix} X_1' & \cdots & X_n' \\ \cdot & \cdot & \cdot \\ \cdot & \cdot & \cdot \\ X_1^{[n]} & \cdots & X_n^{[n]} \end{vmatrix}.$$

Or

$$\begin{vmatrix} X_1' & \cdots & X_n' \\ \cdot & \cdot & \cdot \\ \cdot & \cdot & \cdot \\ X_1^{[n]} & \cdots & X_n^{[n]} \end{vmatrix} = \begin{vmatrix} c_{11} & \cdots & c_{1n} \\ \cdot & \cdot & \cdot \\ \cdot & \cdot & \cdot \\ c_{n1} & \cdots & c_{nn} \end{vmatrix}^{-1} \begin{vmatrix} x_1' & \cdots & x_n' \\ \cdot & \cdot & \cdot \\ \cdot & \cdot & \cdot \\ x_1^{[n]} & \cdots & x_n^{[n]} \end{vmatrix},$$

as was to be proved.

Another extremely simple case arises when we have a single form and a single point:

THEOREM 4. *The system consisting of the form* $f(x_1, \cdots x_n)$ *and the point* $(y_1, \cdots y_n)$ *has as an absolute covariant with regard to linear transformations*

$$f(y_1, \cdots y_n).$$

For let us denote f more explicitly as

$$f(a_1, a_2, \cdots ;\ x_1, \cdots x_n),$$

where $a_1, a_2, \cdots$ are the coefficients of f. If the coefficients after the transformation are $a_1', a_2', \cdots$, we have

$$f(a_1', a_2', \cdots ;\ x_1', \cdots x_n') \equiv f(a_1, a_2, \cdots ;\ x_1, \cdots x_n).$$

This being true for all values of the x's, will be true if the x's are replaced by the y's. But when this is done, the x''s will be replaced by the y''s, since the x's and y's are cogredient. Accordingly

$$f(a_1', a_2', \cdots ;\ y_1', \cdots y_n') \equiv f(a_1, a_2, \cdots ;\ y_1, \cdots y_n),$$

as was to be proved.

The three examples of invariants and covariants which have been given in this section are all polynomials in the coefficients of the forms and the coördinates of the points. Such invariants we shall speak of as *integral rational* invariants and covariants.*

THEOREM 5. *The weight of an integral rational invariant cannot be negative.*†

Let $a_1, a_2, \cdots$; $b_1, b_2, \cdots$; $\cdots$ be the coefficients of the system of forms, and let c_{ij} be the coefficients of the transformation. It is clear that the coefficients $a_1', a_2', \cdots$; $b_1', b_2', \cdots$; $\cdots$ after the transformation are polynomials in the a's, b's, etc., and in the c_{ij}'s. Now let I be an integral rational invariant of weight μ,

$$I(a_1', a_2', \cdots ; \; b_1', b_2', \cdots ; \; \cdots) = c^{\mu}I(a_1, a_2, \cdots ; \; b_1, b_2, \cdots; \; \cdots),$$

where c is the determinant of the transformation. Suppose now that μ were negative, $\mu = -\nu$. Then

$$(5) \qquad c^{\nu}I(a_1', \cdots ; \; b_1', \cdots ; \; \cdots) = I(a_1, \cdots ; \; b_1, \cdots ; \; \cdots).$$

This equality, like the preceding one, is known to hold for all values of the c_{ij}'s for which $c \neq 0$. Hence, since the expressions on both sides of the equality are polynomials in the a's, b's, $\cdots$ and the c_{ij}'s, we infer, by an application of Theorem 5, § 2, that we really have an identity.

Let us now assign to the a's, b's, $\cdots$ any constant values such that $I(a_1, \cdots ; \; b_1, \cdots ; \cdots) \neq 0$. Then $I(a_1', \cdots; \; b_1', \cdots ; \cdots)$ will be a polynomial in the c_{ij}'s alone, which, from (5), cannot be identically zero. The identity (5) thus takes a form which states that the product of two polynomials in the c_{ij}'s is a constant, and since the first of these polynomials, c^{ν}, is of higher degree than zero, this is impossible.

We will agree in future to understand by the terms *invariant* and *covariant*, invariants or covariants (absolute, relative, or arithmetical) *with regard to all non-singular linear transformations*. If we wish to consider invariants or covariants with regard to other sets of transformations, for instance with regard to real linear transformations, this fact will be explicitly mentioned.

---

* All rational invariants and covariants may be formed as the quotients of such as are integral and rational ; cf. Exercises 4, 5, § 78.

† It cannot be zero either ; cf. Theorem 5, § 79.

Finally, let us note the geometric meaning to be associated with the invariants and covariants which have been mentioned in this section. We confine our attention to the case of four variables. The vanishing of the resultant of four linear forms gives a necessary and sufficient condition that the four planes determined by setting the forms equal to zero meet in a point. The vanishing of the covariant of Theorem 3 is a necessary and sufficient condition that the four points lie in a plane. The vanishing of the covariant of Theorem 4 is a necessary and sufficient condition that the point (y_1, y_2, y_3, y_4) lie on the surface $f = 0$. It will be seen that in all cases we are thus led to a projective property; cf. §§ 80, 81.

32. Some Theorems Concerning Linear Forms.

THEOREM 1. *Two systems of n linear forms in n variables are equivalent with regard to non-singular linear transformations if neither resultant is zero.*

Let

$$(1) \quad \begin{cases} a_{11}x_1 + \cdots + a_{1n}x_n \\ \cdot \quad \cdot \quad \cdot \quad \cdot \quad \cdot \\ \cdot \quad \cdot \quad \cdot \quad \cdot \quad \cdot \\ a_{n1}x_1 + \cdots + a_{nn}x_n \end{cases} \qquad (2) \quad \begin{cases} b_{11}x_1 + \cdots + b_{1n}x_n \\ \cdot \quad \cdot \quad \cdot \quad \cdot \quad \cdot \\ \cdot \quad \cdot \quad \cdot \quad \cdot \quad \cdot \\ b_{n1}x_1 + \cdots + b_{nn}x_n \end{cases}$$

be the two systems, whose resultants,

$$a = \begin{vmatrix} a_{11} \cdots a_{1n} \\ \cdot \quad \cdot \\ \cdot \quad \cdot \\ a_{n1} \cdots a_{nn} \end{vmatrix}, \quad b = \begin{vmatrix} b_{11} \cdots b_{1n} \\ \cdot \quad \cdot \\ \cdot \quad \cdot \\ b_{n1} \cdots b_{nn} \end{vmatrix},$$

are, by hypothesis, not zero. Applying the transformations

$$a \begin{cases} x_1' = a_{11}x_1 + \cdots + a_{1n}x_n, \\ \cdot \quad \cdot \quad \cdot \quad \cdot \quad \cdot \quad \cdot \\ \cdot \quad \cdot \quad \cdot \quad \cdot \quad \cdot \quad \cdot \\ x_n' = a_{n1}x_1 + \cdots + a_{nn}x_n, \end{cases} \qquad b \begin{cases} x_1' = b_{11}x_1 + \cdots + b_{1n}x_n, \\ \cdot \quad \cdot \quad \cdot \quad \cdot \quad \cdot \quad \cdot \\ \cdot \quad \cdot \quad \cdot \quad \cdot \quad \cdot \quad \cdot \\ x_n' = b_{n1}x_1 + \cdots + b_{nn}x_n, \end{cases}$$

to (1) and (2) respectively, they are both reduced to the normal form

$$(3) \qquad \begin{cases} x_1', \\ \quad x_2', \\ \quad \cdot \quad \cdot \quad \cdot \quad \cdot \\ \qquad \qquad x_n'. \end{cases}$$

Now, since neither a nor b is zero, the transformations $\mathbf{a}$ and $\mathbf{b}$ have inverses, which when applied to (3) carry it back into (1) and (2) respectively. Hence the transformation $\mathbf{b}^{-1}\mathbf{a}$ carries (1) into (2).

THEOREM 2.   *A system of n linear forms in n variables has no integral rational invariants* * *other than constant multiples of powers of the resultant.*

Let (1) be the given system and a its resultant, and let c be the determinant of a non-singular linear transformation which carries (1) over into

$$
(4) \qquad \begin{cases} a'_{11}x'_1 + \cdots + a'_{1n}x'_n, \\ \ \cdot \quad \cdot \quad \cdot \quad \cdot \quad \cdot \quad \cdot \\ \ \cdot \quad \cdot \quad \cdot \quad \cdot \quad \cdot \quad \cdot \\ a'_{n1}x'_1 + \cdots + a'_{nn}x'_n. \end{cases}
$$

If we call the resultant of (4) a', we have

$$
a' \equiv ac.
$$

Let $I(a_{11}, \cdots a_{nn})$ be any integral rational invariant of the system (1) of weight μ, and write

$$
I' = I(a'_{11}, \cdots a'_{nn}).
$$

Then
$$
I' \equiv c^\mu I.
$$

Now let us assume for a moment that $a \neq 0$, and consider the special transformation which carries over (1) into the normal form (3).   In this special case we have $a' = 1$; hence, as may also be seen directly, $ac = 1$.   Calling the constant value which I' has in this particular case k, we have

$$
k = c^\mu I = a^{-\mu} I,
$$

or
$$
(5) \qquad I = ka^\mu.
$$

This equality, in which k is independent of the coefficients a_{ij}, has been established so far merely for values of the a_{ij}'s for which $a \neq 0$. Since μ is not negative (cf. Theorem 5, § 31), we can now infer that (5) is an identity, by making use of Theorem 5, § 2.   Thus we see that I is merely a constant multiple of a power of the resultant, as was to be proved.

COROLLARY.   *A system of m linear forms in n variables has no integral rational invariants (other than constants) when* $m < n$.

For such an invariant would also be an integral rational invariant of the system of n linear forms obtained by adding $n - m$ new forms to the given system; and hence it would be a constant multiple of a

---

* It has the arithmetical invariant mentioned in Theorem 2, § 30.

power of the resultant of this system. This power must be zero, and hence the invariant must be a mere constant, as otherwise it would involve the coefficients of the added forms, and hence would not be an invariant of the system of original forms.

EXERCISES

1. Prove that if we have two systems of $n + 1$ linear forms in n variables whose matrices are both of rank n, a necessary and sufficient condition that these two systems be equivalent with regard to non-singular linear transformation is that the resultants of the forms of one set taken n at a time be proportional to the resultants of the corresponding forms of the other set.

2. Generalize the preceding theorem.

3. Prove that every integral rational invariant of a system of m linear forms in n variables $(m > n)$ is a homogeneous polynomial in the resultants of these forms taken n at a time.

4. State and prove the theorems analogous to the theorems of the present section, including the three preceding exercises, when the system of linear forms is replaced by a system of points.

33. Cross-ratio and Harmonic Division. Let us consider any four distinct points on a line

$$(1) \qquad (x_1, t_1), (x_2, t_2), (x_3, t_3), (x_4, t_4).$$

We have seen, in § 31, Theorem 3, that each of the six determinants

$$(2) \qquad \begin{array}{ccc} x_1t_2 - x_2t_1, & x_1t_3 - x_3t_1, & x_1t_4 - x_4t_1, \\ x_3t_4 - x_4t_3, & x_4t_2 - x_2t_4, & x_2t_3 - x_3t_2, \end{array}$$

is a covariant of weight -1. The ratio of two of these determinants is therefore an absolute covariant, and we might be tempted, by analogy with the examples of absolute covariants in Exercise 1, § 28, to expect that it might have a geometric meaning. It will be readily seen, however, that this is not the case, for the value of the ratio of two of the determinants (2) will be changed if the two coördinates of one of the four points are multiplied by the same constant, and this does not affect the position of the points.

It is easy, however, to avoid this state of affairs by forming such an expression as the following:*

$$(3) \qquad (1, 2, 3, 4) = \frac{(x_1t_2 - x_2t_1)(x_3t_4 - x_4t_3)}{(x_2t_3 - x_3t_2)(x_4t_1 - x_1t_4)},$$

---

* The reversal of sign of the second factor in the denominator is not essential, but is customary for a reason which will presently be evident.

which is also an absolute covariant of the four points (1), and is called their *cross-ratio* or *anharmonic ratio*.   More accurately it is called the cross-ratio of these four points when taken in the order written in (1).*

In order to determine the geometric meaning of the cross-ratio of four points, let us first suppose the four points to be finite so that $t_1 t_2 t_3 t_4 \neq 0$.   Dividing numerator and denominator of (1, 2, 3, 4) by this product, we find the following expression for the cross-ratio in terms of the non-homogeneous coördinates X_i of the points,

$$(4) \qquad (1, 2, 3, 4) = \frac{(X_1 - X_2)(X_3 - X_4)}{(X_2 - X_3)(X_4 - X_1)}.$$

Finally, denoting the points by P_1, P_2, P_3, P_4, we may write

$$(5) \qquad (1, 2, 3, 4) = \frac{P_1 P_2}{P_3 P_2} \bigg/ \frac{P_1 P_4}{P_3 P_4} = \frac{P_2 P_1}{P_4 P_1} \bigg/ \frac{P_2 P_3}{P_4 P_3}.$$

In words, this formula tells us that the cross-ratio of four finite points is the ratio of the ratio in which the second divides the first and third and the ratio in which the fourth divides the first and third; and that it is also the ratio of the ratios in which the first and third divide the second and fourth.

In this statement, it must be remembered that we have taken the ratio in which C divides the points A, B as AC/BC, so that the ratio is negative if C divides AB internally, positive if it divides it externally.

If we agree that the point at infinity on a line shall be said to divide any two finite points A, B of this line in the ratio $+1$ (and this is a natural convention since the more distant a point the more nearly does it divide AB in the ratio $+1$) it is readily seen, by going back to formula (3), that the first statement following (5) still holds if the second or fourth point is at infinity, while the second statement holds if the first or third is at infinity.   Thus we have in all cases a simple geometric interpretation of the cross-ratio of four distinct points.

The special case in which four points P_1, P_2, P_3, P_4 are so situated that $(1, 2, 3, 4) = -1$ is of peculiar importance.   In this case we have

$$(1, 2, 3, 4) = (1, 4, 3, 2) = (3, 2, 1, 4) = (3, 4, 1, 2) = (2, 1, 4, 3)$$
$$= (2, 3, 4, 1) = (4, 1, 2, 3) = (4, 3, 2, 1) = -1.$$

---

* If these four points are taken in other orders, we get different cross-ratios: (1, 2, 4, 3), (1, 4, 3, 2), etc.   Cf. Exercise 1 at the end of this section

The relation is therefore merely a relation between the two pairs of points P_1, P_3 and P_2, P_4 taken indifferently in either order, and we say that these two pairs of points divide each other harmonically. From the geometric meaning of cross-ratio, we see that, if all four points are finite, the pairs P_1, P_3 and P_2, P_4 divide each other harmonically when, and only when, P_2 and P_4 divide P_1, P_3 internally and externally in the same ratio ; and also when, and only when, P_1 and P_3 divide P_2, P_4 internally and externally in the same ratio. If P_2 or P_4 lies at infinity, the first of these statements alone has a meaning, while if P_1 or P_3 lies at infinity, it is the second statement to which we must confine ourselves.

It is easily seen that the case in which three of the four points, say P_1, P_2, P_3, coincide, while P_4 is any point on the line, may be regarded as a limiting form of two pairs of points which separate one another harmonically. It is convenient to include this case under the term *harmonic division*, and we will therefore lay down the definition:

DEFINITION. *Two pairs of points P_1, P_3 and P_2, P_4 on a line are said to divide one another harmonically if they are distinct and their cross-ratio taken in the order P_1, P_2, P_3, P_4 is -1, and also if at least three of them coincide.*

It will be seen that the property of two pairs of points dividing each other harmonically is a projective property in space of one dimension.

The most important applications of cross-ratio come in geometry of two, three, or more dimensions where the points are not determined as above by two coördinates (or one non-homogeneous coördinate), but by more. Suppose, for instance, we have four distinct finite points on a line in space of three dimensions. Let the points be P_1, P_2, Q_1, Q_2, and suppose the coördinates of P_1, P_2 are (x_1, y_1, z_1, t_1) and (x_2, y_2, z_2, t_2) respectively. Then the coördinates of Q_1, Q_2 may be written

$$(x_1 + \lambda x_2, y_1 + \lambda y_2, z_1 + \lambda z_2, t_1 + \lambda t_2), (x_1 + \mu x_2, y_1 + \mu y_2, z_1 + \mu z_2, t_1 + \mu t_2).$$

Now, let

(6) $$Ax + By + Cz + Dt = 0$$

be any plane through Q_1 but not through P_2, and we have

$$(Ax_1 + By_1 + Cz_1 + Dt_1) + \lambda(Ax_2 + By_2 + Cz_2 + Dt_2) = 0,$$

or, since P_2 does not lie on (6),

$$\frac{Ax_1 + By_1 + Cz_1 + Dt_1}{Ax_2 + By_2 + Cz_2 + Dt_2} = -\lambda.$$

Changing to non-homogeneous coördinates, we have

$$\frac{AX_1 + BY_1 + CZ_1 + D}{AX_2 + BY_2 + CZ_2 + D} = -\lambda\frac{t_2}{t_1}.$$

If P_1M_1 and P_2M_2 are the perpendiculars from P_1 and P_2 on the plane (6), we have

$$\frac{P_1Q_1}{P_2Q_1} = \frac{P_1M_1}{P_2M_2} = \frac{AX_1 + BY_1 + CZ_1 + D}{AX_2 + BY_2 + CZ_2 + D} = -\lambda\frac{t_2}{t_1}.$$

In exactly the same way we get

$$\frac{P_1Q_2}{P_2Q_2} = -\mu\frac{t_2}{t_1}.$$

Consequently $\qquad \dfrac{P_1Q_1}{P_2Q_1} \Big/ \dfrac{P_1Q_2}{P_2Q_2} = \dfrac{\lambda}{\mu}.$

This is the cross-ratio of the four points taken in the order P_1, Q_1, P_2, Q_2.

It is readily seen that if one of the two points Q_1 or Q_2 lies at infinity, all that is essential in the above reasoning remains valid, and the cross-ratio is still λ/μ.

The case in which one of the two points P_1 or P_2 is at infinity may be reduced to the case just considered by writing for the coördinates of Q_1 and Q_2, $(\xi_1, \eta_1, \zeta_1, \tau_1)$ and $(\xi_2, \eta_2, \zeta_2, \tau_2)$. The coördinates of P_1 and P_2 are then

$$\left(\xi_1 - \frac{\lambda}{\mu}\xi_2,\; \eta_1 - \frac{\lambda}{\mu}\eta_2,\; \zeta_1 - \frac{\lambda}{\mu}\zeta_2,\; \tau_1 - \frac{\lambda}{\mu}\tau_2\right),$$

$$(\xi_1 - \xi_2,\; \eta_1 - \eta_2,\; \zeta_1 - \zeta_2,\; \tau_1 - \tau_2).$$

Accordingly, from what has just been proved, we see that the cross-ratio of the four points taken in the order Q_1, P_1, Q_2, P_2 is λ/μ. But this change of order does not change the cross-ratio. Hence in all cases we have the result:

Theorem 1. *The cross-ratio of the four distinct points*

$$P_1 \qquad (x_1, y_1, z_1, t_1),$$

$$P_2 \qquad (x_2, y_2, z_2, t_2),$$

$$Q_1 \qquad (x_1 + \lambda x_2, y_1 + \lambda y_2, z_1 + \lambda z_2, t_1 + \lambda t_2),$$

$$Q_2 \qquad (x_1 + \mu x_2, y_1 + \mu y_2, z_1 + \mu z_2, t_1 + \mu t_2),$$

taken in the order $P_1, Q_1, P_2, Q_2,$ *is* $\lambda/\mu.$

From this theorem we easily deduce the further result :

Theorem 2. *The cross-ratio of four points on a line is invariant with regard to non-singular collineations of space.*[*]

For the four points P_1, P_2, Q_1, Q_2 of Theorem **1** are carried over by a non-singular collineation into the four points

$$P_1' \qquad (x_1', y_1', z_1', t_1'),$$

$$P_2' \qquad (x_2', y_2', z_2', t_2'),$$

$$Q_1' \qquad (x_1' + \lambda x_2', y_1' + \lambda y_2', z_1' + \lambda z_2', t_1' + \lambda t_2'),$$

$$Q_2' \qquad (x_1' + \mu x_2', y_1' + \mu y_2', z_1' + \mu z_2', t_1' + \mu t_2'),$$

whose cross-ratio, when taken in the order $P_1', Q_1', P_2', Q_2',$ is also $\lambda/\mu.$

Theorems similar to Theorems 1 and 2 hold in space of two, and in general in space of n, dimensions and may be proved in the same way.

EXERCISES

1. Denote the six determinants (2) by

$$(1, 2), \qquad (1, 3), \qquad (1, 4), \qquad (3, 4), \qquad (4, 2), \qquad (2, 3),$$

and write

$$A = (1, 2)(3, 4), \qquad B = (1, 3)(4, 2), \qquad C = (1, 4)(2, 3).$$

Prove that six, and only six, cross-ratios can be formed from four points by taking them in different orders, namely the negatives of the six ratios which can be formed from A, B, C taken two and two.

2. Prove that $A + B + C = 0$, and hence show that if λ is one of the cross-ratios of four points, the other five will be

$$\frac{1}{\lambda}, \ 1-\lambda, \ \frac{1}{1-\lambda}, \ \frac{\lambda-1}{\lambda}, \ \frac{\lambda}{\lambda-1}.$$

[*] This also follows from Exercise 5, § 24

3. Prove that the six cross-ratios of four distinct points are all different from each other except in the following two cases:

(α) The case of four harmonic points, where the values of the cross-ratios are $-1, 2, \frac{1}{2}$.

(β) The case known as four equianharmonic points, in which the values of the cross-ratios are $-\frac{1}{2} \pm \frac{1}{2}\sqrt{-3}$.

4. Prove Theorem 2, § 24, by making use of the fact that the cross-ratio of four points on a line is unchanged by non-singular projective transformations of the line.

5. By the cross-ratio of four planes which meet in a line is understood the cross-ratio of the four points in which these planes are met by any line which does not meet their line of intersection.

Justify this definition by proving that if the equations of the four planes are

$$p_1 = 0,\; p_1 + \lambda p_2 = 0,\; p_2 = 0,\; p_1 + \mu p_2 = 0$$

(p_1 and p_2 homogeneous linear polynomials in x, y, z, t), the cross-ratio of the four points in which *any* line which does not meet the line of intersection of the planes is met by the planes is λ / μ.

6. Prove that the cross-ratio of four planes which meet in a line is invariant with regard to non-singular collineations.

34. Plane-Coördinates and Contragredient Variables. If u_1, u_2, u_3, u_4 are constants, and x_1, x_2, x_3, x_4 are the homogeneous coördinates of a point in space, the equation

$$(1) \qquad\qquad u_1x_1 + u_2x_2 + u_3x_3 + u_4x_4 = 0$$

represents a plane. Since the values of the u's determine the position of this plane, the u's may be regarded as coördinates of the plane. We will speak of them as *plane-coördinates*, just as the x's (each set of which determines a point) are called *point-coördinates*. And just as we speak of the point (x_1, x_2, x_3, x_4) so we will speak of the plane (u_1, u_2, u_3, u_4). The u's are evidently analogous to homogeneous coördinates in that if they be all multiplied by the same constant, the plane which they determine is not changed.

Suppose now that we consider the x's as constants and allow the u's to vary, taking on all possible sets of values which, with the fixed set of values of the x's, satisfy (1). This equation will now represent a family of planes, infinite in number, each one of which is determined by a particular set of values of the u's and all of which pass through the fixed point (x_1, x_2, x_3, x_4). The equation (1) may therefore be regarded as the *equation of a point in plane-coördinates*, since it is satisfied by the coördinates of a moving plane which envelops this point, just as when the x's vary and the u's are constant, it is the *equation of a plane in point-coördinates*, since it is satisfied by the coördinates of a moving point whose locus is the plane.*

In the same way, a homogeneous equation of degree higher than the first in the u's will be satisfied by the coördinates of a moving plane which will, in general, envelop a surface. The equation will then be called the equation of this surface in plane-coördinates.†

Let us now subject space to the collineation

$$\mathbf{c} \qquad x'_i = c_{i1}x_1 + c_{i2}x_2 + c_{i3}x_3 + c_{i4}x_4 \qquad (i = 1, 2, 3, 4).$$

We will assume that the determinant c of this transformation is not zero; and we will denote the cofactors in this determinant by C_{ij}. Then the inverse of the transformation $\mathbf{c}$ may be written

$$\mathbf{c}^{-1} \qquad x_i = \frac{C_{1i}}{c}x'_1 + \frac{C_{2i}}{c}x'_2 + \frac{C_{3i}}{c}x'_3 + \frac{C_{4i}}{c}x'_4 \qquad (i = 1, 2, 3, 4).$$

Substituting these expressions, we see that the plane (1) goes over into

(2) $$u'_1 x'_1 + u'_2 x'_2 + u'_3 x'_3 + u'_4 x'_4 = 0,$$

where

$$\mathbf{d} \qquad u'_i = \frac{C_{i1}}{c}u_1 + \frac{C_{i2}}{c}u_2 + \frac{C_{i3}}{c}u_3 + \frac{C_{i4}}{c}u_4 \qquad (i = 1, 2, 3, 4).$$

We thus see that the u's have also suffered a linear transformation, though a different one from the x's, namely, the transformation whose matrix is the conjugate (cf. § 7, Definition 2) of $\mathbf{c}^{-1}$. This transformation $\mathbf{d}$ of the plane-coördinates is merely another way of expressing the collineation which we have commonly expressed by the transformation $\mathbf{c}$ of the point-coördinates. The two sets of variables x and u are called *contragredient variables* according to the following

DEFINITION 1. *Two sets of n variables each are called contragredient if, whenever one is subjected to a non-singular linear transformation, the other is subjected to the transformation which has as its matrix the conjugate of the inverse of the matrix of the first.*

---

* Similarly, in two dimensions, the equation

$$u_1 x_1 + u_2 x_2 + u_3 x_3 = 0$$

represents a *line* in the point-coördinate (x_1, x_2, x_3) if u_1, u_2, u_3 are constants, or a *point* in the line-coördinates (u_1, u_2, u_3) if x_1, x_2, x_3 are constants.

† An example of this will be found in § 53.

Precisely the reasoning used above in the case of four variables establishes here also the theorem :

THEOREM.* *If the two sets of contragredient variables* $x_1, \cdots x_n$ *and* $u_1, \cdots u_n$ *are carried over by a linear transformation into* $x'_1, \cdots x'_n$ *and* $u'_1, \cdots u'_n$, *then*

$$u_1x_1 + u_2x_2 + \cdots + u_nx_n$$

will go over into $\qquad u'_1x'_1 + u'_2x'_2 + \cdots + u'_nx'_n.$

In connection with this subject of contragredient variables it is customary to introduce the conception of *contravariants*, just as the conception of covariants was introduced in connection with the subject of cogredient variables.   For this purpose we lay down the

DEFINITION 2. *If we have a system of forms in* $(x_1, \cdots x_n)$ *and also a number of sets of variables,* $(u'_1, \cdots u'_n), (u''_1, \cdots u''_n), \cdots,$ *contragredient to the* x's, *any rational function of the* u's *and the coefficients of the forms which is unchanged by a non-singular linear transformation of the* x's *except for being multiplied by the* μth *power* (μ *an integer*) *of the determinant of this transformation is called a contravariant of weight* μ.

Thus the theorem that the resultant of n linear forms in n variables is an invariant of weight 1 may, if we prefer, be stated in the form : If we have n sets of n variables each, $(u'_1, \cdots u'_n), \cdots (u_1^{[n]}, \cdots u_n^{[n]})$, each of which is contragredient to the variables $(x_1, \cdots x_n)$, the determinant of the u's is a contravariant of weight 1.†

It will be seen that the conception of contravariant, though sometimes convenient, is unnecessary, since the contragredient variables may always be regarded as the coefficients of linear forms, and, when so regarded, the contravariant is merely an invariant.

Similarly, the still more general conception of *mixed concomitants*, in which, besides the coefficients of forms and the contragredient variables, certain sets of cogredient variables are involved,‡ reduces to the familiar conception of covariants if we regard the contragredient variables as coefficients of linear forms.

---

* This is really a special theorem in the theory of bilinear forms. Cf. the next chapter.

† For other examples of contravariants in which coefficients also occur, see Chap. XII.

‡ An example is $u_1x_1 + u_2x_2 + \cdots + u_nx_n$, the theorem above stating that this is an *absolute* mixed concomitant.

35. Line-Coördinates in Space. A line is determined by two points (y_1, y_2, y_3, y_4), (z_1, z_2, z_3, z_4) which lie on it. It is clear that these eight coördinates are not all necessary to determine the line ; and it will be seen presently that the following six quantities determine the line completely, and may be used as *line-coördinates,*

$$p_{12}, \; p_{13}, \; p_{14}, \; p_{34}, \; p_{42}, \; p_{23},$$

where

(1)
$$p_{ij} = \begin{vmatrix} y_i & y_j \\ z_i & z_j \end{vmatrix}.$$

In other words, the p's are the two-rowed determinants of the matrix

$$\left\| \begin{matrix} y_1 & y_2 & y_3 & y_4 \\ z_1 & z_2 & z_3 & z_4 \end{matrix} \right\|,$$

except that the sign of the determinant obtained by striking out the first and third column has been changed. These six p's are not all zero if, as we assume, the two points y and z are distinct.

These six p's are connected by the relation

(2)
$$p_{12}p_{34} + p_{13}p_{42} + p_{14}p_{23} = 0,^*$$

as may be seen either directly or by expanding the vanishing determinant

$$\begin{vmatrix} y_1 & y_2 & y_3 & y_4 \\ z_1 & z_2 & z_3 & z_4 \\ y_1 & y_2 & y_3 & y_4 \\ z_1 & z_2 & z_3 & z_4 \end{vmatrix}$$

by Laplace's method in terms of the minors of the first two rows.

That the p's may really be used as line-coördinates is shown by the following two theorems:

THEOREM 1. *When a line is given, its line-coördinates p_{ij} are completely determined except for an arbitrary factor different from zero by which they may all be multiplied.*

The definition (1) of the p's shows that they may all be multiplied by an arbitrary factor different from zero without affecting the position of the line, since the y's (and also the z's) may be multiplied by such a factor without affecting the position of the point.

---

* Cf. Exercise 2, § 33.

In order to prove our theorem it is sufficient to show that, if instead of the two points used above for determining the p's we use two other points of the line,

$$(Y_1,\ Y_2,\ Y_3,\ Y_4),\qquad (Z_1,\ Z_2,\ Z_3,\ Z_4),$$

the line-coördinates

$$P_{ij} = \begin{vmatrix} Y_i & Y_j \\ Z_i & Z_j \end{vmatrix}$$

thus determined will be proportional to the p_{ij}'s. Since the points Y and Z are collinear with the distinct points y, z, they are linearly dependent upon them and we may write

$$Y_i = c_1 y_i + c_2 z_i,\qquad Z_i = k_1 y_i + k_2 z_i \qquad (i = 1,\ 2,\ 3,\ 4).$$

Accordingly

$$P_{ij} = \begin{vmatrix} c_1 & c_2 \\ k_1 & k_2 \end{vmatrix} \begin{vmatrix} y_1 & y_2 \\ z_1 & z_2 \end{vmatrix} = K p_{ij},$$

where $K \neq 0$, as Y and Z are distinct points.

THEOREM 2. *Any six constants p_{ij} satisfying the relation (2) and not all zero are the line-coördinates of one, and only one, line.*

That they cannot be the coördinates of more than one line may be seen as follows: Suppose the p_{ij}'s to be the coördinates of a line, and take two distinct points y and z on the line. The coördinates of these points may then be so determined that relations (1) hold. Let us suppose, for definiteness, that $p_{12} \neq 0$.* Now, consider the point whose coördinates are $c_1 y_i + c_2 z_i$. By assigning to c_1 and c_2 first the values $-z_1$ and y_1, then the values $-z_2$ and y_2, we get the two points

(3) $$(0,\ p_{12},\ p_{13},\ p_{14}),\qquad (p_{21},\ 0,\ p_{23},\ p_{24}),$$

where, by definition, $p_{ij} = -p_{ji}$.

These two points are distinct, since for the first of them the first coördinate is zero and the second is not, while for the second the second coördinate is zero and the first is not. These points accordingly determine the line, and since they, in turn, are determined by the p's, we see that the line is uniquely determined by the p's.

---

* By a slight modification of the formulæ this proof will apply to the case in which any one of the p's is assumed different from zero.

It remains, then, merely to show that any set of p_{ij}'s, not all zero which satisfy (2) really determine a line. For this purpose we again assume $p_{12} \neq 0$ * and consider the two points (3) which, as above, are distinct. The line determined by them has as its coördinates

$$p_{12}^2, \quad p_{12}p_{13}, \quad p_{12}p_{14}, \quad -p_{13}p_{42} - p_{14}p_{23}, \quad p_{12}p_{42}, \quad p_{12}p_{23}.$$

Using the relation (2), the fourth of these quantities reduces to $p_{12}p_{34}$, so that, remembering that the coördinates of a line may be multiplied by any constant different from zero, we see that we really have a line whose coördinates are p_{ij}.

In a systematic study of three-dimensional geometry these line-coördinates play as important a part as the point- or plane-coördinates; and in the allied algebraic theories we shall have to consider expressions having the invariant property, into which these line-coördinates enter just as point-coördinates occur in covariants and plane-coördinates in contravariants. We may, if we please, regard these expressions as ordinary covariants, since the line-coördinates are merely functions of the coördinates of two points, but the covariants we get in this way are covariants of a special sort, since the coördinates of the two points occur only in the combinations (1).

As an example, let us consider four points

$$(x_i, \; y_i, \; z_i, \; t_i) \qquad\qquad (i = 1, \, 2, \, 3, \, 4).$$

The determinant of these sixteen coördinates is, by Theorem 3, § 31, a covariant of weight -1. Let us denote by p'_{ij} and p''_{ij} the coördinates of the lines determined by the first two and the last two points respectively. Expanding the four-rowed determinant just referred to by Laplace's method according to the two-rowed determinants of the first two rows, we get

$$(4) \qquad p'_{12}p''_{34} + p''_{12}p'_{34} + p'_{13}p''_{42} + p''_{13}p'_{42} + p'_{14}p''_{23} + p''_{14}p'_{23}.$$

This, then, is an expression having the invariant property and involving only line-coördinates.

Since the vanishing of the four-rowed determinant from which we started gave the condition that the four points lie in a plane, it follows that the vanishing of (4) gives a necessary and sufficient condition that the two lines p' and p'' lie in a plane, or, what amounts to the same thing, that they meet in a point.

---

* By a slight modification of the formulæ, this proof will apply to the case in which any one of the p's is assumed different from zero.

EXERCISES

1. Prove that, if the point-coördinates are subjected to the linear transformation

$$x_i' = c_{i1}x_1 + c_{i2}x_2 + c_{i3}x_3 + c_{i4}x_4 \qquad (i = 1, 2, 3, 4),$$

the line-coördinates will be subjected to the linear transformation

$$p_{ij}' = (c_{i1}c_{j2} - c_{i2}c_{j1})\,p_{12} + (c_{i1}c_{j3} - c_{i3}c_{j1})\,p_{13} + (c_{i1}c_{j4} - c_{i4}c_{j1})\,p_{14} + (c_{i3}c_{j4} - c_{i4}c_{j3})\,p_{34}$$
$$+ (c_{i4}c_{j2} - c_{i2}c_{j4})\,p_{42} + (c_{i2}c_{j3} - c_{i3}c_{j2})\,p_{23}.$$

2. A plane is determined by three points

$$(y_1,\ y_2,\ y_3,\ y_4), \qquad (z_1,\ z_2,\ z_3,\ z_4), \qquad (w_1,\ w_2,\ w_3,\ w_4).$$

Prove that the three-rowed determinants of the matrix of these three points may be used as coördinates of this plane, and that these coördinates are not distinct from the plane-coördinates defined in § 34.

3. A line determined by two of its points may be called a *ray*, and the line-coördinates of the present section may therefore be called ray-coördinates. A line determined as the intersection of two planes may be called an *axis*. If (u_1, u_2, u_3, u_4) and (v_1, v_2, v_3, v_4) are two planes given by their plane-coördinates, discuss the *axis-coördinates* of their intersection,

$$q_{12},\ q_{13},\ q_{14},\ q_{34},\ q_{42},\ q_{23},$$

where $\qquad q_{ij} = u_i v_j - u_j v_i.$

4. Prove that ray-coördinates and axis-coördinates are not essentially different by showing that, for any line, the q's, taken in the order written in Exercise 3, are proportional to the p's taken in the order

$$p_{34},\ p_{42},\ p_{23},\ p_{12},\ p_{13},\ p_{14}.$$

5. A point is determined as the intersection of three planes

$$(u_1,\ u_2,\ u_3,\ u_4), \qquad (v_1,\ v_2,\ v_3,\ v_4), \qquad (w_1,\ w_2,\ w_3,\ w_4).$$

Prove that the three-rowed determinants of the matrix of these planes may be used as coördinates of this point, and that they do not differ from the ordinary point-coördinates.

Hence, show that all covariants may be regarded as invariants.

CHAPTER VIII

BILINEAR FORMS

36. The Algebraic Theory. Before entering on the study of quadratic forms, which will form the subject of the next five chapters, we turn briefly to a very special type of quadratic form in $2n$ variables, known as a bilinear form, and which, as its name implies, forms a natural transition between linear and quadratic forms.

DEFINITION 1. *A polynomial in the $2n$ variables $(x_1, \cdots x_n)$, $(y_1, \cdots y_n)$ is called a bilinear form if each of its terms is of the first degree in the x's and also of the first degree in the y's.*

Thus, for $n = 3$, the most general bilinear form is

$$a_{11}x_1y_1 + a_{12}x_1y_2 + a_{13}x_1y_3$$
$$+ a_{21}x_2y_1 + a_{22}x_2y_2 + a_{23}x_2y_3$$
$$+ a_{31}x_3y_1 + a_{32}x_3y_2 + a_{33}x_3y_3.$$

This may be denoted, for brevity, by $\overset{3}{\underset{1}{\Sigma}} a_{ij}x_iy_j$; and, in general, we may denote the bilinear form in $2n$ variables by

$$(1) \qquad \overset{n}{\underset{1}{\Sigma}} a_{ij}x_iy_j.$$

The matrix

$$\mathbf{a} = \begin{Vmatrix} a_{11} & \cdots & a_{1n} \\ \cdot & \cdot & \cdot \\ \cdot & \cdot & \cdot \\ a_{n1} & \cdots & a_{nn} \end{Vmatrix}$$

is called the matrix of the form (1); its determinant, the determinant of the form; and its rank, the rank of the form.* A bilinear form is called singular when, and only when, its determinant is zero.

---

* It should be noticed that the bilinear form is completely determined when its matrix is given, so there will be no confusion if we speak of the bilinear form $\mathbf{a}$. If two bilinear forms have matrices $\mathbf{a}_1$ and $\mathbf{a}_2$, their sum has the matrix $\mathbf{a}_1 + \mathbf{a}_2$. The bilinear form whose matrix is $\mathbf{a}_1\mathbf{a}_2$ is *not* the product of the two forms, but is sometimes spoken of as their *symbolic product*.

Let us notice that the bilinear form (1) may be obtained by starting from the system of n linear forms in the y's of matrix $\mathbf{a}$, multiplying them respectively by $x_1,\ x_2,\ \cdots\ x_n$, and adding them together. It can also be obtained by starting from the system of n linear forms in the x's whose matrix is the conjugate of $\mathbf{a}$, multiplying them respectively by $y_1,\ y_2,\ \cdots\ y_n$, and adding them together.

Using the first of these two methods, we see (cf. Theorem 1, § 31) that if the y's are subjected to a linear transformation with matrix $\mathbf{d}$, the bilinear form is carried over into a new bilinear form whose matrix is $\mathbf{ad}$. Using the second of the above methods of building up the bilinear form from linear forms, we see that if the x's are subjected to a linear transformation with matrix $\mathbf{c}$, we get a new bilinear form the conjugate of whose matrix is $\mathbf{a'c}$, where accents are used to denote conjugate matrices. The matrix of the form itself is then (cf. Theorem 6, § 22) $\mathbf{c'a}$.*

Combining these two facts, we have

THEOREM 1. *If, in the bilinear form* (1) *with matrix* $\mathbf{a}$, *we subject the x's to a linear transformation with matrix* $\mathbf{c}$ *and the y's to a linear transformation with matrix* $\mathbf{d}$, *we obtain a new bilinear form with matrix* $\mathbf{c'ad}$, *where* $\mathbf{c'}$ *is the conjugate of* $\mathbf{c}$.

Considering the determinants of these matrices, we may say:

THEOREM 2. *The determinant of a bilinear form is multiplied by the product of the determinants of the transformations to which the x's and y's are subjected.*†

We also infer from Theorem 1, in combination with Theorem 7, § 25, the important result: ‡

THEOREM 3. *The rank of a bilinear form is an invariant with regard to non-singular linear transformations of the x's and y's.*

DEFINITION 2. *A bilinear form whose matrix is symmetric is called a symmetric bilinear form.*

---

* These results may also be readily verified without referring to any earlier theorems.

† This theorem tells us that the determinant of a bilinear form is, in a generalized sense, a relative invariant. Such invariants, where the given forms depend on several sets of variables, are known as *combinants*.

‡ This result may also be deduced from Theorem 2, § 30.

THEOREM 4. *A symmetric bilinear form remains symmetric if we subject the x's and the y's to the same linear transformation.*

For if c is the matrix of the transformation to which both the x's and the y's are subjected, the matrix of the transformed form will, by Theorem 1, be $c'ac$. Remembering that a, being symmetric, is its own conjugate, we see, by Theorem 6, § 22, that $c'ac$ is its own conjugate. Hence the transformed form is symmetric.

EXERCISES

1. Prove that a necessary and sufficient condition for the equivalence of two bilinear forms with regard to non-singular linear transformations of the x's and y's is that they have the same rank.

2. Prove that a necessary and sufficient condition that it be possible to factor a bilinear form into the product of two linear forms is that its rank be zero or one.

3. Prove that every bilinear form of rank r can be reduced by non-singular linear transformations of the x's and y's to the normal form

$$x_1 y_1 + x_2 y_2 + \cdots + x_r y_r.$$

4. Do the statements in the preceding exercises remain correct if we confine our attention to real bilinear forms and real linear transformations?

5. Prove that a necessary and sufficient condition that the form

$$x_1 y_1 + x_2 y_2 + \cdots + x_n y_n$$

should be unchanged by linear transformations of the x's and of the y's is that these be contragredient transformations.

37. A Geometric Application. Let (x_1, x_2, x_3) and (y_1, y_2, y_3) be homogeneous coördinates of points in a plane, and let us consider the bilinear equation

$$(1) \qquad \sum_1^3 a_{ij} x_i y_j = 0.$$

If (y_1, y_2, y_3) is a fixed point P, then (1), being linear in the x's, is the equation of a straight line p. The only exception is when the coefficients of (1), regarded as a linear equation in the x's, are all zero, and this cannot happen if the determinant of the form is different from zero. Thus we see that the equation (1) causes one, and only one, line p to correspond to every point P of the plane, provided the bilinear form in (1) is non-singular.

Conversely, if

$$(2) \qquad Ax_1 + Bx_2 + Cx_3 = 0$$

is a line p, there is one, and only one, point P which corresponds to it by means of (1), provided the bilinear form in (1) is non-singular. For if P is the point (y_1, y_2, y_3), the equation of the line corresponding to it is (1), and the necessary and sufficient condition that this line coincide with (2) is

$$a_{11}y_1 + a_{12}y_2 + a_{13}y_3 = \rho A,$$
$$a_{21}y_1 + a_{22}y_2 + a_{23}y_3 = \rho B,$$
$$a_{31}y_1 + a_{32}y_2 + a_{33}y_3 = \rho C,$$

where ρ is a constant, not zero. For a given value of ρ, this set of equations has one, and only one, solution (y_1, y_2, y_3), since the determinant a is not zero, while a change in ρ merely changes all the y's in the same ratio. Hence,

THEOREM. *If the bilinear equation* (1) *is non-singular, it establishes a one-to-one correspondence between the points and lines of the plane.*

This correspondence is called a *correlation*.

EXERCISES

1. Discuss the singular correlations of the plane, considering separately the cases in which the rank of the bilinear form is 2 and 1.

2. Examine the corresponding equation in three dimensions, that is, the equation obtained by equating to zero a bilinear form in which $n = 4$, and discuss it for all possible suppositions as to the rank of the form.

3. Show that a necessary and sufficient condition for three or more lines, which correspond to three or more given points by means of a non-singular correlation, to be concurrent is that the points be collinear.

4. Show that the cross-ratio of any four concurrent lines is the same as that of the four points to which they correspond by means of a non-singular correlation.

5. Let P be any point in a plane and p the line corresponding to it by means of a non-singular correlation. Prove that a necessary and sufficient condition for the lines corresponding to the points of p to pass through P is that the bilinear form be symmetrical.

6. State and prove the corresponding theorem for points and planes in space of three dimensions, showing that here it is necessary and sufficient that the form be *symmetrical* or *skew-symmetrical*.*

---

* The correlation given by a symmetric bilinear equation is known as a *reciprocation*. By reference to the formulæ of the next chapter, it will be readily seen that in this case every point corresponds, in the plane, to its polar with regard to a fixed conic ; in space, to its polar plane with regard to a fixed quadric surface. The skew-symmetric bilinear equation gives rise in the plane merely to a very special singular correlation. In space, however, it gives an important correlation which is in general non-singular and is known as a *null-system*. Cf. any treatment of line geometry, where, however, the subject is usually approached from another side.

CHAPTER IX

GEOMETRIC INTRODUCTION TO THE STUDY OF QUADRATIC FORMS

38. Quadric Surfaces and their Tangent Lines and Planes. If x_1, x_2, x_3 are homogeneous coördinates in a plane, we see, by reference to § 4, that the equation of any conic may be written

$$a_{11}x_1^2 + a_{22}x_2^2 + a_{33}x_3^2 + 2a_{12}x_1x_2 + 2a_{13}x_1x_3 + 2a_{23}x_2x_3 = 0.$$

Similarly, in space of three dimensions, the equation of any quadric surface may be written

$$a_{11}x_1^2 + a_{22}x_2^2 + a_{33}x_3^2 + a_{44}x_4^2 + 2a_{12}x_1x_2 + 2a_{13}x_1x_3 + 2a_{14}x_1x_4$$
$$+ 2a_{34}x_3x_4 + 2a_{42}x_4x_2 + 2a_{23}x_2x_3 = 0.$$

This form may be made still more symmetrical if, besides the coefficients $a_{12}, a_{13}, a_{14}, a_{34}, a_{42}, a_{23}$, we introduce the six other constants $a_{21}, a_{31}, a_{41}, a_{43}, a_{24}, a_{32}$, defined by the general formula

$$a_{ij} = a_{ji}.$$

The equation of the quadric surface may then be written

$$a_{11}x_1^2 \;\;\;+ a_{12}x_1x_2 + a_{13}x_1x_3 + a_{14}x_1x_4$$
$$+ a_{21}x_2x_1 + a_{22}x_2^2 \;\;\;+ a_{23}x_2x_3 + a_{24}x_2x_4$$
$$+ a_{31}x_3x_1 + a_{32}x_3x_2 + a_{33}x_3^2 \;\;\;+ a_{34}x_3x_4$$
$$+ a_{41}x_4x_1 + a_{42}x_4x_2 + a_{43}x_4x_3 + a_{44}x_4^2 \;\;\;= 0,$$

or for greater brevity

$$(1) \qquad \sum_1^4 a_{ij}x_ix_j = 0.$$

DEFINITION 1. — *The matrix of the sixteen a's taken in the order written above is called the matrix of the quadric surface (1), the determinant of this matrix is called the discriminant of the quadric surface, its rank is called the rank of the quadric surface, and if the discriminant vanishes, the quadric surface is said to be singular.*

A fundamental problem is the following: If $(y_1,\ y_2,\ y_3,\ y_4)$ and $(z_1,\ z_2,\ z_3,\ z_4)$ are two points, in what points does the line yz meet the surface (1)?

The coördinates of any point on yz, other than y, may be written

$$(z_1 + \lambda y_1,\ z_2 + \lambda y_2,\ z_3 + \lambda y_3,\ z_4 + \lambda y_4).$$

A necessary and sufficient condition for this point to lie on (1) is

$$\sum_1^4 a_{ij}(z_i + \lambda y_i)(z_j + \lambda y_j) = 0,$$

or expanded,

$$(2) \qquad \sum_1^4 a_{ij}z_iz_j + 2\lambda \sum_1^4 a_{ij}y_iz_j + \lambda^2 \sum_1^4 a_{ij}y_iy_j = 0.$$

If the point y does not lie on (1), this is a quadratic equation in λ. To each root of this equation corresponds one point where the line meets the quadric. Thus we see that every line through a point y which does not lie on a quadric surface, meets this surface either in two, and only two, distinct points, or in only one point.

On the other hand, if y does lie on (1), the equation (2) reduces to an equation of the first degree, provided $\Sigma a_{ij}y_iz_j \neq 0$. In this case, also, the line meets the surface in two, and only two, distinct points, viz., the point y and the point corresponding to the root of the equation of the first degree (2).

Finally, if $\Sigma a_{ij}y_iy_j = \Sigma a_{ij}y_iz_j = 0$, the first member of equation (2) reduces to a constant, so that (2) is either satisfied by no value of λ, in which case the line meets the surface at the point y only, or by all values of λ (if $\Sigma a_{ij}z_iz_j = 0$), in which case every point on the line is also a point on the surface.

Combining the preceding results we may say:

THEOREM 1. *If a quadric surface and a straight line are given, one of the following three cases must occur :*

(1) The line meets the quadric in two, and only two, points, in which case the line is called a secant.

*(2) The line meets the quadric in one, and only one, point, in which case it is called a tangent.**

(3) Every point of the line is a point of the quadric. In this case the line is called a ruling of the quadric.†

---

* We shall presently distinguish between true tangents and pseudo-tangents.

† Also called a *generator*, because, as will presently appear, the whole surface may be generated by the motion of such a line.

That all these three cases are possible is shown by simple examples; for instance, in the case of the surface

$$y^2 + z^2 - zt = 0,$$

the three coördinate axes illustrate the three cases.

We shall often find it convenient to say that a tangent line meets the quadric in *two coincident points*.

From the proof we have given of Theorem 1, we can also infer the further result:

THEOREM 2. *If (y_1, y_2, y_3, y_4) is a point on the quadric* (1), *then if*

$$(3) \qquad \overset{4}{\underset{1}{\Sigma}} a_{ij} x_i y_j \equiv 0,*$$

every line through y is either a tangent or a ruling of (1), *otherwise every line through y which lies in the plane*

$$(4) \qquad \overset{4}{\underset{1}{\Sigma}} a_{ij} x_i y_j = 0$$

is a tangent or ruling of (1), *while every other line through y is a secant.*

A theorem of fundamental importance, which follows immediately from this, is:

THEOREM 3. *If there exists a point y on the quadric* (1) *such that the identity* (3) *is fulfilled, then* (1) *is a cone with y as a vertex; and, conversely, if* (1) *is a cone with y as a vertex, then the identity* (3) *is fulfilled.*

We pass now to the subject of tangent planes, which we define as follows:

DEFINITION 2. *A plane p is said to be tangent to the quadric* (1) *at one of its points P, if every line of p which passes through P is either a tangent or a ruling of* (1).

It will be seen that, according to this definition, if (1) is a cone, every plane through a vertex of (1) is tangent to (1) at this vertex. We have thus included among the tangent planes, planes which in ordinary geometric parlance would not be called tangent. The same objection applies to our definition of tangent lines. We therefore now introduce the distinction between true tangent lines or planes and pseudo-tangent lines or planes.

DEFINITION 3. *A line or plane which touches a quadric surface at a point which is not a vertex is called a true tangent; all other tangent lines and planes are called pseudo-tangents.*

---

* It should be noticed that, on account of the relation $a_{ij} = a_{ji}$, $\Sigma a_{ij} x_i y_j \equiv \Sigma a_{ij} y_i x_j$.

EXERCISES

1. Prove that if P is a point on a quadric surface S, which is not a vertex, and p the tangent plane at this point, one of the following three cases must occur :

(a) Two, and only two, lines of p are rulings of S, and these rulings intersect at P.

(b) One, and only one, line of p is a ruling of S, and this ruling passes through P.

(c) Every line of p is a ruling of S.

2. Prove that

(a) When case (a) of Exercise 1 occurs, the quadric surface is not a cone ; and, conversely, if the quadric surface is not a cone, case (a) will always occur.

(b) If case (b) of Exercise 1 occurs, p is tangent to S at every point of the ruling which lies in p.

(c) If case (b) of Exercise 1 occurs, S is a cone with one, and only one, vertex, and this vertex is on the ruling which lies in p; and conversely, if S is a cone with one, and only one, vertex, case (b) will always occur.

(d) If case (c) of Exercise 1 occurs, there is a line l in p every point of which (but no other point) is a vertex of S; and S consists of two planes one of which is p, while the other intersects it in l.

39. Conjugate Points and Polar Planes. Two points are commonly said to be conjugate with regard to a quadric surface

$$\text{(1)} \qquad \sum_{1}^{4} a_{ij} x_i x_j = 0,$$

when they are divided harmonically by the points where the line connecting them meets the surface. In order to include all limiting cases, we frame the definition as follows:

DEFINITION. *Two distinct points are said to be conjugate with regard to the surface* (1) *if*

(a) *The line joining them is a tangent or a secant to* (1), *and the points are divided harmonically by the points where this line meets* (1); *or*

(b) *The line joining them is a ruling of* (1).

Two coincident points are called conjugate if they both lie on (1).

Let the coördinates of the points be $(y_1,\, y_2,\, y_3,\, y_4)$ and $(z_1,\, z_2,\, z_3,\, z_4)$, and let us look first at the case in which the points are distinct and neither of them lies on (1), and in which the line connecting them is a secant of (1). The points of intersection of the line yz with (1) may therefore be written

$$(z_1 + \lambda_i y_1,\, z_2 + \lambda_i y_2,\, z_3 + \lambda_i y_3,\, z_4 + \lambda_i y_4) \qquad (i = 1,\, 2),$$

where λ_1 and λ_2 are the roots of Equation (2), § 38. A necessary and sufficient condition for harmonic division is that the cross-ratio λ_1 / λ_2 have the value -1 ; that is $\lambda_1 + \lambda_2 = 0$; or, referring back to Equation (2), § 38,

$$(2) \qquad \overset{4}{\underset{1}{\Sigma}} a_{ij} y_i z_j = 0.$$

We leave it for the reader to show that in all other cases in which y and z are conjugate this relation (2) is fulfilled ; and that, conversely, whenever this condition is fulfilled, the points are conjugate. That is:

THEOREM 1. *A necessary and sufficient condition that the points y, z be conjugate with regard to (1) is that (2) be fulfilled.*

This theorem enables us at once to write down the equation of the locus of the point x conjugate to a fixed point y, namely,

$$(3) \qquad \overset{4}{\underset{1}{\Sigma}} a_{ij} x_i y_j = 0.$$

Except when the first member of this equation vanishes identically, this locus is therefore a plane called the *polar plane* of the point y. We saw in the last section that the first member of (3) vanishes identically when (1) is a cone and y is a vertex. This is the only case in which it vanishes identically ; for, if y is any point, not a vertex, on a quadric surface, (3) represents the tangent plane at that point; while if y is not on (1), the first member of (3) can clearly not vanish identically, since it does not vanish when the x's are replaced by the y's. Hence the theorem :

THEOREM 2. *If (1) is not a cone, every point y has a definite polar plane (3) ; if (1) is a cone, every point except its vertices has a definite polar plane (3), while for the vertices the first member of (3) is identically zero.*

We note that the property that a plane is the polar of a given point with regard to a quadric surface is a projective property, since a collineation of space evidently carries over two conjugate points into points conjugate with regard to the transformed surface.

THEOREM 3. *If two points P_1 and P_2 are so situated that the polar plane of P_1 passes through P_2, then, conversely, the polar plane of P_2 will pass through P_1.*

For, from the hypothesis, it follows that P_1 and P_2 are conjugate points, and from this the conclusion follows.

40. Classification of Quadric Surfaces by Means of their Rank. Theorem 2 of the last section may be stated by saying that a necessary and sufficient condition that the quadric surface

$$(1) \qquad \overset{4}{\underset{1}{\Sigma}} a_{ij}x_i x_j = 0$$

be a cone and that $(y_1,\ y_2,\ y_3,\ y_4)$ be its vertex (or one of its vertices) is that

$$(2) \qquad \overset{4}{\underset{1}{\Sigma}} a_{ij}x_i y_j \equiv 0.$$

This identity (2) is equivalent to the four equations

$$(3) \qquad \begin{cases} a_{11}y_1 + a_{12}y_2 + a_{13}y_3 + a_{14}y_4 = 0, \\ a_{21}y_1 + a_{22}y_2 + a_{23}y_3 + a_{24}y_4 = 0, \\ a_{31}y_1 + a_{32}y_2 + a_{33}y_3 + a_{34}y_4 = 0, \\ a_{41}y_1 + a_{42}y_2 + a_{43}y_3 + a_{44}y_4 = 0. \end{cases}$$

A necessary and sufficient condition for this set of equations to have a common solution other than $(0,\ 0,\ 0,\ 0)$ is that the determinant of their coefficients be zero. We notice that this determinant is the discriminant a of the quadric surface. Hence,

THEOREM 1. *A necessary and sufficient condition for a quadric surface to be a cone is that its discriminant vanish.*

If, then, the rank of the quadric surface is *four*, the surface is not a cone.

If the rank is *three*, the set of equations (3) has one, and, except for multiples of this, only one, solution. Hence in this case the surface is an ordinary cone with a single vertex.

If the rank is *two*, equations (3) have two linearly independent solutions (cf. §18), on which all other solutions are linearly dependent. Hence in this case the surface is a cone with a whole line of vertices.

If the rank is *one*, equations (3) have three linearly independent solutions on which all other solutions are linearly dependent. Hence we have a cone with a whole plane of vertices.

If the rank is *zero* we have, strictly speaking, no quadric surface; but the locus of (1) may be regarded as a cone, every point in space being a vertex.

It is clear that the property of a quadric surface being a cone is a projective property ; and the same is true of the property of a point being a vertex of a cone. Hence from the classification we have just given we infer

THEOREM 2. *The rank of a quadric surface is unchanged by non-singular collineations.*

EXERCISES

1. DEFINITION. *If a plane p is the polar of a point P with regard to a quadric surface, then P is called a pole of p.*

Prove that if the quadric surface is non-singular, every plane has one, and only one, pole.

2. Prove that if the quadric surface is a cone, a plane which does not pass through a vertex has no pole.

What can be said here about planes which do pass through a vertex?

41. Reduction of the Equation of a Quadric Surface to a Normal Form. Since cross-ratio is invariant under a non-singular collineation, a quadric surface S, a point P, not on S, and its polar plane with regard to S, are carried over by any non-singular collineation into a quadric surface S', a point P', and its polar plane with regard to S'. A point (y_1, y_2, y_3, y_4) not on the quadric surface $\sum_1^4 a_{ij}x_ix_j = 0$, cannot be on its own polar plane $\sum_1^4 a_{ij}x_iy_j = 0$ as we see by replacing the x's in this last equation by the y's. Now transform by a collineation so that this point becomes the origin and its polar plane the plane at infinity.* The quadric surface will now be a central quadric with center at the origin, since, if any line be drawn through the origin, the two points in which this line meets the surface are divided harmonically by the origin and the point at infinity on this line.

The equation of the polar plane of the point (y_1', y_2', y_3', y_4') with regard to the transformed quadric

$$\sum_1^4 a_{ij}'x_i'x_j' = 0$$

is

$$\sum_1^4 a_{ij}'x_i'y_j' = 0,$$

---

* Such a collineation can obviously be determined in an infinite number of ways by means of the theorem that there exists a collineation which carries over any five linearly independent points into any five linearly independent points ; cf. Exercises 2, 3, § 24.

which reduces to the simple form

$$a'_{14}x'_1 + a'_{24}x'_2 + a'_{34}x'_3 + a'_{44}x'_4 = 0$$

when the point is the origin $(0, 0, 0, 1)$. For this equation to represent the plane at infinity, we must have

$$a'_{14} = a'_{24} = a'_{34} = 0, \; a'_{44} \neq 0.$$

Hence the quadric surface becomes

$$\begin{aligned}
& a'_{11}\, x'^2_1 && + a'_{12}\, x'_1\, x'_2 + a'_{13}\, x'_1\, x'_3 \\
& + a'_{21}\, x'_2\, x'_1 && + a'_{22}\, x'^2_2 && + a'_{23}\, x'_2\, x'_3 \\
& + a'_{31}\, x'_3\, x'_1 && + a'_{32}\, x'_3\, x'_2 + a'_{33}\, x'^2_3 \\
& && && + a'_{44}\, x'^2_4 = 0.
\end{aligned}$$

A slightly different reduction can be performed by transforming the point (y_1, y_2, y_3, y_4) to the point at infinity on the x_1-axis and its polar plane to the x_2x_3-plane. It is easy to see that we thus get rid of the terms containing x_1 except the square term.

Similarly we can get rid of the terms containing x_2 and x_3. Thus we see that *any quadric surface can be reduced by a collineation to a form where its equation contains no term in x_i except the term in x_i^2 whose coefficient then is not zero.*

According as we take for i the values 1, 2, 3, 4, we get thus four different normal forms for the equation of our quadric surface, and inasmuch as each of these forms can be obtained in a great variety of ways, the question naturally arises whether we cannot perform all four reductions simultaneously. That this can, in general, be done may be seen as follows: let y be a point not on the quadric surface, and z any point on the polar plane of y, but not on the quadric surface. Its polar plane contains y. Let w be any point on the intersection of the polar planes of y and z, but not on the quadric surface. Then its polar plane passes through y and z. These three polar planes meet in some point u, and it is readily seen that the four points y, z, w, u do not lie on a plane. The tetrahedron $yzwu$ is called a *polar* or *self-conjugate* tetrahedron of the quadric surface, since it has the property that any vertex is the pole of the opposite face.

If we transform the four points, y, z, w, u to the origin and the points at infinity on the three axes, the effect will be the same as that of the separate transformations above, that is, the equation of the quadric surface will be reduced to the form

$$a'_{11}\, x'^2_1 + a'_{22}\, x'^2_2 + a'_{33}\, x'^2_3 + a'_{44}\, x'^2_4 = 0.$$

We have tacitly assumed that it is possible to find points y, z, w, constructed as indicated above, and not lying on the quadric surface. We leave it for the reader to show that, if the quadric surface is not a cone, this will always be possible in an infinite number of ways. A cone, however, has no self-conjugate tetrahedron, and in this case the above reduction is impossible.

EXERCISES

1. Prove that if the discriminant of a quadric surface is zero, the equation of the surface can always be reduced, by a suitable collineation, to a form in which the coördinate x_4 does not enter.

[SUGGESTION. Show, by using the results of this chapter, that if the vertex of a quadric cone is at the origin, $a_{14} = a_{24} = a_{34} = a_{44} = 0$.]

2. Show that, provided the cone has a finite vertex, the collineation of Exercise 1 may be taken in the form

$$x_1' = x_1 + a\,x_4,$$
$$x_2' = x_2 + \beta\,x_4,$$
$$x_3' = x_3 + \gamma\,x_4,$$
$$x_4' = x_4.$$

[SUGGESTION. Use non-homogeneous coördinates.]

CHAPTER X

QUADRATIC FORMS

42. The General Quadratic Form and its Polar. The general quadratic form in n variables is

$$(1) \qquad \sum_1^n a_{ij}x_ix_j \equiv a_{11}x_1^2 \quad + a_{12}x_1x_2 + \dots + a_{1n}x_1x_n$$
$$+ a_{21}x_2x_1 + a_{22}x_2^2 \quad + \dots + a_{2n}x_2x_n$$
$$\cdot \quad \cdot \quad \cdot \quad \cdot \quad \cdot \quad \cdot \quad \cdot \quad \cdot$$
$$\cdot \quad \cdot \quad \cdot \quad \cdot \quad \cdot \quad \cdot \quad \cdot \quad \cdot$$
$$+ a_{n1}x_nx_1 + a_{n2}x_nx_2 + \dots + a_{nn}x_n^2,$$

where $a_{ij} = a_{ji}$.* The bilinear form $\sum_1^n a_{ij}y_iz_j$ is called the *polar form* of (1). Subjecting (1) to the linear transformation

$$\mathbf{c} \qquad \begin{cases} x_1 = c_{11}x_1' + \dots + c_{1n}x_n', \\ \cdot \quad \cdot \quad \cdot \quad \cdot \quad \cdot \quad \cdot \quad \cdot \\ \cdot \quad \cdot \quad \cdot \quad \cdot \quad \cdot \quad \cdot \quad \cdot \\ x_n = c_{n1}x_1' + \dots + c_{nn}x_n', \end{cases}$$

we get a new quadratic form

$$(2) \qquad \sum_1^n a_{ij}'x_i'x_j'.$$

The polar form of (2) is $\Sigma a_{ij}'y_i'z_j'$. If we transform the y's and z's of the polar form of (1) by the same transformation $\mathbf{c}$, we get a new bilinear form $\sum_1^n \bar{a}_{ij}y_i'z_j'$. We will now prove that $\bar{a}_{ij} = a_{ij}'$.

We have the identities

$$(3) \qquad \sum_1^n a_{ij}x_ix_j \equiv \sum_1^n a_{ij}'x_i'x_j',$$

$$(4) \qquad \sum_1^n a_{ij}y_iz_j \equiv \sum_1^n \bar{a}_{ij}y_i'z_j'.$$

---

* It should be clearly understood that this restriction is a matter of convenience, not of necessity. If it were not made, the quadratic form would be neither more nor less general.

Each of these we may regard as identities in the x'''s, y'''s, z' s, the x's, y's, z's being merely abbreviations for certain polynomials in the corresponding primed letters. The last written identity reduces, when we let $y_i' = z_i' = x_i'$ $(i = 1, 2, \ldots n)$, to

$$\sum_1^n a_{ij} x_i x_j \equiv \sum_1^n \bar{a}_{ij} x_i' x_j'.$$

Combining this with (3) gives

$$\sum_1^n \bar{a}_{ij} x_i' x_j' \equiv \sum_1^n a_{ij}' x_i' x_j'.$$

Hence $\qquad \qquad \bar{a}_{ii} = a_{ii}'$ and $\bar{a}_{ij} + \bar{a}_{ji} = a_{ij}' + a_{ji}'.$

We have assumed that $a_{ij}' = a_{ji}'$, these being merely the coefficients of a certain quadratic form, and we proved, in Theorem 4, § 36, that $\bar{a}_{ij} = \bar{a}_{ji}$. Hence we infer that $\bar{a}_{ij} = a_{ij}'.$

From this fact and from (4) we get at once the further result:

$$\sum_1^n a_{ij}' y_i' z_j' \equiv \sum_1^n a_{ij} y_i z_j.$$

That is:

THEOREM. *The polar form*

$$\sum_1^n a_{ij} y_i z_j$$

is an absolute covariant of the system composed of the quadratic form

$$\sum_1^n a_{ij} x_i x_j$$

and the two points $\qquad (y_1, \ldots y_n), \; (z_1, \ldots z_n).$

43. The Matrix and the Discriminant of a Quadratic Form.

DEFINITION. *The matrix*

$$a = \left\| \begin{array}{ccc} a_{11} & \cdots & a_{1n} \\ \cdot & \cdot \; \cdot & \cdot \\ \cdot & \cdot \; \cdot & \cdot \\ a_{n1} & \cdots & a_{nn} \end{array} \right\|$$

is called the matrix of the quadratic form

$$(1) \qquad \qquad \sum_1^n a_{ij} x_i x_j.$$

The determinant of **a** *is called the discriminant of* (1); *and the rank of* **a**, *the rank of* (1). *If the discriminant vanishes,* (1) *is called singular.*

The matrix of (1) is the matrix of its polar form. Moreover, as was shown in the last section, if the x's in (1) are subjected to a linear transformation, and the y's and z's in the polar of (1) are subjected to the same transformation, the matrix of the new quadratic form will be the same as the matrix of the new bilinear form. But we saw, in Theorem 1, §36, how the matrix of a bilinear form is changed by linear transformations of the variables. Thus we have the theorem:

THEOREM 1. *If in the quadratic form* (1) *with matrix* a *we subject the x's to a linear transformation with matrix* c, *we obtain a new quadratic form with matrix* c′ac, *where* c′ *is the conjugate of* c.

From this there follow at once, precisely as in §36, the further results:

THEOREM 2. *The rank of a quadratic form is not changed by non-singular linear transformation.*

THEOREM 3. *The discriminant of a quadratic form is a relative invariant of weight two.*

44. Vertices of Quadratic Forms.

DEFINITION. *By a vertex of the quadratic form*

$$(1) \qquad \sum_1^n a_{ij} x_i x_j,$$

we understand a point $(c_1, \cdots c_n)$ *where the c's are not all zero, such that*

$$(2) \qquad \sum_1^n a_{ij} x_i c_j \equiv 0.$$

A quadratic form clearly vanishes at all of its vertices.

It is merely another way of stating this definition when we say:

THEOREM 1. *A necessary and sufficient condition that* $(c_1, \cdots c_n)$ *be a vertex of* (1) *is that it be a solution, not consisting exclusively of zeros, of the system of equations*

$$(3) \qquad \begin{aligned} a_{11}c_1 + \cdots + a_{1n}c_n &= 0, \\ &\ \cdots \cdots \cdots \\ a_{n1}c_1 + \cdots + a_{nn}c_n &= 0. \end{aligned}$$

Since the resultant of (3) is the discriminant of (1), we may add:

THEOREM 2. *A necessary and sufficient condition for a quadratic form to have a vertex is that its discriminant be zero; and if the rank of the form is r, it has n − r linearly independent vertices, and every point linearly dependent on these is a vertex.*

In particular, we note that if the discriminant of a quadratic form is zero and if the cofactors of the elements of this determinant are denoted in the ordinary way by A_{ij}, then $(A_{i1}, \cdots A_{in})$ is a vertex, provided all these A's are not zero.

The following identity is of great importance (cf. formula (2), § 38),

$$(4) \qquad \sum_1^n a_{ij}(z_i + \lambda y_i)(z_j + \lambda y_j) \equiv \sum_1^n a_{ij}z_iz_j + 2\lambda \sum_1^n a_{ij} z_iy_j + \lambda^2 \sum_1^n a_{ij}y_iy_j.$$

This may be regarded as an identity in all the letters involved.

If $(c_1, \cdots c_n)$ is a vertex of the quadratic form $\sum_1^n a_{ij}x_ix_j$, and these c's are substituted in (4) in place of the y's, the last two terms of the second member of this identity are zero, and we have

$$(5) \qquad \sum_1^n a_{ij}(z_i + \lambda c_i)(z_j + \lambda c_j) \equiv \sum_1^n a_{ij}z_iz_j;$$

and conversely, if (5) holds, $(c_1, \cdots c_n)$ is a vertex; for subtracting (5) from (4), after substituting the c's for the y's in (4), we have

$$2\lambda \sum_1^n a_{ij}z_ic_j + \lambda^2 \sum_1^n a_{ij} c_ic_j \equiv 0,$$

and, this being an identity in λ as well as in the z's, we have

$$\sum_1^n a_{ij}z_ic_j \equiv 0.$$

Thus we have proved the following theorem:

THEOREM 3. *A necessary and sufficient condition that $(c_1, \cdots c_n)$ be a vertex of the quadratic form (1) is that $z_1, \cdots z_n$ and λ being independent variables, the identity (5) be fulfilled.*

<h3 style="text-align:center">EXERCISES</h3>

1. Prove that if $(c_1, \ldots c_n)$ is a vertex of (1), and $(y_1, \ldots y_n)$ is any point at which the quadratic form vanishes, then the quadratic form vanishes at every point linearly dependent on c and y.

2. State and prove a converse to 1.

45. Reduction of a Quadratic Form to a Sum of Squares. If in the quadratic form

$$(1) \qquad \phi(x_1, \cdots x_n) \equiv \sum_1^n a_{ij} x_i x_j$$

the coefficient a_{ii} is not zero, we may simplify the form by the following transformation due to Lagrange.

The difference

$$\phi(x_1, \cdots x_n) - \frac{1}{a_{ii}}(a_{i1}x_1 + \cdots + a_{in}x_n)^2$$

is evidently independent of x_i. Denoting it by ϕ_1, we have

$$\phi \equiv \frac{1}{a_{ii}}(a_{i1}x_1 + \cdots + a_{in}x_n)^2 + \phi_1.$$

If, then, we perform the non-singular linear transformation

$$(2) \qquad \begin{cases} x_1' = a_{i1}x_1 + a_{i2}x_2 + \cdots + a_{in}x_n \\ x_2' = \qquad\qquad\quad x_2 \\ \;\cdot\;\cdot\;\cdot\;\cdot\;\cdot\;\cdot\;\cdot\;\cdot\;\cdot \\ x_i' = \qquad x_1 \\ \;\cdot\;\cdot\;\cdot\;\cdot\;\cdot\;\cdot\;\cdot\;\cdot\;\cdot \\ x_n' = \qquad\qquad\qquad\qquad\qquad x_n, \end{cases}$$

the quadratic form ϕ is reduced to the form

$$(3) \qquad \frac{1}{a_{ii}} x_1'^2 + \phi_1(x_2', \cdots x_n'),$$

in which all the terms in x_1' are wanting except the term in $x_1'^2$.

It will be seen that this reduction can in general be performed in a variety of ways. It becomes impossible only when the coefficients of all the square terms in the original quadratic form are zero.

Unless, in the new quadratic form ϕ_1, the coefficients of all the square terms are zero, we can apply the same reduction to this form by subjecting the variables $x_2', \cdots x_n'$ to a suitable non-singular linear transformation. This transformation may also be regarded as a non-singular linear transformation of all the x''s : $(x_1', x_2', \cdots x_n')$ if we write $x_1'' = x_1'$. We thus reduce (3) to the form

$$(4) \qquad \frac{1}{a_{ii}} x_1''^2 + \frac{1}{a_{jj}'} x_2''^2 + \phi_2(x_3'', \cdots x_n'').$$

Applying this reduction now to ϕ_2, and proceeding as before, we see that by a number of successive non-singular transformations the form ϕ can finally be reduced to the form:

$$(5) \qquad c_1 x_1^2 + c_2 x_2^2 + \cdots + c_n x_n^2.$$

These successive transformations can now be combined into a single non-singular linear transformation, and we are thus led to the

THEOREM. *Every quadratic form in n variables can be reduced to the form (5) by a non-singular linear transformation.*

The proof of this theorem is not yet complete ; for if at any stage of the reduction the quadratic form ϕ_i has the peculiarity that all its square terms are wanting, the next step in the reduction will be impossible by the method we have used. Before considering this point, we will illustrate the method of reduction by a numerical case.

Example.

$$\phi \equiv \left\{ \begin{array}{l} 2\,x_1^2 + \ \ x_1 x_2 + 8\,x_1 x_3 \\ +\ \ x_2 x_1 - 3\,x_2^2 \ \ +9\,x_2 x_3 \\ +8\,x_3 x_1 + 9\,x_3 x_2 + 2\,x_3^2 \end{array} \right\} \equiv \tfrac{1}{2}(2\,x_1 + x_2 + 8\,x_3)^2 + \phi_1$$

where

$$\phi_1 \equiv -\tfrac{1}{2}(x_2 + 8\,x_3)^2 - 3\,x_2^2 + 18\,x_2 x_3 + 2\,x_3^2 \equiv \left\{ \begin{array}{l} -\tfrac{7}{2}\,x_2^2 \ \ + \ \ 5\,x_2 x_3 \\ +5\,x_3 x_2 - 30\,x_3^2 \end{array} \right\}$$

$$\equiv -\tfrac{2}{7}\left(-\tfrac{7}{2}\,x_2 + 5\,x_3\right)^2 - \tfrac{160}{7}\,x_3^2.$$

Accordingly, by means of the non-singular linear transformation

$$\begin{cases} x_1' = 2\,x_1 + x_2 \ \ + 8\,x_3, \\ x_2' = \qquad\ -\tfrac{7}{2}\,x_2 + 5\,x_3, \\ x_3' = \qquad\qquad\qquad x_3, \end{cases}$$

the form ϕ reduces to $\quad \tfrac{1}{2}\,x_1'^2 - \tfrac{2}{7}\,x_2'^2 - \tfrac{160}{7}\,x_3'^2.$

We have given here merely *one* method of reduction. Three differ ent methods were open to us at the first step and two at the second.

We proceed now to complete the proof of the general theorem. Let us suppose that the coefficients of all the square terms in ϕ are zero,* but that $a_{12} \neq 0$. Then

$$\phi(x_1, \cdots x_n) \equiv 2\,a_{12} x_1 x_2 + 2\,x_1(a_{13} x_3 + \cdots + a_{1n} x_n)$$

$$+ 2\,x_2(a_{23} x_3 + \cdots + a_{2n} x_n) + \sum_{3}^{n} a_{ij} x_i x_j$$

$$\equiv \frac{2}{a_{12}}(a_{12} x_2 + a_{13} x_3 + \cdots + a_{1n} x_n)(a_{21} x_1 + a_{23} x_3 + \cdots + a_{2n} x_n)$$

$$+ \phi_1,$$

where $\quad \phi_1 \equiv -\dfrac{2}{a_{12}}(a_{13} x_3 + \cdots + a_{1n} x_n)(a_{23} x_3 + \cdots + a_{2n} x_n) + \sum_{3}^{n} a_{ij} x_i x_j.$

---

* This method may be used whenever $a_{11} = a_{22} = 0$ whether all the other coef ficients a_{ii} are zero or not.

The non-singular linear transformation

$$\begin{cases} x_1' = \phantom{a_{21}x_1} a_{12}x_2 + a_{13}x_3 + \cdots + a_{1n}x_n \\ x_2' = a_{21}x_1 \phantom{+a_{12}x_2} + a_{23}x_3 + \cdots + a_{2n}x_n \\ x_3' = x_3 \\ \cdot \ \cdot \ \cdot \ \cdot \ \cdot \ \cdot \ \cdot \ \cdot \ \ \cdot \ \cdot \\ \cdot \ \cdot \ \cdot \ \cdot \ \cdot \ \cdot \ \cdot \ \cdot \ \ \cdot \ \cdot \\ x_n' = x_n \end{cases}$$

thus reduces ϕ to the form

$$\frac{2}{a_{12}} x_1' x_2' + \phi_1(x_3', \cdots x_n').$$

The further non-singular transformation

$$\begin{cases} x_1'' = x_1' + x_2', \\ x_2'' = x_1' - x_2', \\ x_3'' = x_3', \\ \cdot \ \cdot \ \cdot \ \cdot \ \cdot \ \cdot \\ \cdot \ \cdot \ \cdot \ \cdot \ \cdot \ \cdot \\ x_n'' = x_n' \end{cases}$$

reduces ϕ to the form

$$\frac{1}{2\,a_{12}} x_1''^2 - \frac{1}{2\,a_{12}} x_2''^2 + \phi_1(x_3'', \cdots x_n'').$$

The above reduction was performed on the supposition that $a_{12} \neq 0$. It is clear, however, that only a slight change in notation would be necessary to carry through a similar reduction if $a_{12} = 0$ but $a_{ij} \neq 0$. The only case to which the reduction does not apply is, therefore, the one in which all the coefficients of the quadratic form are zero, a case in which no further reduction is necessary or possible.

We thus see that whenever Lagrange's reduction fails, the method last explained will apply, and thus our theorem is completely established.

EXERCISES

1. Given a quadratic form in which $n = 5$ and $a_{ij} = |\,i - j\,|$. Reduce to the form (5).

2. Reduce the quadratic form

$$9\,x^2 - 6\,y^2 - 8\,z^2 + 6xy - 14\,xz + 18\,xw + 8\,yz + 12\,yw - 4\,zw$$

to the form (5).

3. Prove that if $(y_1, \ldots y_n)$ is any point at which a given quadratic form is not zero, a linear transformation can be found (and that in an infinite number of ways) which carries this point into the point $(0, \cdots 0, 1)$ and its polar into kx_n; and show that this linear transformation eliminates from the quadratic form all terms in x_n except the term in x_n^2 which then has a coefficient not zero.

4. Prove that the transformations described in Exercise 3 are the only ones which have the effect there described.

5. Show how the two methods of reduction explained in this section come as special cases under the transformation of Exercise 3.

46. A Normal Form, and the Equivalence of Quadratic Forms. In the method of reduction explained in the last section, it may happen that, after we have taken a number of steps, and thus reduced ϕ to the form

$$c_1 x_1^2 + \cdots + c_k x_k^2 + \phi_k (x_{k+1}, \cdots x_n),$$

the form ϕ_k is identically zero. In this case no further reduction would be necessary and the form (5) of the last section to which ϕ is reduced would have the peculiarity that $c_{k+1} = c_{k+2} = \cdots = c_n = 0$, while all the earlier c's are different from zero. It is easy to see just when this case will occur.

For this purpose, consider the matrix

$$\left\| \begin{array}{cccc} c_1 & 0 & \cdots & 0 \\ 0 & c_2 & \cdots & 0 \\ \cdot & \cdot & \cdot & \cdot \\ \cdot & \cdot & \cdot & \cdot \\ 0 & 0 & \cdots & c_n \end{array} \right\|$$

of the reduced form (5) of § 45. It is clear that the rank of this matrix is precisely equal to the number of c's different from zero; and, since the rank of this reduced form is the same as that of the original form, we have the result :

THEOREM 1. *A necessary and sufficient condition that it be possible to reduce a quadratic form by means of a non-singular linear transformation to the form*

(1) $$c_1 x_1^2 + \cdots + c_r x_r^2,$$

where none of the c's are zero, is that the rank of the quadratic form be r.

This form (1) involves r coefficients $c_1, \cdots c_r$. That the values of these coefficients, apart from the fact that none of them are zero, are immaterial will be seen if we consider the effect on (1) of the transformation

$$(2) \quad \begin{cases} x_1 = \sqrt{\dfrac{k_1}{c_1}}\, x_1', \\[2mm] \cdot \quad \cdot \quad \cdot \quad \cdot \quad \cdot \\ \cdot \quad \cdot \quad \cdot \quad \cdot \quad \cdot \\ x_r = \sqrt{\dfrac{k_r}{c_r}}\, x_r', \\[2mm] x_{r+1} = x_{r+1}', \\ \cdot \quad \cdot \quad \cdot \quad \cdot \quad \cdot \\ \cdot \quad \cdot \quad \cdot \quad \cdot \quad \cdot \\ x_n = x_n', \end{cases}$$

where $k_1, \cdots k_r$ are arbitrarily given constants none of which, however, is to be zero. The transformation (2) is non-singular, and reduces (1) to the form

$$(3) \qquad k_1 x_1'^2 + \cdots + k_r x_r'^2.$$

Thus we have proved

THEOREM 2. *A quadratic form of rank r can be reduced by means of a non-singular linear transformation to the form* (3), *where the values of the constants $k_1, \cdots k_r$ may be assigned at pleasure provided none of them are zero.*

If, in particular, we assign to all the k's the value 1, we get

THEOREM 3. *Every quadratic form of rank r can be reduced to the normal form*

$$(4) \qquad x_1^2 + \cdots + x_r^2$$

by means of a non-singular linear transformation.

From this follows

THEOREM 4. *A necessary and sufficient condition that two quadratic forms be equivalent with regard to non-singular linear transformations is that they have the same rank.*

That this is a necessary condition is evident from the fact that the rank is an invariant. That it is a sufficient condition follows from the fact that, if the ranks are the same, both forms can be reduced to the same normal form (4).

The normal form (4) has no special advantage, except its symmetry, over any other form which could be obtained from (3) by assigning to the k's particular numerical values.   Thus, for instance, a normal form which might be used in place of (4) is

$$x_1^2 + \cdots + x_{r-1}^2 - x_r^2.$$

This form would have the advantage, in geometrical work, of giving rise to a real locus.

Finally we note that the transformations used in this section are not necessarily real, even though the form we start with be real.

EXERCISE

Apply the results of this section to the study of quadric surfaces.

47. Reducibility.   A quadratic form is called *reducible* when it is identically equal to the product of two linear forms, that is, when

$$(1) \qquad \sum_1^n a_{ij} x_i x_j \equiv (b_1 x_1 + b_2 x_2 + \cdots + b_n x_n)(c_1 x_1 + c_2 x_2 + \cdots + c_n x_n).$$

Let us seek a necessary and sufficient condition that this be the case. We begin by supposing the identity (1) to hold, and we consider in succession the case in which the two factors in the right-hand member of (1) are linearly independent, and that in which they are proportional.   In the first case the b's are not all proportional to the corresponding c's, and by a mere change of notation we may insure b_1, b_2 not being proportional to c_1, c_2.   This being done, the transformation

$$\begin{cases} x_1' = b_1 x_1 + b_2 x_2 + \cdots + b_n x_n \\ x_2' = c_1 x_1 + c_2 x_2 + \cdots + c_n x_n \\ x_3' = \qquad\qquad\qquad x_3 \\ \cdot \quad \cdot \quad \cdot \quad \cdot \quad \cdot \quad \cdot \quad \cdot \quad \cdot \quad \cdot \\ \cdot \quad \cdot \quad \cdot \quad \cdot \quad \cdot \quad \cdot \quad \cdot \quad \cdot \quad \cdot \\ x_n' = \qquad\qquad\qquad x_n \end{cases}$$

is non-singular and carries our quadratic form over into the form

$$x_1' x_2'.$$

The matrix of this form is readily seen to be of rank 2, hence the original form was of rank 2.

Turning now to the case in which the two factors in (1) are proportional to each other, we see that (1) may be written

$$\sum_1^n a_{ij}x_ix_j \equiv C(b_1x_1 + \cdots + b_nx_n)^2 \qquad \text{where } C \neq 0.$$

Unless all the b's are zero (in which case the rank of the quadratic form is zero) we may without loss of generality suppose $b_1 \neq 0$, in which case the linear transformation

$$\begin{cases} x_1' = b_1x_1 + \cdots + b_nx_n \\ x_2' = \qquad\qquad x_2 \\ \cdot \quad \cdot \quad \cdot \quad \cdot \quad \cdot \quad \cdot \quad \cdot \\ x_n' = \qquad\qquad x_n \end{cases}$$

will be non-singular and will reduce the quadratic form to

$$Cx_1'^2,$$

which is of rank 1.

Thus we have shown that if a quadratic form is reducible, its rank is 0, 1, or 2. We wish now, conversely, to prove that every quadratic form whose rank has one of these values is reducible.

A quadratic form of rank zero is obviously reducible.

A form of rank 1 can be reduced by a non-singular linear transformation to the form $x_1'^2$, that is,

$$\sum_1^n a_{ij}x_ix_j \equiv x_1'^2.$$

If here we substitute for x_1' its value in terms of the x's, it is clear that the form is reducible.

A form of rank 2 can be reduced to the form $x_1'^2 + x_2'^2$, that is,

$$\sum_1^n a_{ij}x_ix_j \equiv x_1'^2 + x_2'^2 \equiv (x_1' + \sqrt{-1}\,x_2')(x_1' - \sqrt{-1}\,x_2').$$

Here again, replacing x_1' and x_2' by their values in terms of the x's, the reducibility of the form follows. Hence,

THEOREM. *A necessary and sufficient condition that a quadratic form be reducible is that its rank be not greater than 2.*

48. **Integral Rational Invariants of a Quadratic Form.** We have seen that the discriminant a of a quadratic form is an invariant of weight 2. Any integral power of a, or more generally, any constant multiple of such a power, will therefore also be an invariant. We will now prove conversely the

THEOREM. *Every integral rational invariant of a quadratic form is a constant multiple of some power of the discriminant.*

Let us begin by assuming that the quadratic form

$$\text{(1)} \qquad \overset{n}{\underset{1}{\Sigma}} a_{ij} x_i x_j$$

is non-singular, and let c be the determinant of a linear transformation which carries it over into the normal form

$$\text{(2)} \qquad x_1'^2 + x_2'^2 + \cdots + x_n'^2.$$

Let $I(a_{11}, \cdots a_{nn})$ be any integral rational invariant of (1) of weight μ, and denote by k the value of this invariant when formed from (2). It is clear that k is a constant, that is, independent of the coefficients a_{ij} of (1).   Then

$$k = c^\mu I.$$

Moreover, the discriminant a being of weight 2, and having for (2) the value 1, we have

$$1 = c^2 a.$$

Raising the last two equations to the powers 2 and μ respectively, we get

$$k^2 = c^{2\mu} I^2, \qquad 1 = c^{2\mu} a^\mu.$$

From which follows

$$\text{(3)} \qquad I^2 = k^2 a^\mu.$$

This formula has been established so far for all values of the coefficients a_{ij} for which $a \neq 0$.   That it is really an identity in the a_{ij}'s is seen at once by a reference to Theorem 5, § 2.   The polynomial on the right-hand side of (3) is of degree μ in a_{11};[*] hence we see that μ must be an even number, since I^2 is of even degree in a_{11}.   Letting $\mu = 2\nu$, we infer from (3) (cf. Exercise 1, § 2) that one or the other of the identities

$$I \equiv ka^\nu, \qquad I \equiv - ka^\nu$$

must hold, and either of these identities establishes our theorem.

A comparison of the result of this section with Theorem 4, § 46 will bring out clearly the essential difference between the two conceptions of a *complete system of invariants* mentioned in § 29. It will be seen that the rank of a quadratic form is in itself a complete system of invariants for this form in the sense of Definition 2, § 29 ; while the discriminant of the form is in itself a complete system in the sense of the footnote appended to this definition.

---

[*] We assume here that $k \neq 0$, as otherwise the truth of the theorem would be obvious.

49. A Second Method of Reducing a Quadratic Form to a Sum of Squares. By the side of Lagrange's method of reducing a quadratic form to a sum of squares, there are many other methods of accomplishing the same result, one of the most useful of which we proceed to explain. It depends on the following three theorems. The proof of the first of these theorems is due to Kronecker and establishes, in a remarkably simple manner, the fact that any quadratic form of rank r can be written in terms of r variables only, a fact which has already been proved by another method in Theorem 1, § 46.

THEOREM 1. *If the rank of the quadratic form*

$$(1) \qquad \phi(x_1, \cdots x_n) \equiv \sum_1^n a_{ij} x_i x_j$$

is $r > 0$, and if the variables $x_1, \cdots x_n$ are so numbered that the r-rowed determinant in the upper left-hand corner of its matrix is not zero, *new variables $x_1', \cdots x_n'$ can be introduced by means of a non-singular linear transformation such that*

$$x_i' = x_i \qquad\qquad (i = r + 1, \cdots n),$$

and such that (1) reduces to the form

$$\sum_1^r a_{ij} x_i' x_j'.$$

This, it will be noticed, is a quadratic form in r variables in which the coefficients, so far as they go, are the same as in the given form (1).

In order to prove this theorem, we begin by finding a vertex $(c_1, \cdots c_n)$ of the form (1) by means of Equations (3), § 44. Since the r-rowed determinant which stands in the upper left-hand corner of the matrix of these equations is not zero, the values of $c_{r+1}, \cdots c_n$ may be chosen at pleasure, and the other c's are then completely determined. If we let $c_{r+1} = c_{r+2} = \cdots = c_{n-1} = 0$, $c_n = 1$, we get a vertex

$$(c_1, \cdots c_r, 0, \cdots 0, 1).$$

Using this vertex in the identity (5), § 44, we have

$$\phi(x_1 + \lambda c_1, \cdots x_r + \lambda c_r, x_{r+1}, \cdots x_{n-1}, x_n + \lambda) \equiv \phi(x_1, \cdots x_n).$$

If we let $\lambda = -x_n$, this identity reduces to

$$\phi(x_1 - c_1 x_n, \cdots x_r - c_r x_n, x_{r+1}, \cdots x_{n-1}, 0) \equiv \phi(x_1, \cdots x_n).$$

---

* That such an arrangement is possible is evident from Theorem 3, § 20.

Accordingly, if we perform the non-singular linear transformation [*]

$$\begin{cases} x_i' = x_i - c_i x_n & (i = 1, \cdots r), \\ x_i' = x_i & (i = r+1, \cdots n), \end{cases}$$

the quadratic form (1) reduces to

$$\phi(x_1', \cdots x_{n-1}', 0) \equiv \sum_1^{n-1} a_{ij} x_i' x_j'.$$

This, being a quadratic form in $n-1$ variables of rank r and so arranged that the r-rowed determinant which stands in the upper left-hand corner of its matrix is not zero, can be reduced, by the method just explained, to the form

$$\sum_1^{n-2} a_{ij} x_i'' x_j'',$$

where the linear transformation used is non-singular and such that

$$x_i'' = x_i' \qquad (i = r+1, \cdots n-1).$$

By adding the formula $\quad x_n'' = x_n',$

we may regard this as a non-singular linear transformation in the n variables. This transformation may then be combined with the one previously used, thus giving a non-singular transformation in which

$$x_i'' = x_i, \qquad (i = r+1, \cdots n),$$

and such that it reduces (1) to the form

$$\sum_1^{n-2} a_{ij} x_i'' x_j''.$$

Proceeding in this way step by step, our theorem is at last proved.

In the next two theorems we denote by A_{ij} in the usual way the cofactor of a_{ij} in the discriminant a of the quadratic form (1).

THEOREM 2. *If $A_{nn} \neq 0$, new variables $x_1', \cdots x_n'$ can be introduced by a non-singular transformation in such a way that*

$$x_n' = x_n$$

and that (1) takes the form

$$\sum_1^{n-1} a_{ij} x_i' x_j' + \frac{a}{A_{nn}} x_n'^2.$$

---

[*] This transformation should be compared with Exercise 2, § 41.

To prove this we consider the quadratic form

$$\sum_{1}^{n} a_{ij}\, x_i x_j - \frac{a}{A_{nn}}\, x_n^2.$$

Its discriminant is

$$\begin{vmatrix} a_{11} & \cdots & a_{1,\,n-1} & a_{1n} \\ \cdot & \cdot \cdot \cdot \cdot \cdot \cdot & \cdot \cdot \cdot & \cdot \\ \cdot & \cdot \cdot \cdot \cdot \cdot \cdot & \cdot \cdot \cdot & \cdot \\ a_{n-1,\,1} & \cdots & a_{n-1,\,n-1} & a_{n-1,\,n} \\ a_{n1} & \cdots & a_{n,\,n-1} & a_{nn} - \dfrac{a}{A_{nn}} \end{vmatrix} = a - A_{nn}\frac{a}{A_{nn}} = 0.$$

Hence by means of a non-singular transformation of the kind used in the last theorem, an essential point being that $x_n' = x_n$, we get

$$\sum_{1}^{n} a_{ij}\, x_i x_j - \frac{a}{A_{nn}}\, x_n^2 \equiv \sum_{1}^{n-1} a_{ij}\, x_i' x_j',$$

or

$$\sum_{1}^{n} a_{ij}\, x_i x_j \equiv \sum_{1}^{n-1} a_{ij}\, x_i' x_j' + \frac{a}{A_{nn}}\, x_n'^2.$$

THEOREM 3. *If*

$$A_{nn} = A_{n-1,n-1} = 0, \qquad A_{n,n-1} \neq 0,$$

new variables $x_1', \cdots x_n'$ can be introduced by a non-singular transforma-tion in such a way that

$$x_{n-1}' = x_{n-1}, \qquad x_n' = x_n,$$

and that (1) *takes the form*

$$\sum_{1}^{n-2} a_{ij}\, x_i' x_j' + \frac{2\,a}{A_{n,\,n-1}}\, x_n' x_{n-1}'.$$

Let us denote by B the determinant obtained by striking out the last two rows and columns of a. Then (cf. Corollary 3, § 11) we have

$$(2) \qquad aB = \begin{vmatrix} A_{n-1,\,n-1} & A_{n-1,\,n} \\ A_{n,\,n-1} & A_{nn} \end{vmatrix} = -A_{n,\,n-1}^2 \neq 0.$$

Consider, now, the quadratic form

$$(3) \qquad \sum_{1}^{n} a_{ij} x_i x_j - \frac{2\,a}{A_{n,\,n-1}}\, x_n x_{n-1}.$$

Its discriminant is

$$(4) \quad \begin{vmatrix} a_{11} & \cdots & a_{1,\,n-1} & & a_{1n} \\ \cdot & \cdots & \cdots & \cdots & \cdot \\ \cdot & \cdots & \cdots & \cdots & \cdot \\ a_{n-1,\,1} & \cdots & a_{n-1,\,n-1} & & a_{n-1,\,n} - \dfrac{a}{A_{n,\,n-1}} \\ a_{n1} & \cdots & a_{n,\,n-1} - \dfrac{a}{A_{n,\,n-1}} & & a_{nn} \end{vmatrix}$$

$$= a - A_{n,\,n-1} \frac{a}{A_{n,\,n-1}} - A_{n,\,n-1} \frac{a}{A_{n,\,n-1}} - B \left(\frac{a}{A_{n,\,n-1}} \right)^2,$$

which has the value zero, as we see by making use of (2).   Not only does the determinant (4) vanish, but its principal minors obtained by striking out its last row and column and its next to the last row and column are zero, being A_{nn} and $A_{n-1,\,n-1}$ respectively.   The minor obtained by striking out the last two rows and columns from (4) is B, and, by (2), this is not zero.   Thus we see (cf. Theorem 1, § 20) that the determinant (4) is of rank $n-2$.   Hence, by Theorem 1, we can reduce (3) by a non-singular linear transformation in which $x'_{n-1} = x_{n-1}$, $x'_n = x_n$ to the form

$$\sum_1^n a_{ij} x_i x_j - \frac{2\,a}{A_{n,\,n-1}} x_n x_{n-1} \equiv \sum_1^{n-2} a_{ij} x'_i x'_j.$$

Hence

$$\sum_1^n a_{ij} x_i x_j \equiv \sum_1^{n-2} a_{ij} x'_i x'_j + \frac{2\,a}{A_{n,\,n-1}} x'_n x'_{n-1}.$$

COROLLARY.   *Under the conditions of Theorem 3, the quadratic form* (1) *can be reduced to the form*

$$\sum_1^{n-2} a_{ij} x'_i x'_j + \frac{2a}{A_{n,\,n-1}} \left(x'^2_{n-1} - x'^2_n \right)$$

by a non-singular linear transformation.

To see this we have merely first to perform the reduction of Theorem 3, and then to follow this by the additional non-singular transformation

$$\begin{cases} x'_i = x''_i & (i = 1, 2, \cdots n - 2). \\ x'_{n-1} = x''_{n-1} - x''_n, \\ x'_n = x''_{n-1} + x''_n. \end{cases}$$

Having thus established these three theorems, the method of reducing a quadratic form completely is obvious. If the form (1) is singular, we begin by reducing it by Theorem 1 to

$$\sum_{1}^{r} a_{ij} x_i x_j,$$

where r is the rank of the form. Unless all the principal $(r-1)$-rowed minors of the discriminant of this form are zero, the order of the variables $x_1, \cdots x_r$ can be so arranged that the reduction of Theorem 2 is possible, a reduction which may be regarded as a non-singular linear transformation of all n variables. If all the principal $(r-1)$-rowed minors are zero, there will be at least one of the cofactors A_{ij} which is not zero, and, by a suitable rearrangement of the order of the variables, this may be taken as $A_{r,\,r-1}$. The reduction of Theorem 3, Corollary, will then be possible. Proceeding in this way, we finally reach the result, precisely as in Theorem 1, § 46, that a quadratic form of rank r can always be reduced by a non-singular linear transformation to the form

$$c_1 x_1^2 + \cdots + c_r x_r^2.$$

It may be noticed that the arrangement of the transformation of this section is in a certain sense precisely the reverse of that of §45, inasmuch as we here leave at each step the *coefficients* of the unreduced part of the form unchanged, but change the *variables* which enter into this part; while in §45 we change the *coefficients* of the unreduced part, but leave the *variables* in it unchanged.

CHAPTER XI

REAL QUADRATIC FORMS

50. The Law of Inertia. We come now to the study of real quadratic forms and the effect produced on them by real linear transformations.

We notice, here, to begin with, that the only operations involved in the last chapter are rational operations (*i.e.* addition, subtraction, multiplication, and division) with the single exception of the radicals which come into formula (2), § 46. In particular the reduction of § 45 (or the alternative reduction of § 49) involves only rational operations. Consequently, since rational operations performed on real quantities give real results, we have

THEOREM 1. *A real quadratic form of rank r can be reduced by means of a real non-singular linear transformation to the form*

$$(1) \qquad c_1 x_1'^2 + c_2 x_2'^2 + \cdots + c_r x_r'^2$$

where $c_1, \cdots c_r$ are real constants none of which are zero.

As we saw in the last chapter, this reduction can be performed in a variety of ways, and the values of the coefficients $c_1, \cdots c_r$ in the reduced form will be different for the different reductions. The signs of these coefficients, apart from the order in which they occur, will not depend on the particular reduction used, as is stated in the following important theorem discovered independently by Jacobi and Sylvester and called by the latter the Law of Inertia of Quadratic Forms:

THEOREM 2. *If a real quadratic form of rank r is reduced by two real non-singular linear transformations to the forms* (1) *and*

$$(2) \qquad k_1 x_1''^2 + k_2 x_2''^2 + \cdots + k_r x_r''^2,$$

respectively, then the number of positive c's in (1) *is equal to the number of positive k's in* (2).

In order to prove this, let us suppose that the x''s and x'''s have been so numbered that the first μ of the c's and the first ν of the k's are positive while all the remaining c's and k's are negative. Our

144

theorem will be established if we can show that $\mu = \nu$. If this is not the case, one of the two integers μ and ν must be the greater, and it is merely a matter of notation to assume that $\mu > \nu$. We will prove that this assumption leads to a contradiction.

If we regard the x''s and x'''s simply as abbreviations for certain linear forms in the x's, (1) and (2) are both of them identically equal to the original quadratic form, and hence to each other. This identity may be written

$$(3) \qquad c_1 x_1'^2 + \cdots + c_\mu x_\mu'^2 - |c_{\mu+1}| x_{\mu+1}'^2 - \cdots - |c_r| x_r'^2$$

$$\equiv k_1 x_1''^2 + \cdots + k_\nu x_\nu''^2 - |k_{\nu+1}| x_{\nu+1}''^2 - \cdots - |k_r| x_r''^2.$$

Let us now consider the system of homogeneous linear equations in $(x_1, \cdots x_n)$,

$$(4) \qquad x_1'' = 0, \cdots x_\nu'' = 0, \; x_{\mu+1}' = 0, \cdots x_n' = 0.$$

We have here $\nu + n - \mu < n$ equations. Hence, by Theorem 3, Corollary 1, §17, we can find a solution of these equations in which all the unknowns are not zero. Let $(y_1, \cdots y_n)$ be such a solution and denote by y_i', y_i'' the values of x_i', x_i'' when the constants $y_1, \cdots y_n$ are substituted in them for the variables $x_1, \cdots x_n$. Substituting the y's for the x's in (3) gives

$$c_1 y_1'^2 + \cdots + c_\mu y_\mu'^2 = - |k_{\nu+1}| y_{\nu+1}''^2 - \cdots - |k_r| y_r''^2.$$

The expression on the left cannot be negative, and that on the right cannot be positive, hence they must both be zero; and this is possible only if

$$y_1' = \cdots = y_\mu' = 0.$$

But by (4) we also have $\quad y_{\mu+1}' = \cdots = y_n' = 0.$

That is, $(y_1, \cdots y_n)$ is a solution, not composed exclusively of zeros, of the system of n homogeneous linear equations in n unknowns,

$$x_1' = 0, \; x_2' = 0, \cdots x_n' = 0.$$

The determinant of these equations must therefore be zero, that is, the linear transformation which carries over the x's into the x''s must be a singular transformation. We are here led to a contradiction, and our theorem is proved.

We can thus associate with every real quadratic form two in·tegers P and N, namely, the number of positive and negative coefficients respectively which we get when we reduce the form by any real non-singular linear transformation to the form (1). These two numbers are evidently arithmetical invariants of the quadratic form with regard to real non-singular linear transformations, since two real quadratic forms which can be transformed into one another by means of such a transformation can obviously be reduced to the *same* expression of form (1).*

The two arithmetical invariants P and N which we have thus arrived at, and the arithmetical invariant r which we had before, are not independent since we have the relation

$$(5) \qquad\qquad P + N = r.$$

One of the invariants P and N is therefore superfluous and either might be dispensed with. It is found more convenient, however, to use neither P nor N, but their difference,

$$(6) \qquad\qquad s = P - N,$$

which is called the *signature* of the quadratic form.

DEFINITION. *By the signature of a real quadratic form is understood the difference between the number of positive and the number of negative coefficients which we obtain when we reduce the form by any real non-singular linear transformation to the form* (1).

Since the integers P and N used above were arithmetical invariants, their difference s will also be an arithmetical invariant. It should be noticed, however, that s is not necessarily a *positive* integer. We have thus proved

THEOREM 3. *The signature of a quadratic form is an arithmetical invariant with regard to real non-singular linear transformations.*

EXERCISES

1. Prove that the rank r and the signature s of a quadratic form are either both even or both odd ; and that $-r \leqq s \leqq r.$

2. Prove that any two integers r and s (r positive or zero) satisfying the conditions of Exercise 1 may be the rank and signature respectively of a quadratic form.

* P is sometimes called the *index of inertia* of the quadratic form.

3. Prove that a necessary and sufficient condition that a real quadratic form of rank r and signature s be factorable into two real linear factors is that

$$\text{either} \qquad r < 2;$$
$$\text{or} \qquad r = 2, s = 0.$$

4. A quadratic form of rank r shall be said to be *regularly arranged* (cf. § 20, Theorem 4) if the x's are so numbered that no two consecutive A's are zero in the set

$$A_0 = 1, A_1 = a_{11}, A_2 = \begin{vmatrix} a_{11} & a_{12} \\ a_{21} & a_{22} \end{vmatrix}, \dots A_r = \begin{vmatrix} a_{11} & \cdots & a_{1r} \\ \cdot & \cdot & \cdot \\ \cdot & \cdot & \cdot \\ a_{r1} & \cdots & a_{rr} \end{vmatrix},$$

and that $A_r \neq 0$. Prove that if the form is real and any one of these A's is zero, the two adjacent A's have opposite signs.

[SUGGESTION. In this exercise and the following ones, the work of § 49 should be consulted.]

5. Prove that the signature of a regularly arranged real quadratic form is equal to the number of permanences minus the number of variations of sign in the sequence of the A's, if the A's which are zero are counted as positive or as negative at pleasure.

6. Defining the expression sgn x (read *signum x*) by the equations

$$\text{sgn } x = +1 \qquad x > 0,$$
$$\text{sgn } x = \quad 0 \qquad x = 0,$$
$$\text{sgn } x = -1 \qquad x < 0,$$

show that the signature of a regularly arranged real quadratic form of rank r is

$$\text{sgn } (A_0 A_1) + \text{sgn } (A_1 A_2) + \cdots + \text{sgn } (A_{r-1} A_r).$$

51. Classification of Real Quadratic Forms. We saw in the last section that a real quadratic form has two invariants with regard to real non-singular linear transformations, — its rank and its signature. The main result to be established in the present section (Theorem 2) is that these two invariants form a complete system.

If in § 46 the c's and k's are real, the transformation (2) will be real when, but only when, each c has the same sign as the corresponding k. All that we can infer from the reasoning of that section now is, therefore, that if a real quadratic form of rank r can be reduced by a real non-singular linear transformation to the form

$$c_1 x_1^2 + \cdots + c_r x_r^2,$$

it can also be reduced by a real non-singular linear transformation to the form

$$k_1 x_1^2 + \cdots + k_r x_r^2,$$

where the k's are arbitrarily given real constants, not zero, subject to the condition that each k has the same sign as the corresponding c. Using the letters P and N for the number of positive and negative c's respectively, the transformation can be so arranged that the first P c's are positive, the last N negative. Accordingly the first P k's can be taken as $+1$, the last N as -1. From equations (5) and (6) of § 50, we see that P and N may be expressed in terms of the rank and signature of the form by the formulæ

$$(1) \qquad P = \frac{r+s}{2}, \qquad N = \frac{r-s}{2}.$$

Thus we have the theorem :

THEOREM 1. *A real quadratic form of rank r and signature s can be reduced by a real non-singular linear transformation to the normal form*

$$(2) \qquad x_1^2 + \cdots + x_P^2 - x_{P+1}^2 - \cdots - x_r^2$$

where P is given by (1).

We are now able to prove the fundamental theorem :

THEOREM 2. *A necessary and sufficient condition that two real quadratic forms be equivalent with regard to real non-singular linear transformations is that they have the same rank and the same signature.*

That this is a necessary condition is evident from the invariance of rank and signature. That it is sufficient follows from the fact that if the two forms have the same rank and signature, they can both be reduced to the same normal form (2).

DEFINITION. *All real quadratic forms, equivalent with regard to real non-singular linear transformations to a given form, and therefore to each other, are said to form a class.*[*]

Thus, for instance, since every real non-singular quadratic form in four variables can be reduced to one or the other of the five normal forms,

$$(3) \qquad \begin{cases} x_1^2 + x_2^2 + x_3^2 + x_4^2, \\ x_1^2 + x_2^2 + x_3^2 - x_4^2, \\ x_1^2 + x_2^2 - x_3^2 - x_4^2, \\ x_1^2 - x_2^2 - x_3^2 - x_4^2, \\ - x_1^2 - x_2^2 - x_3^2 - x_4^2, \end{cases}$$

[*] This term may be used in a similar manner whenever the conception of equivalence is involved.

we see that all such forms belong to one or the other of five classes characterized by the values

$$s = 4, 2, 0, -2, -4, \qquad r = 4.$$

If, however, as is the case in many problems in geometry, we are concerned not with quadratic forms, but with the equations obtained by equating these forms to zero, the number of classes to be distinguished will be reduced by about one half, since two equations are the same if their first members differ merely in sign.

Thus there are only three classes of non-singular quadric surfaces with real equations, whose normal forms are obtained by equating the first three of the forms (3) to zero. These equations written in non-homogeneous coördinates are

$$X^2 + Y^2 + Z^2 = -1,$$
$$X^2 + Y^2 + Z^2 = 1,$$
$$X^2 + Y^2 - Z^2 = 1.$$

The first of these represents an imaginary sphere, the second a real sphere, and the third an unparted hyperboloid generated by the revolution of a rectangular hyperbola about its conjugate axis. It may readily be proved that this last surface may also be generated by the revolution of either of the lines

$$Y = 1, \ X = \pm Z$$

about the axis of Z. We may therefore say:

THEOREM 3. *There are three, and only three, classes of non-singular quadric surfaces with real equations. In the first the surfaces are imaginary; in the second real, but their rulings are imaginary; in the third they are real, and the rulings through their real points are real.*[*]

This classification is complete from the point of view we have adopted of regarding quadric surfaces as equivalent if one can be transformed into the other by a real non-singular collineation. The more familiar classification does not adopt this projective view, but distinguishes in our second class between ellipsoids, biparted hyperboloids, and elliptic paraboloids; and in the third class between unparted hyperboloids and hyperbolic paraboloids.

---

[*] If, as here, we consider not real quadratic forms, but real homogeneous quadratic equations we must use, not s, but $|s|$ as an invariant. In place of $|s|$ we may use what is known as the *characteristic* of the quadratic form, that is the smaller of the two integers P N. This characteristic is simply $\frac{1}{2}(r - |s|)$.

EXERCISES

1. Prove that there are $\frac{1}{2}(n+1)(n+2)$ classes of real quadratic forms in n variables.

2. Give a complete classification of singular quadric surfaces with real equations from the point of view of the present section.

52. Definite and Indefinite Forms.

DEFINITION. *By an indefinite quadratic form is understood a real quadratic form such that, when it is reduced to the normal form (2), § 51, by a real non-singular linear transformation, both positive and negative signs occur. All other real quadratic forms are called definite;* and we distinguish between positive and negative definite forms according as the terms in the normal form are all positive or all negative.*

In other words, a real quadratic form of rank r and signature s is definite if $s = \pm r$, otherwise it is indefinite.†

The names *definite* and *indefinite* have been given on account of the following fundamental property:

THEOREM 1. *An indefinite quadratic form is positive for some real values of the variables, negative for others. A positive definite form is positive or zero for all real values of the variables; a negative definite form, negative or zero.*

The part of this theorem which relates to definite forms follows directly from the definition. To prove the part concerning indefinite forms, suppose the form reduced by a real non-singular linear transformation to the normal form

$$(1) \qquad x_1'^2 + \cdots + x_\beta'^2 - x_{\beta+1}'^2 - \cdots - x_r'^2$$

Regarding the x's as abbreviations for certain real linear forms in the x's, let us consider the system of $n - P$ homogeneous linear equations

$$(2) \qquad x_{P+1}' = 0, \ x_{P+2}' = 0, \ \cdots x_n' = 0.$$

Since these equations are real, and their number is less than the number of unknowns, they have a real solution not consisting

---

* Some writers reserve the name *definite* for non-singular forms, and call the singular definite forms *semidefinite*.

† Otherwise stated, the condition for a definite form is that the *characteristic* be zero. Cf. the footnote to Theorem 3, § 51.

exclusively of zeros. Let $(y_1, \cdots y_n)$ be such a solution. This solution cannot satisfy all the equations

$$(3) \qquad x_1' = 0, \cdots x_P' = 0,$$

for equations (2) and (3) together form a system of n homogeneous linear equations in n unknowns whose determinant is not zero, since it is the determinant of the linear transformation which reduces the given quadratic form to the normal form (1). Accordingly, if we substitute $(y_1, \cdots y_n)$ for the variables $(x_1, \cdots x_n)$ in the given quadratic form, this form will have a positive value, as we see from the reduced form (1).

Similarly, by choosing for the x's a real solution of the equations

$$x_1' = 0, \cdots x_P' = 0, \qquad x_{r+1}' = 0, \cdots x_n' = 0,$$

which does not consist exclusively of zeros, we see that the quadratic form takes on a negative value.

We pass now to some theorems which will be better appreciated by the reader if he considers their geometrical meaning in the case $n = 4$.

THEOREM 2. *If an indefinite quadratic form is positive at the real point $(y_1, \cdots y_n)$ and negative at the real point $(z_1, \cdots z_n)$, then there are two real points linearly dependent on these two, but linearly independent of each other, at which the quadratic form is zero, and neither of which is a vertex of the form.*

The condition that the quadratic form

$$(4) \qquad \sum_1^n a_{ij} x_i x_j$$

vanish at the point $(y_1 + \lambda z_1, \cdots y_n + \lambda z_n)$ is

$$\sum_1^n a_{ij} y_i y_j + 2\lambda \sum_1^n a_{ij} y_i z_j + \lambda^2 \sum_1^n a_{ij} z_i z_j = 0.$$

This quadratic equation in λ has two real distinct roots, since, from our hypothesis that (4) is positive at y and negative at z, it follows that

$$\left(\sum_1^n a_{ij} y_i z_j \right)^2 - \left(\sum_1^n a_{ij} y_i y_j \right) \left(\sum_1^n a_{ij} z_i z_j \right) > 0.$$

Let us call these roots λ_1 and λ_2. Then the points

$$(5) \qquad (y_1 + \lambda_1 z_1, \cdots y_n + \lambda_1 z_n), \qquad (y_1 + \lambda_2 z_1, \cdots y_n + \lambda_2 z_n)$$

are two real points linearly dependent on the points y and z at which
(4) vanishes.

Next notice that

$$(6) \qquad \begin{vmatrix} y_i + \lambda_1 z_i & y_j + \lambda_1 z_j \\ y_i + \lambda_2 z_i & y_j + \lambda_2 z_j \end{vmatrix} = \begin{vmatrix} 1 & \lambda_1 \\ 1 & \lambda_2 \end{vmatrix} \cdot \begin{vmatrix} y_i & y_j \\ z_i & z_j \end{vmatrix}.$$

Since the points y and z are linearly independent, the integers i, j
can be so chosen that the last determinant on the right of (6) is not
zero. Then the determinant on the left of (6) is not zero; and,
consequently, the points (5) are linearly independent.

In order, finally, to prove that neither of the points (5) is a
vertex, denote them for brevity by

$$(Y_1, \cdots Y_n), \qquad (Z_1, \cdots Z_n).$$

Letting $\lambda_1 - \lambda_2 = 1/\mu$, we have

$$z_i = \mu Y_i - \mu Z_i \qquad\qquad (i = 1, 2, \cdots n).$$

Therefore

$$(7) \qquad \sum_1^n a_{ij} z_i z_j = \mu^2 \sum_1^n a_{ij} Y_i Y_j - 2\,\mu^2 \sum_1^n a_{ij} Y_i Z_j + \mu^2 \sum_1^n a_{ij} Z_i Z_j.$$

Since the points Y and Z have been so determined that (4) vanishes
at them, the first and last terms on the right of (7) are zero. If
either Y or Z were a vertex, the middle term would also be zero;
but this is impossible since the left-hand member of (7) is, by
hypothesis, negative. Thus our theorem is proved.

For the sake of completeness we add the corollary, whose truth
is at once evident :

COROLLARY. *The only points linearly dependent on y and z at
which the quadratic form vanishes are points linearly dependent on one
or the other of the points referred to in the theorem; and none of these
are vertices.*

We come now to a theorem of fundamental importance in the
theory of quadratic forms.

THEOREM 3. *A necessary and sufficient condition that a real
quadratic form be definite is that it vanish at no real points except its
vertices and the point* $(0, 0, \cdots 0)$.

Suppose, first, that we have a real quadratic form which vanishes at no real points except its vertices and the point $(0, 0, \cdots 0)$. If it were indefinite, we could (Theorem 1) find two real points y, z, at one of which it is positive, at the other, negative. Hence (Theorem 2) we could find two real points linearly dependent on y and z, at which the quadratic form vanishes. Neither of these will be the point $(0, 0, \cdots 0)$, since, by Theorem 2, they are linearly independent. Moreover, they are neither of them vertices. Thus we see that the form must be definite, and the sufficiency of the condition is established.

It remains to be proved that a definite form can vanish only at its vertices and at the point $(0, 0, \cdots 0)$.

Suppose (4) is definite and that $(y_1, \cdots y_n)$ is any real point at which it vanishes. Then,

$$\sum_1^n a_{ij}(x_i + \lambda y_i)(x_j + \lambda y_j) \equiv \sum_1^n a_{ij}x_i x_j + 2\lambda \sum_1^n a_{ij}x_i y_j.$$

If y were neither a vertex nor the point $(0, 0, \cdots 0)$, $\Sigma a_{ij}x_i y_j$ would not vanish identically, and we could find a real point $(z_1, \cdots z_n)$ such that

$$k = \sum_1^n a_{ij}z_i y_j \neq 0.$$

If we let

$$c = \sum_1^n a_{ij}z_i z_j,$$

we have

(8)
$$\sum_1^n a_{ij}(z_i + \lambda y_i)(z_j + \lambda y_j) = c + 2\lambda k.$$

For a given real value of λ, the left-hand side of this equation is simply the value of the quadratic form (4) at a certain real point. Accordingly, for different values of λ it will not change sign, while the right-hand side of (8) has opposite signs for large positive and large negative values of λ. Thus the assumption that y was neither a vertex nor the point $(0, 0, \cdots 0)$ has led to a contradiction; and our theorem is proved.

COROLLARY. *A non-singular definite quadratic form vanishes, for real values of the variables, only when its variables are all zero.*

As a simple application of the last corollary we will prove

THEOREM 4. *In a non-singular definite form, none of the coefficients of the square terms can be zero.*

For suppose the form (4) were definite and non-singular; and that $a_{ii} = 0$. Then the form would vanish at the point

$$x_1 = \cdots = x_{i-1} = x_{i+1} = \cdots = x_n = 0, \ x_i = 1;$$

and this is impossible, since this is not the point $(0, 0, \cdots 0)$.

EXERCISES

1. Definition. *By an orthogonal transformation* * *is understood a linear transformation which carries over the variables* $(x_1, \cdots x_n)$ *into the variables* $(x'_1, \cdots x_n)$ *in such a way that*

$$x_1^2 + x_2^2 + \cdots + x_n^2 \equiv x_1'^2 + x_2'^2 + \cdots + x_n'^2.$$

Prove that every orthogonal transformation is non-singular, and, in particular, that its determinant must have the value $+1$ or -1.

2. Prove that all orthogonal transformations in n variables form a group; and that the same is true of all orthogonal transformations in n variables of determinant $+1$.

3. Prove that a necessary and sufficient condition that a linear transformation be orthogonal is that it leave the "distance"

$$\sqrt{(y_1 - z_1)^2 + (y_2 - z_2)^2 + \cdots + (y_n - z_n)^2}$$

between every pair of points $(y_1, \cdots y_n)$, $(z_1, \cdots z_n)$ invariant.

4. Prove that if $n = 3$, and if x_1, x_2, x_3 be interpreted as non-homogeneous rectanglar coördinates in space, an orthogonal transformation represents either a rigid displacement which leaves the origin fixed, or such a displacement combined with reflection in a plane through the origin.

Show that the first of these cases will occur when the determinant of the transformation is $+1$, the second when this determinant is -1.

5. If the coefficients of a linear transformation are denoted in the usual way by c_{ij}, prove that a necessary and sufficient condition that the transformation be orthogonal is that

$$c_{1i}^2 + c_{2i}^2 + \cdots + c_{ni}^2 = 1 \qquad (i = 1, 2, \cdots n),$$

$$c_{1i}c_{1j} + c_{2i}c_{2j} + \cdots + c_{ni}c_{nj} = 0 \qquad \begin{cases} i = 1, 2, \cdots n \\ j = 1, 2, \cdots n \end{cases} i \neq j.$$

Show that these will still be necessary and sufficient conditions for an orthogonal transformation if the two subscripts of every c be interchanged.†

---

* The matrix of such a transformation is called an orthogonal matrix, and its determinant an orthogonal determinant.

† We have here $\frac{1}{2} n (n + 1)$ relations between the n^2 coefficients of the transformation. This suggests that it should be possible to express all the coefficients in terms of

$$n^2 - \frac{n(n+1)}{2} = \frac{n(n-1)}{2}$$

of them, or if we prefer in terms of $\frac{1}{2} n(n-1)$ other parameters. For Cayley's discussion of this question cf. Pascal's book, *Die Determinanten*, § 47. Cayley's formulæ, however, do not include all orthogonal transformations except as limiting cases.

CHAPTER XII

THE SYSTEM OF A QUADRATIC FORM AND ONE OR MORE LINEAR FORMS

53. Relations of Planes and Lines to a Quadric Surface. If the plane

$$(1) \qquad u_1 x_1 + u_2 x_2 + u_3 x_3 + u_4 x_4 = 0$$

is a true tangent plane to the quadric surface

$$(2) \qquad \overset{4}{\underset{1}{\Sigma}} a_{ij} x_i x_j = 0,$$

there will be a point (y_1, y_2, y_3, y_4) (namely the point of contact) lying in (1) and such that its polar plane

$$(3) \qquad \overset{4}{\underset{1}{\Sigma}} a_{ij} x_i y_j = 0$$

coincides with (1). From elementary analytic geometry we know that a necessary and sufficient condition that two equations of the first degree represent the same plane is that their coefficients be proportional. Accordingly, from the coincidence of (1) and (3), we deduce the equations

$$(4) \qquad \begin{cases} a_{11} y_1 + a_{12} y_2 + a_{13} y_3 + a_{14} y_4 - \rho u_1 = 0, \\ a_{21} y_1 + a_{22} y_2 + a_{23} y_3 + a_{24} y_4 - \rho u_2 = 0, \\ a_{31} y_1 + a_{32} y_2 + a_{33} y_3 + a_{34} y_4 - \rho u_3 = 0, \\ a_{41} y_1 + a_{42} y_2 + a_{43} y_3 + a_{44} y_4 - \rho u_4 = 0. \end{cases}$$

From the fact that the point y lies on (1), we infer the further relation

$$(5) \qquad u_1 y_1 + u_2 y_2 + u_3 y_3 + u_4 y_4 = 0.$$

These equations (4) and (5) have been deduced on the supposition that (1) is a true tangent plane to (2). They still hold if it is a pseudo-tangent plane; for then the quadric must be a cone, and a vertex of this cone must lie on (1). Taking the point y as this vertex, equation (5) is fulfilled. Moreover, since now the first member of (3) is identically zero, equations (4) will also be fulfilled if we let $\rho = 0$. Thus we have shown in all cases, that if (1) is a tangent plane to (2), there exist five constants, y_1, y_2, y_3, y_4, ρ, of which the first four are not all zero, and which satisfy equations (4) and (5). Hence

$$(6) \qquad \begin{vmatrix} a_{11} & a_{12} & a_{13} & a_{14} & u_1 \\ a_{21} & a_{22} & a_{23} & a_{24} & u_2 \\ a_{31} & a_{32} & a_{33} & a_{34} & u_3 \\ a_{41} & a_{42} & a_{43} & a_{44} & u_4 \\ u_1 & u_2 & u_3 & u_4 & 0 \end{vmatrix} = 0.$$

Conversely, if this last equation is fulfilled, there exist five constants, y_1, y_2, y_3, y_4, ρ, not all zero, and which satisfy equations (4) and (5). We can go a step farther and say that y_1, y_2, y_3, y_4 cannot all be zero, as otherwise, from equations (4) and the fact that the u's are not all zero, ρ would also be zero. Thus we see that if equation (6) is fulfilled, there exists a point (y_1, y_2, y_3, y_4) in the plane (1) whose coördinates, together with a certain constant ρ, satisfy (4). If $\rho = 0$, this shows that the quadric is a cone with y as a vertex, and hence that (1) is at least a pseudo-tangent plane. If $\rho \neq 0$, equations (4) show us that the polar plane (3) of y coincides with the plane (1). Moreover we see, either geometrically, or by multiplying equations (4) by y_1, y_2, y_3, y_4 respectively and adding, that the point y lies on the quadric; so that, in this case, (1) is a true tangent plane.

We have thus established the theorem:

THEOREM 1. *Equation* (6) *is a necessary and sufficient condition that the plane* (1) *be tangent to the quadric* (2).

It will be seen that this theorem gives us no means of distinguishing between true and pseudo-tangent planes of quadric cones. In the case of non-singular quadrics, pseudo-tangent planes are impossible, and therefore equation (6) may, in this case, be regarded as the equation of the quadric in plane-coördinates.

In the case of a quadric surface of rank 3, that is, of a cone with a single vertex, the coördinates (u_1, u_2, u_3, u_4) of every plane through this vertex satisfy equation (6), so that in this case this equation represents a single point, and not the quadric cone.*

* In fact a cone cannot be represented by a single equation in plane-coördinates.

If the rank of (2) is less than 3, the coördinates of every plane in space should satisfy (6), since every such plane passes through a vertex and is therefore a tangent plane. This fact may be verified by noticing that equation (6) may also be written

$$\sum_1^4 A_{ij}u_iu_j = 0,$$

where the A's are the cofactors in the discriminant of (2) according to our usual notation.

We pass now to the condition that a straight line touch the quadric (2). This line we will determine as the intersection of the two planes (1) and

(7)
$$v_1x_1 + v_2x_2 + v_3x_3 + v_4x_4 = 0.$$

If the line of intersection of these planes is a true tangent to (2), there will be a point (y_1, y_2, y_3, y_4), namely the point of contact, lying upon it, and such that its polar plane (3) contains the line. It must therefore be possible to write the equation of this polar plane in the form

(8)
$$\sum_1^4 (\mu u_i + \nu v_i) x_i = 0 ;$$

and, in fact, by properly choosing the constants μ and ν, the coefficients of (8) may be made not merely proportional, but equal to the coefficients of (3):

(9)
$$\begin{cases} a_{11}y_1 + a_{12}y_2 + a_{13}y_3 + a_{14}y_4 - \mu u_1 - \nu v_1 = 0, \\ a_{21}y_1 + a_{22}y_2 + a_{23}y_3 + a_{24}y_4 - \mu u_2 - \nu v_2 = 0, \\ a_{31}y_1 + a_{32}y_2 + a_{33}y_3 + a_{34}y_4 - \mu u_3 - \nu v_3 = 0, \\ a_{41}y_1 + a_{42}y_2 + a_{43}y_3 + a_{44}y_4 - \mu u_4 - \nu v_4 = 0. \end{cases}$$

Since the point y lies on the line of intersection of the planes (1) and (7), we also have the relations

(10)
$$\begin{cases} u_1y_1 + u_2y_2 + u_3y_3 + u_4y_4 = 0, \\ v_1y_1 + v_2y_2 + v_3y_3 + v_4y_4 = 0. \end{cases}$$

Since the six equations (9) and (10) are satisfied by six constants $y_1, y_2, y_3, y_4, \mu, \nu$ not all zero, we infer finally the relation

(11)
$$\begin{vmatrix} a_{11} & a_{12} & a_{13} & a_{14} & u_1 & v_1 \\ a_{21} & a_{22} & a_{23} & a_{24} & u_2 & v_2 \\ a_{31} & a_{32} & a_{33} & a_{34} & u_3 & v_3 \\ a_{41} & a_{42} & a_{43} & a_{44} & u_4 & v_4 \\ u_1 & u_2 & u_3 & u_4 & 0 & 0 \\ v_1 & v_2 & v_3 & v_4 & 0 & 0 \end{vmatrix} = 0.$$

We have deduced this equation on the supposition that the line of intersection of (1) and (7) is a true tangent to (2). We leave it to the reader to show that (11) holds if this line is a pseudo-tangent, and also if it is a ruling of (2).

We also leave it for him to show that if (11) holds, the line of intersection of (1) and (7) will be either a true tangent, a pseudo-tangent, or a ruling, and thus to establish the theorem:

THEOREM 2. *A necessary and sufficient condition that the line of intersection of the planes (1) and (7) be either a tangent or a ruling of (2) is that equation (11) be fulfilled.*

On expanding the determinant in (11), it will be seen that it is a quadratic form in the six line-coördinates q_{ij} (cf. Exercise 3, § 35). Equation (11) may therefore be regarded as the equation of the quadric surface in line-coördinates if the surface is not a cone, or is a cone with a single vertex. If the rank of (2) is 2, so that the quadric consists of two planes, (11) is the equation of the line of intersection of these planes. While if the rank is 1 or 0, (11) is identically fulfilled.

EXERCISES

1. Two planes are said to be conjugate with regard to a non-singular quadric surface if each passes through the pole of the other.

Prove that if (2) is a non-singular quadric, a necessary and sufficient condition that the planes (1) and (7) be conjugate with regard to it is the vanishing of the determinant

$$\begin{vmatrix} a_{11} & a_{12} & a_{13} & a_{14} & u_1 \\ a_{21} & a_{22} & a_{23} & a_{24} & u_2 \\ a_{31} & a_{32} & a_{33} & a_{34} & u_3 \\ a_{41} & a_{42} & a_{43} & a_{44} & u_4 \\ v_1 & v_2 & v_3 & v_4 & 0 \end{vmatrix} = -\sum_{1}^{4} A_{ij} u_i v_j.$$

How must this definition of conjugate planes be extended in order that this theorem be true for singular quadrics also?

2. Prove that if (2) is a non-singular quadric, a necessary and sufficient condition that the point of intersection of three planes lie on (2) is the vanishing of the seven-rowed determinant formed by bordering the discriminant of (2) with the coefficients of the three planes.

3. Admitting it to be obvious geometrically that a necessary and sufficient condition that a line touch a non-singular quadric is that the two tangent planes which can be passed through this line should coincide, prove that, if (2) is non-singular, a necessary and sufficient condition that the line of intersection of (1) and (7) touch (2) is

$$\left(\sum_{1}^{4} A_{ij} u_i u_j\right)\left(\sum_{1}^{4} A_{ij} v_i v_j\right) - \left(\sum_{1}^{4} A_{ij} u_i v_j\right)^2 = 0.$$

4. Show algebraically that the condition of Exercise 3 is equivalent to (11).

54. The Adjoint Quadratic Form and Other Invariants. Passing now to the case of n variables, we begin by considering the system consisting of a quadratic form and a single linear form

$$(1) \qquad \sum_1^n a_{ij} x_i x_j,$$

$$(2) \qquad \sum_1^n u_i x_i.$$

The geometrical considerations of the last section suggest that we form the expression

$$(3) \qquad \sum_1^n A_{ij} u_i u_j \equiv - \begin{vmatrix} a_{11} & \cdots & a_{1n} & u_1 \\ \cdot & \cdot & \cdot & \cdot \\ \cdot & \cdot & \cdot & \cdot \\ a_{n1} & \cdots & a_{nn} & u_n \\ u_1 & \cdots & u_n & 0 \end{vmatrix}.$$

This, it will be seen, is a quadratic form in the variables $(u_1, \cdots u_n)$ whose matrix is the adjoint of the matrix of (1). We will speak of (3) as the *adjoint* of (1).

The invariance of (3) is at once suggested by the fact that in the case $n = 4$ the vanishing of (3) gave a necessary and sufficient condition for a projective relation. In fact we will prove the theorem:

THEOREM 1. *The adjoint form* (3) *is an invariant of weight two of the pair of forms* (1), (2).

Inasmuch as the u's are, as we saw in § 34, contragredient to the x's, we may also call (3) a contravariant (cf. Definition 2, § 34).

In order to prove this theorem we must subject the x's to a linear transformation,

$$(4) \qquad \begin{cases} x_1 = c_{11} x_1' + \cdots + c_{1n} x_n' \\ \cdot \quad \cdot \quad \cdot \quad \cdot \quad \cdot \quad \cdot \quad \cdot \quad \cdot \\ \cdot \quad \cdot \quad \cdot \quad \cdot \quad \cdot \quad \cdot \quad \cdot \quad \cdot \\ x_n = c_{n1} x_1' + \cdots + c_{nn} x_n', \end{cases}$$

whose determinant we will call c. Let us denote by a_{ij}' and u_i' respectively the coefficients of the quadratic and linear form into which this transformation carries (1) and (2).

Let us now introduce an auxiliary variable t, and consider the quadratic form in $x_1, \cdots x_n, t$,

$$(5) \qquad \sum_1^n a_{ij} x_i x_j + 2 t (u_1 x_1 + \cdots + u_n x_n).$$

The discriminant of this form is precisely the determinant in (3), that is, the negative of the adjoint of (1).

Let us now perform on the variables $x_1, \cdots x_n, t$ the linear transformation given by formulæ (4) and the additional formula

$$(6) \qquad\qquad t = t'.$$

The determinant of this transformation is c, and it carries over the form (5) into

$$\sum_1^n a'_{ij} x'_i x'_j + 2\, t'(u'_1 x'_1 + \cdots + u'_n x'_n).$$

From the fact that the discriminant of (5) is an invariant of weight 2, we infer the relation we wished to obtain:

$$\begin{vmatrix} a'_{11} & \cdots & a'_{1n} & u'_1 \\ \cdot & \cdot & \cdot & \cdot \\ \cdot & \cdot & \cdot & \cdot \\ a'_{n1} & \cdots & a'_{nn} & u'_n \\ u'_1 & \cdots & u'_n & 0 \end{vmatrix} \equiv c^2 \begin{vmatrix} a_{11} & \cdots & a_{1n} & u_1 \\ \cdot & \cdot & \cdot & \cdot \\ \cdot & \cdot & \cdot & \cdot \\ a_{n1} & \cdots & a_{nn} & u_n \\ u_1 & \cdots & u_n & 0 \end{vmatrix}.$$

The method just used admits of immediate extension to the proof of the following more general theorem:

THEOREM 2. *If we have a system consisting of a quadratic form in n variables and p linear forms, the $(n+p)$-rowed determinant formed by bordering the discriminant of the quadratic form by p rows and p columns each of which consists of the coefficients of one of the linear forms is an invariant of weight 2.*

We leave the details of the proof of this theorem to the reader.

If the discriminant a of the quadratic form (1) is not zero, we may form a new quadratic form whose matrix is the inverse of the matrix of (1). This quadratic form, which is known as the *inverse* or *reciprocal* of (1), is simply the adjoint of (1) divided by the discriminant a. We will prove the following theorem concerning it:

THEOREM 3. *If the quadratic form (1) is non-singular, it will be carried over into its inverse by the non-singular transformation*

$$(7) \qquad\qquad x'_i = a_{i1} x_1 + \cdots + a_{in} x_n \qquad\qquad (i = 1, 2, \cdots n).$$

For we have

$$\sum_1^n a_{ij} x_i x_j \equiv \sum_1^n x_i x'_i.$$

But from (7) we have

$$x_i = \frac{A_{i1}}{a} x'_1 + \cdots + \frac{A_{in}}{a} x'_n$$

and therefore

$$\sum_1^n a_{ij} x_i x_j \equiv \sum_1^n \frac{A_{ij}}{a} x'_i x'_j,$$

as was to be proved.

It will be noticed that if (1) is a real quadratic form, the transformation (7) is real; and from this follows

THEOREM 4. *A real non-singular quadratic form and its inverse have the same signature.*

EXERCISES

1. Given a quadratic form $\Sigma a_{ij}x_i x_j$ and two linear forms $\Sigma u_i x_i$, $\Sigma v_i x_i$. Prove that

$$\sum_1^n A_{ij}u_i v_j \equiv - \begin{vmatrix} a_{11} & \cdots & a_{1n} & u_1 \\ \cdot & \cdot & \cdot & \cdot & \cdot & \cdot \\ \cdot & \cdot & \cdot & \cdot & \cdot & \cdot \\ a_{n1} & \cdots & a_{nn} & u_n \\ v_1 & \cdots & v_n & 0 \end{vmatrix}$$

is an invariant of the system of weight 2.

2. Generalize the theorem of Exercise 1 to the case in which we have more than two linear forms.

3. Prove that if a first quadratic form is transformed into a second by the linear transformation of matrix **c**, then the adjoint of the first will be transformed into the adjoint of the second by the linear transformation whose matrix is the conjugate of the adjoint of **c**.

4. Prove a similar theorem for bilinear forms.

5. State and prove a theorem for bilinear forms analogous to Theorem 3.

55. The Rank of the Adjoint Form. Suppose the discriminant a of the quadratic form $\sum_1^n a_{ij}x_i x_j$ is of rank r, and that the discriminant A of its adjoint $\sum_1^n A_{ij}u_i u_j$ is of rank R. Then, if $r < n-1$, all the $(n-1)$-rowed determinants of a are zero; but these are the elements of A, hence $R = 0$. If $r = n-1$, at least one of the elements of A is not zero, and all two-rowed determinants of A are zero (since by § 11 each of them contains a as a factor), hence $R = 1$. If $r = n$, $R = n$; for if R were less than n we should have $A = 0$, and therefore $a = 0$ (since $A = a^{n-1}$). But this is impossible, since by hypothesis $r = n$. We have then:

THEOREM 1. *If the rank of a quadratic form in n variables and of its adjoint are r and R respectively, then*

$$\begin{aligned} &\textit{if } r = n, & R &= n, \\ &\textit{if } r = n-1, & R &= 1, \\ &\textit{if } r < n-1, & R &= 0. \end{aligned}$$

Let us consider further the case $r = n - 1$. Here we have seen that $R = 1$, that is, that the adjoint is the square of a linear form,

$$\sum_{1}^{n} A_{ij} u_i u_j \equiv (\sum_{1}^{n} c_i u_i)^2 \equiv \sum_{1}^{n} c_i c_j u_i u_j.$$

Comparing coefficients, we see that

$$A_{ij} = c_i c_j.$$

All the c's cannot be zero, as otherwise we should have $R = 0$. Let $c_\lambda \neq 0$. Then since

$$A_{\lambda\lambda} = c_\lambda^2 \neq 0$$

we see that not all the quantities $(A_{\lambda 1}, \cdots A_{\lambda n})$ are zero. Accordingly (cf. § 44) the point $(A_{\lambda 1}, A_{\lambda 2}, \cdots A_{\lambda n})$, and therefore also the point $(c_1, \cdots c_n)$, is a vertex of the original quadratic form. Thus we have the theorem :

THEOREM 2. *If the rank of a quadratic form in n variables is $n - 1$, its adjoint is the square of a linear form, and the coefficients of this linear form are the coördinates of a vertex of the original form.*

Since, in the case we are considering, all the vertices of the quadratic form are linearly dependent on any one, this theorem completely determines the linear form in question except for a constant factor.

CHAPTER XIII

PAIRS OF QUADRATIC FORMS

56. Pairs of Conics. We will give in this section a short geometrical introduction to the study of pairs of quadratic forms, confining
ourselves, for the sake of brevity, to two dimensions.

Let u and v be two conics which we will assume to be so situated
that they intersect in four, and only four, distinct points, A, B, C, D.
Consider all conics through these four points. These conics, we will

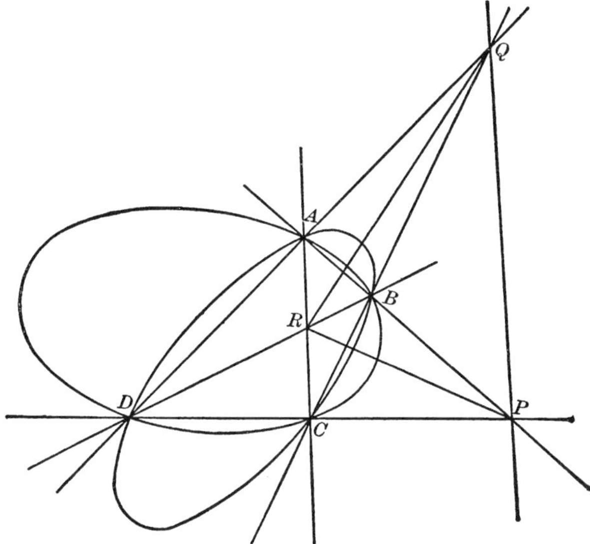

say, form a *pencil*. It is obvious that there are three and only three
singular conics (*i.e.* conics which consist of pairs of lines) in this
pencil, namely, the three pairs of lines AB, CD; BC, DA; AC, BD
Let us call the "vertices" of these conics P, Q, and R respectively.

From the harmonic properties of the complete quadrilateral* we
see that the secants PAB and PCD are divided harmonically by the

* Cf. any book on modern geometry.

line QR. Accordingly QR is the polar of P with regard to every conic of the pencil. In a similar manner PR is the polar of Q, and PQ the polar of R with regard to every conic of the pencil. Thus, we see that the triangle PQR is a self-conjugate triangle (see § 41) with regard to every conic of the pencil. Accordingly, if we perform a collineation which carries over P, Q, R into the origin and the points at infinity on the axes of x and y, the equation of every conic of the pencil will be reduced to a form in which only the square terms enter. We are thus led to the result:

THEOREM. *If two conics intersect in four and only four distinct points, there exists a non-singular collineation which reduces their equations to the normal form*

$$\begin{cases} A_1 x_1^2 + A_2 x_2^2 + A_3 x_3^2 = 0, \\ B_1 x_1^2 + B_2 x_2^2 + B_3 x_3^2 = 0. \end{cases}$$

If we wish to carry through this reduction analytically, we shall write the equations of the two conics u and v in the forms

$$(1) \qquad \sum_1^3 a_{ij} x_i x_j = 0, \qquad \sum_1^3 b_{ij} x_i x_j = 0.$$

The pencil of conics may then be written

$$(2) \qquad \sum_1^3 (a_{ij} - \lambda b_{ij}) x_i x_j = 0,$$

or rather, to be accurate, this equation will represent for different values of λ all the conics of the pencil except the conic v. The singular conics of the pencil will be obtained by equating the discriminant of (2) to zero,

$$(3) \qquad \begin{vmatrix} a_{11} - \lambda b_{11} & a_{12} - \lambda b_{12} & a_{13} - \lambda b_{13} \\ a_{21} - \lambda b_{21} & a_{22} - \lambda b_{22} & a_{23} - \lambda b_{23} \\ a_{31} - \lambda b_{31} & a_{32} - \lambda b_{32} & a_{33} - \lambda b_{33} \end{vmatrix} = 0.$$

This equation we will call the λ-equation of the two conics When expanded, it takes the form

$$(4) \qquad -\Delta' \lambda^3 + \Theta' \lambda^2 - \Theta \lambda + \Delta = 0,$$

where Δ, Δ' are the discriminants of u and v respectively, and

$$\Theta = \begin{vmatrix} a_{11} & a_{12} & b_{13} \\ a_{21} & a_{22} & b_{23} \\ a_{31} & a_{32} & b_{33} \end{vmatrix} + \begin{vmatrix} a_{11} & b_{12} & a_{13} \\ a_{21} & b_{22} & a_{23} \\ a_{31} & b_{32} & a_{33} \end{vmatrix} + \begin{vmatrix} b_{11} & a_{12} & a_{13} \\ b_{21} & a_{22} & a_{23} \\ b_{31} & a_{32} & a_{33} \end{vmatrix},$$

while Θ' can be obtained from Θ by an interchange of the letters a and b. It can readily be proved (cf. the next section) that the coefficients Θ and Θ' as well as Δ and Δ' are invariants of weight two.

Except when the discriminant Δ' of v is zero, the equation (4) is of the third degree, and its three roots, which in the case we have considered must evidently be distinct, give, when substituted in (2), the three singular conics of the pencil.

We will not stop here to show how the theory of any two conics, where no restriction as to the number of points of intersection is made, can be deduced from equation (3).* This will follow in Chapter XXII as an application of the method of elementary divisors. Our only object in this section has been to give a geometrical basis for the appreciation of the following sections.

57. Invariants of a Pair of Quadratic Forms. Their λ-Equation. We consider the pair of quadratic forms

$$\phi(x_1, \cdots x_n) \equiv \sum_1^n a_{ij} x_i x_j,$$

$$\psi(x_1, \cdots x_n) \equiv \sum_1^n b_{ij} x_i x_j,$$

and form from them the *pencil* of quadratic forms

$$\phi - \lambda\psi \equiv \sum_1^n (a_{ij} - \lambda b_{ij}) x_i x_j.$$

The discriminant of this pencil,

$$\begin{vmatrix} a_{11} - \lambda b_{11} & \cdots & a_{1n} - \lambda b_{1n} \\ \cdot & \cdots & \cdot \\ \cdot & \cdots & \cdot \\ a_{n1} - \lambda b_{n1} & \cdots & a_{nn} - \lambda b_{nn} \end{vmatrix} \equiv F(\lambda),$$

is a polynomial in λ which is in general of degree n, and which may be written

$$F(\lambda) \equiv \Theta_0 - \Theta_1\lambda + \cdots + (-1)^n \Theta_n \lambda^n.$$

* An elementary discussion of the λ-equation of two conics (*l'équation en λ*) is regularly given in French text-books on analytic geometry. See, for instance, Briot et Bouquet, *Leçons de Géométrie analytique*, 14th ed., p. 349, or Niewenglowski, *Cours de Géométrie analytique*, Vol. I, p. 459.

The coefficients of this polynomial are themselves polynomials in the a_{ij}'s and b_{ij}'s, Θ_0 and Θ_n being merely the discriminants of ϕ and ψ respectively, while Θ_k is the sum of all the different determinants which can be formed by replacing k columns of the discriminant of ϕ by the corresponding columns of the discriminant of ψ.

THEOREM 1. *The coefficients $\Theta_0, \cdots \Theta_n$ of $F(\lambda)$ are integral rational invariants of weight two of the pair of quadratic forms ϕ, ψ.*[*]

In order to prove this, let us consider a linear transformation of determinant c which carries over ϕ and ψ into ϕ' and ψ' respectively, where

$$\phi' \equiv \sum_1^n a'_{ij}\, x'_i\, x'_j,$$

$$\psi' \equiv \sum_1^n b'_{ij}\, x'_i\, x'_j.$$

Let us denote by Θ'_i the polynomial in the a'_{ij}'s and b'_{ij}'s obtained by putting accents to the a's and b's in Θ_i. Our theorem will then be proved if we can establish the identities

$$\Theta'_i \equiv c^2 \Theta_i \qquad\qquad (i = 0,\, 1,\, \cdots n)$$

This follows at once from the fact that $F(\lambda)$, being the discriminant of $\phi - \lambda\psi$, is an invariant of weight two, so that if we denote by $F'(\lambda)$ the discriminant of $\phi' - \lambda\psi'$, we have

$$F'(\lambda) \equiv c^2\, F(\lambda).$$

This being an identity in λ as well as in the a's and b's, we can equate the coefficients of like powers of λ on the two sides, and this gives precisely the identities we wished to establish.[†]

The equation

$$F(\lambda) = 0$$

we will call the λ-equation of the pair of forms ϕ, ψ. Since, as we have seen, F is merely multiplied by a constant different from zero when ϕ and ψ are subjected to a non-singular linear transformation,

---

[*] Cf. Exercise 13, § 90.

[†] The method by which we have here arrived at invariants of the system of two quadratic forms will be seen to be of very general application. If we have an integral rational invariant I of weight μ of a single form of the kth degree in n variables, we can find a large number of invariants of the system $\phi_1, \phi_2, \cdots \phi_p$ of p forms of the kth degree in n variables by forming the invariant I for the form $\lambda_1\phi_1 + \cdots + \lambda_p\phi_p$. This will be a polynomial in the λ's, each of whose coefficients is seen, precisely as above, to be an integral rational invariant of the systems of ϕ's of weight μ.

the roots of the λ-equation will not be changed by such a transformation. These roots, however, are irrational functions of the Θ's and hence of the a's and b's. We may therefore state the result:

THEOREM 2. *The roots of the λ-equation of a pair of quadratic forms are absolute irrational invariants of this pair of forms with regard to non-singular linear transformations.*

It is clear that the multiplicity of any root of the λ-equation will not be changed by a non-singular linear transformation. Hence

THEOREM 3. *The multiplicities of the roots of the λ-equation are arithmetical invariants of the pair of quadratic forms with regard to non-singular linear transformations.*

If
$$\phi \equiv a_1 x_1^2 + \cdots + a_n x_n^2,$$
$$\psi \equiv \ \ x_1^2 + \cdots + x_n^2,$$

the roots of the λ-equation are $a_1, \cdots a_n$. This example shows that the absolute invariants of Theorem 2 may have any values, and also that the arithmetical invariants of Theorem 3 are subject to no other restriction than the obvious one of being positive integers whose sum is n.

58. Reduction to Normal Form when the λ-Equation has no Multiple Roots. Although our main concern in this section is with the case in which the λ-equation has no multiple roots, we begin by establishing a theorem which applies to a much more general case.

THEOREM 1. *If λ_1 is a simple root of the λ-equation of the pair ϕ, ψ of quadratic forms in n variables, then, by a non-singular linear transformation, ϕ and ψ can be reduced respectively to the forms*

$$(1) \qquad \begin{cases} \lambda_1 c_1 z_1^2 + \phi_1(z_2, \cdots z_n) \\ \ \ c_1 z_1^2 + \psi_1(z_2, \cdots z_n) \end{cases}$$

where c_1 is a constant not zero and ϕ_1, ψ_1 are quadratic forms in the $n - 1$ variables $z_2, \cdots z_n$.

To prove this, we will consider the pencil of forms

$$(2) \qquad \phi - \lambda \psi \equiv \phi - \lambda_1 \psi + (\lambda_1 - \lambda)\psi.$$

Since λ_1 is a root of the λ-equation of the pair of forms ϕ, ψ, the form $\phi - \lambda_1\psi$ is singular, and can therefore, by a suitable non-singular linear transformation, be written in a form in which one of the variables, say x_1', does not enter, $\phi - \lambda_1\psi \equiv \phi'(x_2', \cdots x_n').$

If this transformation reduces ψ to ψ', we have

$$(3) \qquad \phi - \lambda\psi \equiv \phi'(x_2', \cdots x_n') + (\lambda_1 - \lambda)\psi'(x_1', \cdots x_n').$$

The discriminant of the quadratic form which stands here on the right cannot contain $\lambda_1 - \lambda$ as a factor more than once, since λ_1 is, by hypothesis, not a multiple root of the λ-equation of ϕ and ψ. It follows from this that the coefficient of $x_1'^2$ in the quadratic form ψ' cannot be zero, for otherwise the discriminant of the right-hand side of (3) would have a zero in the upper left-hand corner, and $\lambda_1 - \lambda$ would be a factor of all the elements of its first row and also of its first column ; so that it would contain the factor $(\lambda_1 - \lambda)^2$.

Since the coefficient of $x_1'^2$ in ψ' is not zero, we can by Lagrange's reduction (Formulæ (2), (3), §45) obtain a non-singular linear transformation of the form

$$\begin{cases} z_1 = \gamma_1 x_1' + \gamma_2 x_2' + \cdots + \gamma_n x_n' \\ z_2 = \qquad\quad\ x_2' \\ \cdot \quad \cdot \quad \cdot \quad \cdot \quad \cdot \quad \cdot \quad \cdot \quad \cdot \\ \cdot \quad \cdot \quad \cdot \quad \cdot \quad \cdot \quad \cdot \quad \cdot \quad \cdot \\ z_n = \qquad\qquad\qquad\qquad x_n' \end{cases}$$

which reduces ψ' to the form

$$c_1 z_1^2 + \psi_1(z_2, \cdots z_n) \qquad\qquad (c_1 \neq 0).$$

This transformation carries over the second member of (3) into

$$\phi'(z_2, \cdots z_n) + (\lambda_1 - \lambda)\psi_1(z_2, \cdots z_n) + (\lambda_1 - \lambda)c_1 z_1^2.$$

Combining these two linear transformations and writing

$$\phi'(z_2, \cdots z_n) + \lambda_1\psi_1(z_2, \cdots z_n) \equiv \phi_1(z_2, \cdots z_n),$$

we have thus obtained a non-singular linear transformation which effects the reduction,

$$\phi(x_1, \cdots x_n) - \lambda\psi(x_1, \cdots x_n) \equiv \phi_1(z_2, \cdots z_n) - \lambda\psi_1(z_2, \cdots z_n) + (\lambda_1 - \lambda)c_1 z_1^2.$$

If, here, we equate the coefficients of λ on both sides, and the terms independent of λ, we see that we have precisely the reduction of the forms ϕ, ψ to the forms (1); and the theorem is proved.

Let us now assume that the form ψ is non-singular, thus insuring that the λ-equation be of degree n. We will further assume that the roots $\lambda_1, \lambda_2, \cdots \lambda_n$ of this equation are all distinct. We can then, by the theorem just proved, reduce the forms ϕ, ψ to the forms (1) by a non-singular linear transformation. The λ-equation of these two forms is seen to differ from the λ-equation of the pair of forms in $(n-1)$ variables ϕ_1, ψ_1 only by the presence of the extra factor $\lambda_1 - \lambda$. Accordingly the λ-equation of the pair of forms ϕ_1, ψ_1 has as its roots $\lambda_2, \cdots \lambda_n$ and these are all simple roots. We may therefore apply the reduction of Theorem 1 to the two forms ϕ_1, ψ_1 and thus by a non-singular linear transformation of $z_2, \cdots z_n$ reduce them to the forms

$$\lambda_2 c_2 z_2'^2 + \phi_2(z_3', \cdots z_n'),$$
$$c_2 z_2'^2 + \psi_2(z_3', \cdots z_n').$$

This linear transformation may, by means of the additional formula

$$z_1' = z_1,$$

be regarded as a non-singular linear transformation of $z_1, \cdots z_n$ which carries over ϕ, ψ into the forms

$$\lambda_1 c_1 z_1'^2 + \lambda_2 c_2 z_2'^2 + \phi_2(z_3', \cdots z_n'),$$
$$c_1 z_1'^2 + c_2 z_2'^2 + \psi_2(z_3', \cdots z_n').$$

Proceeding in this way, we establish the theorem:

THEOREM 2. *If ϕ, ψ are quadratic forms in $(x_1, \cdots x_n)$ of which the second is non-singular, and if the roots $\lambda_1, \cdots \lambda_n$ of their λ-equation are all distinct, there exists a non-singular linear transformation which carries over ϕ and ψ into*

$$\lambda_1 c_1 x_1'^2 + \lambda_2 c_2 x_2'^2 + \cdots + \lambda_n c_n x_n'^2,$$
$$c_1 x_1'^2 + c_2 x_2'^2 + \cdots + c_n x_n'^2$$

respectively, where $c_1, \cdots c_n$ are constants all different from zero.

Since none of the c's are zero, the linear transformation

$$x_i'' = \sqrt{c_i}\, x_i' \qquad\qquad (i = 1, 2, \cdots n)$$

is non-singular. Performing this transformation, we get the further result:

THEOREM 3. *Under the same conditions as in Theorem 2, ϕ and ψ may be reduced by means of a non-singular linear transformation to the normal forms*

$$\lambda_1 x_1^2 + \lambda_2 x_2^2 + \cdots + \lambda_n x_n^2,$$
$$x_1^2 + x_2^2 + \cdots + x_n^2.$$

From this we infer at once

THEOREM 4. *If in the two pairs of quadratic forms ϕ, ψ and ϕ', ψ' the forms ψ and ψ' are both non-singular, and if the λ-equations of these two pairs of forms have no multiple roots, a necessary and sufficient condition for the equivalence of the two pairs of forms is that these two λ-equations have the same roots; or, what amounts to the same thing, that the invariants Θ_0, Θ_1, $\cdots \Theta_n$ of the first pair of forms be proportional to the invariants Θ_0', Θ_1', $\cdots \Theta_n'$ of the second.*

EXERCISE

Prove that, under the conditions of Theorem 3, the reduction to the normal form can be performed in essentially only one way, the only possible variation consisting in a change of sign of some of the x's in the normal form.

59. Reduction to Normal Form when ψ is Definite and Non-singular. We now consider the case of two real quadratic forms ϕ, ψ of which ψ is definite and non-singular. Our main problem is to reduce this pair of forms to a normal form by means of a real linear transformation. For this purpose we begin by proving

THEOREM 1. *The λ-equation of a pair of real quadratic forms ϕ, ψ can have no imaginary root if the form ψ is definite and non-singular.*

For, if possible, let $\alpha + \beta i$ (α and β real) be an imaginary root of this λ-equation, so that $\beta \neq 0$. Then $\phi - \alpha \psi - i\beta \psi$ will be a singular quadratic form, and can therefore be reduced by a non-singular linear transformation

$$
\begin{cases}
x_1' = (p_{11} + iq_{11})x_1 + \cdots + (p_{1n} + iq_{1n})x_n, \\
 \cdot \quad \cdot \quad \cdot \quad \cdot \quad \cdot \quad \cdot \quad \cdot \quad \cdot \quad \cdot \quad \cdot \quad \cdot \quad \cdot \\
 \cdot \quad \cdot \quad \cdot \quad \cdot \quad \cdot \quad \cdot \quad \cdot \quad \cdot \quad \cdot \quad \cdot \quad \cdot \quad \cdot \\
x_n' = (p_{n1} + iq_{n1})x_1 + \cdots + (p_{nn} + iq_{nn})x_n
\end{cases}
$$

to the sum of k squares, where $k < n$,

(1) $$\phi - \alpha\psi - i\beta\psi \equiv x_1'^2 + x_2'^2 + \cdots + x_k'^2.$$

    Let

(2) $$y_l = p_{l1}x_1 + \cdots + p_{ln}x_n,$$

(3) $$z_l = q_{l1}x_1 + \cdots + q_{ln}x_n,$$

so that $$x_l' = y_l + iz_l.$$

By equating the coefficients of i on the two sides of (1) we thus get

$$(4) \qquad -\beta\psi \equiv 2\,y_1 z_1 + 2\,y_2\,z_2 + \cdots + 2\,y_k z_k.$$

Let us now determine $x_1, \cdots x_n$ so as to make the right-hand side of (4) vanish, for instance by means of the equations

$$y_1 = y_2 = \cdots = y_k = 0.$$

A reference to (2) shows that we have here a system of k real homogeneous linear equations in n unknowns, so that real values of $x_1, \cdots x_n$ not all zero can be found satisfying these equations. For these values of the variables, we see from (4) that ψ vanishes; but this is impossible (cf. the Corollary of Theorem 3, § 52), since ψ is by hypothesis non-singular and definite.

THEOREM 2. *If ψ is a non-singular definite quadratic form and ϕ is any real quadratic form, the pair of forms ϕ, ψ can be reduced by a real non-singular linear transformation to the normal form*

$$(5) \qquad \begin{cases} \phi \equiv \pm\,(\lambda_1 x_1'^2 + \cdots + \lambda_n x_n'^2), \\ \psi \equiv \pm\,(\quad x_1'^2 + \cdots + \quad x_n'^2), \end{cases}$$

where $\lambda_1, \cdots \lambda_n$ are the roots of the λ-equation, and the upper or lower sign is to be used in both cases according as ψ is a positive or a negative form.

The proof of this theorem is very similar to the proof of Theorem 2, § 58. We must first prove, as in Theorem 1, § 58, that ϕ, ψ can be reduced by a real non-singular linear transformation to the forms

$$(6) \qquad \begin{cases} \lambda_1 c_1 z_1^2 + \phi_1\,(z_2, \cdots z_n) \\ \quad c_1 z_1^2 + \psi_1(z_2, \cdots z_n). \end{cases} \qquad (c_1 \neq 0)$$

To prove this, we consider the pencil of forms

$$\phi - \lambda\psi \equiv \phi - \lambda_1\psi + (\lambda_1 - \lambda)\psi.$$

Since λ_1 is real by Theorem 1, $\phi - \lambda_1\psi$ is a real singular quadratic form, and can therefore by a real non-singular linear transformation be reduced to a form in which one of the variables does not enter,

$$\phi - \lambda_1\psi \equiv \phi'(x_2', \cdots x_n').$$

If this transformation reduces ψ to ψ', we have

$$(7) \qquad \phi - \lambda\psi \equiv \phi'(x_2', \cdots x_n') + (\lambda_1 - \lambda)\psi'(x_1', \cdots x_n').$$

At this point comes the essential difference between the case we are now considering and the case considered in § 58, as λ_1 may now be a multiple root of the discriminant of the right-hand side of (7). We need, then, a different method for showing that the coefficient of $x_1'^2$ in ψ' is not zero. For this purpose it is sufficient to notice that ψ, and therefore also ψ', is a non-singular definite form, and that accordingly, by Theorem 4, § 52, the coefficient of none of the square terms in ψ' can be zero.

Having thus shown that the coefficient of $x_1'^2$ in ψ' is not zero, we can apply Lagrange's reduction to ψ', and thus complete the reduction of the forms ϕ, ψ to the forms (6) precisely as in the proof of Theorem 1, § 58, noticing that the transformation we have to deal with is real.

In (6), ϕ_1, ψ_1 are real quadratic forms in the $n-1$ variables $z_2, \cdots z_n$. Moreover, since

$$\psi(x_1, \cdots x_n) \equiv c_1 z_1^2 + \psi_1(z_2, \cdots z_n)$$

is non-singular and definite, it follows that the same is true of ψ_1. For, if ψ_1 were either singular or indefinite, we could find values of $z_2, \cdots z_n$ not all zero and such that $\psi_1 = 0$; and these values together with the value $z_1 = 0$ would make $\psi = 0$. This, however, is impossible by the Corollary of Theorem 3, § 52.

The λ-equation of the two forms ϕ_1, ψ_1 evidently differs from the λ-equation of ϕ, ψ only by the absence of the factor $\lambda - \lambda_1$. The roots of the λ-equation of ϕ_1, ψ_1 are therefore $\lambda_2, \cdots \lambda_n$, so that if we reduce ϕ_1 and ψ_1 by the method already used for ϕ, ψ (we have just seen that ϕ_1, ψ_1 satisfy all the conditions imposed on ϕ, ψ), we get

$$\phi_1(z_2, \cdots z_n) \equiv \lambda_2 c_2 z_2'^2 + \phi_2(z_3', \cdots z_n'),$$

$$\psi_1(z_2, \cdots z_n) \equiv c_2 z_2'^2 + \psi_2(z_3', \cdots z_n').$$

Proceeding in this way, we finally reduce ϕ, ψ by a real non-singular linear transformation to the forms

$$(8) \qquad \begin{cases} \phi \equiv \lambda_1 c_1 y_1^2 + \cdots + \lambda_n c_n y_n^2, \\ \psi \equiv c_1 y_1^2 + \cdots + c_n y_n^2. \end{cases}$$

Since ψ is definite, the constants $c_1, \cdots c_n$ are all positive or all negative according as ψ is a positive or a negative form. By means of the further non-singular real linear transformation

$$x_i' = \sqrt{|c_i|}\, y_i \qquad\qquad (i = 1, 2, \cdots n),$$

the forms (8) may be reduced to the forms (5), and our theorem is proved.

EXERCISES

1. If ϕ is a real quadratic form in n variables of rank r, prove that it can be reduced by a real orthogonal transformation in n variables to the form

$$c_1 x_1^2 + c_2 x_2^2 + \cdots + c_r x_r^2.$$

Cf. Exercises, § 52.

2. Show that the determinant of the orthogonal transformation of Exercise 1 may be taken at pleasure as $+1$ or -1.

3. Discuss the metrical classification of real quadric surfaces along the following lines:

Assume the equation in non-homogeneous rectangular coördinates, and show that by a transformation to another system of rectangular coördinates having the same origin the equation can be reduced to a form where the terms of the second degree have one or the other of the five forms (the A's being positive constants)

$$A_1 x_1^2 + A_2 x_2^2 + A_3 x_3^2,$$
$$A_1 x_1^2 + A_2 x_2^2 - A_3 x_3^2,$$
$$A_1 x_1^2 + A_2 x_2^2,$$
$$A_1 x_1^2 - A_2 x_2^2,$$
$$A_1 x_1^2.$$

Then simplify each of the non-homogeneous equations thus obtained by further transformations of coördinates; thus getting finally the standard forms of the equations of ellipsoids, hyperboloids, paraboloids, cones, cylinders, and planes which are discussed in all elementary text-books of solid analytic geometry.

CHAPTER XIV

SOME PROPERTIES OF POLYNOMIALS IN GENERAL

60. Factors and Reducibility. In the present section we will introduce certain conceptions of fundamental importance in our subsequent work.

DEFINITION 1. *By a factor or divisor of a polynomial f in any number of variables is understood a polynomial ϕ which satisfies an identity of the form*
$$f \equiv \phi\psi,$$
ψ being also a polynomial.

It will be noticed that every constant different from zero is a factor of every polynomial; that every polynomial is a factor of a polynomial which vanishes identically; while a polynomial which is a mere constant, different from zero, has no factors other than constants.

We note also that a polynomial in $x_1, \cdots x_n$ which is not identically zero cannot have as a factor a polynomial which actually contains any other variables.

The conception of reducibility, which we have already met in a special case (§47), we define as follows:

DEFINITION 2. *A polynomial is said to be reducible if it is identically equal to the product of two polynomials neither of which is a constant.*

In dealing with real polynomials, a narrower determination of the conception of reducibility is usually desirable. We consider, then, what we will call *reducibility in the domain of reals*, a conception which we define as follows:

DEFINITION 3. *A real polynomial is said to be reducible in the domain of reals if it is identically equal to the product of two other real polynomials neither of which is a constant.*

In many branches of algebra still another modification of the conception of reducibility plays an important part. In order to explain this, we first lay down the following definition:

DEFINITION 4. *A set of numbers is said to form a domain of rationality if, when a and b are any numbers of the set, $a + b$, $a - b$, ab, and, so far as $b \neq 0$, a/b are also numbers of the set.*

Thus all numbers, real and imaginary, form a domain of rationality, and the same is true of all real numbers. The simplest of all domains of rationality, apart from the one which contains only the single number zero, is what is known as the natural domain, that is all rational numbers, positive and negative. A more complicated domain of rationality would be the one consisting of all numbers of the form $a + b\sqrt{-1}$, where a and b are not merely real, but rational. These illustrations, which might be multiplied indefinitely, should suffice to make the scope of the above definition clear.*

DEFINITION 5. *A polynomial all of whose coefficients lie in a domain of rationality R is said to be reducible in this domain if it is identically equal to a product of two polynomials, neither of which is a constant, whose coefficients also lie in this domain.*

It will be noticed that Definitions 2 and 3 are merely the special cases of this definition in which the domain of rationality is the domain of all numbers, and the domain of all reals respectively. To illustrate these three definitions, we note that the polynomial $x^2 + 1$ is reducible according to Definition 2, since it is identically equal to $(x + \sqrt{-1})(x - \sqrt{-1})$. It is, however, not reducible in the domain of reals, nor in the natural domain. On the other hand, $x^2 - 2$ is reducible in the domain of reals, but not in the natural domain. Finally, $x^2 - 4$ is reducible in the natural domain.

Leaving these modifications of the conception of reducibility, we close this section with the following two definitions:

DEFINITION 6. *Two polynomials are said to be relatively prime if they have no common factor other than a constant.*

---

*By $R(a_1, a_2, \cdots a_n)$ is understood the domain of rationality consisting of all numbers which can be obtained from the given numbers $a_1, \cdots a_n$ by the rational processes (addition, subtraction, multiplication, and division). In this notation the natural domain would be most simply denoted by $R(1)$; the domain last mentioned in the text by $R(1, \sqrt{-1})$ or, even more simply, by $R(\sqrt{-1})$. This notation would not apply to all cases (*e.g.* the real domain) except by the use of an infinite number of arguments.

DEFINITION 7. *Two methods of factoring a polynomial shall be said to be not essentially different if there are the same number n of factors in each case, and these factors can be so arranged that the kth factors are proportional for all values of k, from 1 to n inclusive.*

EXERCISES

1. Prove that every polynomial in (x, y) is irreducible if it is of the form

$$f(x) + y,$$

where $f(x)$ is a polynomial in x alone.

Would this also be true for polynomials of the form

$$f(x) + y^2 ?$$

2. If f, ϕ, ψ are polynomials in any number of variables which satisfy the relation

$$f \equiv \phi \psi,$$

and if the coefficients of f and ϕ lie in a certain domain of rationality, prove that the coefficients of ψ will lie in the same domain provided $\phi \not\equiv 0$.

61. The Irreducibility of the General Determinant and of the Symmetrical Determinant.

THEOREM 1. *The determinant*

$$D = \begin{vmatrix} a_{11} & a_{12} & \cdots & a_{1n} \\ a_{21} & a_{22} & \cdots & a_{2n} \\ \cdot & \cdot & \cdot & \cdot \\ a_{n1} & a_{n2} & \cdots & a_{nn} \end{vmatrix}$$

is an irreducible polynomial if its n^2 elements are regarded as independent variables.

For suppose it were reducible, and let

$$D \equiv f(a_{11}, \cdots a_{nn}) \, \phi(a_{11}, \cdots a_{nn}),$$

where neither f nor ϕ is a constant. Expanding D according to the elements of the first row, we see that it is of the first degree in a_{11}. Hence one of the two polynomials f and ϕ must be of the first degree in a_{11}, the other of the zeroth degree. Precisely the same reasoning shows that if a_{ij} is any element of D, one of the polynomials f and ϕ will be of the first degree in a_{ij}, while the other will not involve this variable.

Let us denote by f that one of the two polynomials which involves a_{ii}, any element of the principal diagonal of D. Then ϕ does not involve any element of the ith row or the ith column.   For if it did, since f is of the first degree in a_{ii} and ϕ is of the zeroth, their product D would involve terms containing products of the form $a_{ii}a_{ij}$ or $a_{ii}a_{ji}$, which, from the definition of a determinant, is impossible. Consequently, if either one of the polynomials f and ϕ contains any element of the principal diagonal of D, it must contain all the elements standing in the same row and all those standing in the same column with this one, and none of these can occur in the other polynomial.

Now suppose f contains a_{ii} and that ϕ contains any other element of the principal diagonal, say a_{jj}.   Then a_{ij} and a_{ji} can be in neither f nor ϕ, which is impossible.   Hence, if f contains any one of the elements in the principal diagonal, it must contain all the others, and hence *all* the elements, and our theorem is proved.

THEOREM 2.   *The symmetrical determinant*

$$D = \begin{vmatrix} a_{11} & \cdots & a_{1n} \\ \cdot & \cdot & \cdot \\ a_{n1} & \cdots & a_{nn} \end{vmatrix} \qquad (a_{ij} = a_{ji})$$

is an irreducible polynomial if its $\frac{1}{2}n(n+1)$ *elements be regarded as independent variables.*

The proof given for the last theorem holds, almost word for word, in this case also, the only difference being that while D is of the first degree in each of the elements of its principal diagonal, it is of the *second* degree in each of the other elements.   The slight changes in the proof made necessary by this difference are left to the reader.

EXERCISES

1. The general bordered determinant

$$\begin{vmatrix} a_{11} & \cdots & a_{1n} & u_{11} & \cdots & u_{1p} \\ \cdot & & \cdot & \cdot & & \cdot \\ a_{n1} & \cdots & a_{nn} & u_{n1} & \cdots & u_{np} \\ v_{11} & \cdots & v_{1n} & 0 & \cdots & 0 \\ \cdot & & \cdot & \cdot & & \cdot \\ v_{p1} & \cdots & v_{pn} & 0 & \cdots & 0 \end{vmatrix}$$

is irreducible if $p < n$, the a's, u's, and v's being regarded as independent variables.

2. The symmetrical bordered determinant obtained from the determinant in Exercise 1 by letting $a_{ij} = a_{ji}$, $u_{ij} = v_{ji}$ is irreducible if $p < n$.

3. If for certain values of i and j, but not for all, $a_{ij} = a_{ji}$, but if the a's are otherwise independent, can we still say that

$$\begin{vmatrix} a_{11} & \cdots & a_{1n} \\ \cdot & \cdot & \cdot \\ \cdot & \cdot & \cdot \\ a_{n1} & \cdots & a_{nn} \end{vmatrix}$$

is irreducible ?

4. Prove that a skew-symmetric determinant (cf. Exercises 2, 3, §20) is always reducible by showing that, when it is of even order, it is a perfect square.

[SUGGESTION. Use Corollary 3, § 11, and Theorem 6 and Exercise 1, § 76.]

Does this theorem require any modification if the elements are real and we consider reducibility in the domain of reals?

62. Corresponding Homogeneous and Non-Homogeneous Polynomials.   It is often convenient to consider side by side two polynomials, one homogeneous and the other non-homogeneous, which bear to one another the same relation as the first members of the equations of a plane curve or of a surface in homogeneous and non-homogeneous coördinates respectively. Such polynomials we will speak of as *corresponding* to one another according to the following definition:

DEFINITION. *If we have a non-homogeneous polynomial of the kth degree in any number of variables* $(x_1, \cdots x_{n-1})$ *and form a new polynomial by multiplying each term of the old by the power of a new variable* x_n *necessary to bring up the degree of this term to* k, *the homogeneous polynomial thus formed shall be said to* **correspond** *to the given non-homogeneous polynomial.*

Thus the two polynomials

(1) $2x^3 + 3x^2y - 5xz^2 - yz + 2z^2 + x - 3y - 9,$

(2) $2x^3 + 3x^2y - 5xz^2 - yzt + 2z^2t + xt^2 - 3yt^2 - 9t^3,$

correspond to each other.

It may be noticed that if $\phi(x_1, \cdots x_{n-1})$ is the non-homogeneous polynomial of degree k, the corresponding homogeneous polynomial may be written

$$x_n^k \, \phi\left(\frac{x_1}{x_n}, \frac{x_2}{x_n}, \cdots \frac{x_{n-1}}{x_n}\right).$$

To every non-homogeneous polynomial there corresponds one, and only one, homogeneous polynomial. Conversely, however, to a homogeneous polynomial in n variables there correspond in general

n different non-homogeneous polynomials which are obtained by setting one of the variables equal to 1.   For instance, in the example given above, to (2) corresponds not only (1) but also

$$(3) \qquad 2 \quad + 3y \quad - 5\,z^2 \ - yzt + 2z^2t + t^2 \ - 3yt^2 - 9t^3,$$

$$(4) \qquad 2x^3 + 3x^2 \ - 5xz^2 - zt \ + 2z^2t + xt^2 - 3t^2 \ - 9t^3,$$

$$(5) \qquad 2x^3 + 3x^2y - 5x \ - yt \ + 2t \ + xt^2 - 3yt^2 - 9t^3.$$

It should be noticed, however, that if one of the variables enters into every term of a homogeneous polynomial, the result of setting this variable equal to unity is to give, not a *corresponding* non-homogeneous polynomial, but a polynomial of lower degree.   In fact, in the extreme case in which every variable enters into every term of the homogeneous polynomial, there is no corresponding non-homogeneous polynomial; as, for instance, in the case of the polynomial

$$x^2yz + xy^2z + xyz^2.$$

THEOREM 1.   *If one of two corresponding polynomials is reducible, then the other is, also, and the factors of each polynomial correspond to the factors of the other.*

For let $\phi(x_1, \cdots x_n)$ be a homogeneous polynomial of degree $(k + l)$, and suppose it can be factored into two factors of degrees k and l, respectively,

$$(6) \qquad \phi_{k+l}(x_1, \cdots x_n) \equiv \psi_k(x_1, \cdots x_n)\,\chi_l(x_1, \cdots x_n).$$

Now suppose the corresponding non-homogeneous polynomial in question is the one formed by setting $x_n = 1$.   We have

$$(7) \qquad \phi_{k+l}(x_1, \cdots x_{n-1}, 1) \equiv \psi_k(x_1, \cdots x_{n-1}, 1)\,\chi_l(x_1, \cdots x_{n-1}, 1).$$

Since by hypothesis the degree of the polynomial on the left is unchanged by this operation, neither of the factors on the right-hand side of (6) can have had its degree reduced, hence neither of the factors on the right of (7) is a constant.   Our non-homogeneous polynomial is therefore reducible; and moreover the two factors on the right of (7), being of degrees k and l respectively, are precisely the two functions corresponding to the two factors on the right of (6).

Now, let $\Phi_{k+l}(x_1, \cdots x_{n-1})$ be a non-homogeneous polynomial, and suppose

$$\Phi_{k+l}(x_1, \cdots x_{n-1}) \equiv \Psi_k(x_1, \cdots x_{n-1}) \, \mathrm{X}_l(x_1, \cdots x_{n-1}),$$

where the subscripts denote the degrees of the polynomials. Let $\phi_{k+l}, \psi_{k+l}, \chi_{k+l}$ be the homogeneous polynomials corresponding to Φ, Ψ, X. Then when $x_n \neq 0$,

$$\Phi_{k+l}\left(\frac{x_1}{x_n}, \cdots \frac{x_{n-1}}{x_n}\right) = \Psi_k\left(\frac{x_1}{x_n}, \cdots \frac{x_{n-1}}{x_n}\right) \mathrm{X}_l\left(\frac{x_1}{x_n}, \cdots \frac{x_{n-1}}{x_n}\right).$$

Multiplying this equation by x_n^{k+l} we have

$$\phi_{k+l}(x_1, \cdots x_{n-1}, x_n) = \psi_k(x_1, \cdots x_{n-1}, x_n) \, \chi_l(x_1, \cdots x_{n-1}, x_n),$$

an equation which holds whenever $x_n \neq 0$, and, therefore, by Theorem 5, § 2, is an identity. Thus our theorem is proved.

As a simple illustration of the way in which this theorem may be applied we mention the condition for reducibility of a non-homogeneous quadratic polynomial in any number of variables. By applying the test of § 47 to the corresponding homogeneous polynomial we obtain at once a test for the reducibility of any non-homogeneous quadratic polynomial.

THEOREM 2. *If f and ϕ are non-homogeneous polynomials, and F, Φ are the corresponding homogeneous polynomials, a necessary and sufficient condition that F and Φ be relatively prime is that f and ϕ be relatively prime.*

For if f and ϕ have a common factor ψ which is not a constant, the homogeneous polynomial Ψ which corresponds to ψ is, by Theorem 1, a common factor of F and Φ, and is clearly not a constant. Conversely, if Ψ is a common factor of F and Φ which is not a constant, f and ϕ will have, by Theorem 1, a common factor which corresponds to Ψ and which therefore cannot be a mere constant.

63. Division of Polynomials. We will consider first two polynomials in one variable:

$$(1) \qquad \begin{cases} f(x) \equiv a_0 x^n + a_1 x^{n-1} + \cdots + a_n, \\ \phi(x) \equiv b_0 x^m + b_1 x^{m-1} + \cdots + b_m. \end{cases}$$

We learn in elementary algebra how to divide f by ϕ, getting a quotient $Q(x)$ and a remainder $R(x)$. What is essential here is contained in the following theorem:

THEOREM 1. *If f and ϕ are two polynomials in x of which ϕ is not identically zero, there exists one, and only one, pair of polynomials, Q and R, which satisfy the identity*

$$(2) \qquad f(x) \equiv Q(x)\,\phi(x) + R(x),$$

and such that either $R \equiv 0$, or the degree of R is less than the degree of ϕ.*

We begin by proving that *at least* one pair of polynomials Q, R exists which satisfies the conditions of the theorem.

If f is of lower degree than ϕ (or if $f \equiv 0$), the truth of this statement is obvious, for we may then let $Q \equiv 0$, $R \equiv f$.

Suppose, then, that f is of at least as high degree as ϕ. Writing f and ϕ in the form (1), we may assume

$$a_0 \neq 0, \quad b_0 \neq 0, \quad n \geqq m.$$

LEMMA. *If ϕ is not of higher degree than f, there exist two polynomials Q_1 and R_1 which satisfy the identity*

$$f(x) \equiv Q_1(x)\,\phi(x) + R_1(x),$$

and such that either $R_1 \equiv 0$, or the degree of R_1 is less than the degree of f.

The truth of this lemma is obvious if we let

$$Q_1(x) \equiv \frac{a_0}{b_0} x^{n-m}.$$

These two polynomials Q_1 and R_1 will serve as the polynomials Q and R of our theorem if $R_1 \equiv 0$, or if the degree of R_1 is less than the degree of ϕ. If not, apply the lemma again to the two functions R_1 and ϕ, getting

$$R_1(x) \equiv Q_2(x)\,\phi(x) + R_2(x),$$

where R_2 is either identically zero or is of lower degree than R_1. We may then write,

$$f(x) \equiv [\,Q_1(x) + Q_2(x)\,]\,\phi(x) + R_2(x).$$

If $R_2 \equiv 0$, or if the degree of R_2 is less than the degree of ϕ, we may take for the polynomials Q, R of our theorem, $Q_1 + Q_2$ and R_2. If not, we apply our lemma again to R_2 and ϕ. Proceeding in this way

---

* It will be remembered that, according to the definition we have adopted, a polynomial which vanishes identically has no degree.

we get a series of polynomials R_1, R_2, $\cdots$ whose degrees are constantly decreasing. We therefore, after a certain number of steps, reach a polynomial R_i which is either identically zero or of degree less than ϕ. Combining the identities obtained up to this point, we have

$$f(x) \equiv [Q_1(x) + \cdots + Q_i(x)] \, \phi(x) + R_i(x),$$

an identity which proves the part of our theorem which states that at least one pair of polynomials Q, R of the kind described exists.[*]

Suppose now that besides the polynomials Q, R of the theorem there existed a second pair of polynomials Q', R' satisfying the same conditions. Subtracting from (2) the similar identity involving Q', R', we have

$$(3) \qquad\qquad 0 \equiv (Q - Q') \, \phi + (R - R').$$

From this we infer, as was to be proved,

$$Q \equiv Q', \qquad R \equiv R'.$$

For if Q and Q' were not identical, the first term on the right of (3) would be of at least the mth degree, while the second involves no power of x as high as m.

Turning now to polynomials in several variables:

$$(4) \quad \begin{cases} f(x_1, \cdots x_k) \equiv a_0(x_2, \cdots x_k)x_1^n + a_1(x_2, \cdots x_k)x_1^{n-1} + \cdots + a_n(x_2, \cdots x_k), \\ \phi(x_1, \cdots x_k) \equiv b_0(x_2, \cdots x_k)x_1^m + b_1(x_2, \cdots x_k)x_1^{m-1} + \cdots + b_m(x_2, \cdots x_k), \end{cases}$$

the ordinary method of dividing f by ϕ would give us as quotient and remainder, not polynomials, but fractional rational functions. In order to avoid this, we state our theorem in the following form:

THEOREM 2. *If f and ϕ are polynomials in $(x_1, \cdots x_k)$ of which ϕ is not identically zero, there exist polynomials Q, R, P, of which the last is not identically zero and does not involve the variable x_1, which satisfy the identity,*

$$(5) \quad P(x_2, \cdots x_k)f(x_1, \cdots x_k) \equiv Q(x_1, \cdots x_k)\phi(x_1, \cdots x_k) + R(x_1, \cdots x_k),$$

and such that either $R \equiv 0$, or the degree in x_1 of R is less than the degree in x_1 of ϕ.

The proof of this theorem follows the same lines as the proof of Theorem 1.

---

[*] The reader should notice that the process just used is merely the ordinary process of long division.

If f is of lower degree in x_1 than ϕ (or if $f \equiv 0$), the truth of the theorem is obvious, for we may then let $P \equiv 1$, $Q \equiv 0$, $R \equiv f$.

Suppose, then, that f is of at least as high degree in x_1 as ϕ. Writing f and ϕ in the form (4), we may assume

$$a_0 \not\equiv 0, \quad b_0 \not\equiv 0, \quad n \geqq m.$$

LEMMA. *If ϕ is not of higher degree in x_1 than f, there exist two polynomials Q_1, R_1 which satisfy the identity,*

$$b_0(x_2, \cdots x_k) f(x_1, \cdots x_k) \equiv Q_1(x_1, \cdots x_k) \phi(x_1, \cdots x_k) + R_1(x_1, \cdots x_k),$$

and such that either $R_1 \equiv 0$, or the degree of R_1 in x_1 is less than the degree of f in x_1.

The truth of this lemma is obvious if we let

$$Q_1 \equiv a_0(x_2, \cdots x_k)\, x_1^{n-m}\,.$$

The polynomials Q_1, R_1, b_0 will serve as the polynomials Q, R, P of our theorem if $R_1 \equiv 0$, or if the degree of R_1 in x_1 is less than the degree of ϕ in x_1. If not, apply the lemma again to the two functions R_1 and ϕ, getting

$$b_0(x_2, \cdots x_k) R_1(x_1, \cdots x_k) \equiv Q_2(x_1, \cdots x_k) \phi(x_1, \cdots x_k) + R_2(x_1, \cdots x_k),$$

where R_2 is either identically zero or is of lower degree in x_1 than R_1. We may then write $\quad b_0^2 f \equiv (b_0 Q_1 + Q_2)\phi + R_2.$

If $R_2 \equiv 0$, or if the degree of R_2 in x_1 is less than the degree of ϕ in x_1, we may take for the polynomials Q, R, P of our theorem the functions $b_0 Q_1 + Q_2$, R_2, b_0^2. If not, we apply our lemma again to R_2 and ϕ. Proceeding in this way, we get a series of polynomials R_1, R_2, $\cdots$ whose degrees in x_1 are constantly decreasing. We therefore, after a certain number of steps, reach a polynomial R_i which is either identically zero, or of degree in x_1 less than ϕ. Combining the identities obtained up to this point, we have

$$b_0^i f \equiv (b_0^{i-1} Q_1 + b_0^{i-2} Q_2 + \cdots + Q_i)\phi + R_i,$$

an identity which proves our theorem, and which also establishes the additional result:

COROLLARY. *The polynomial P whose existence is stated in our theorem may be taken as a power of b_0.*

We note that it would obviously not be correct to add to the statement of Theorem 2 the further statement that there is only one set of polynomials Q, R, P, since the identity (5) may be multiplied by any polynomial in $(x_2, \cdots x_n)$ without changing its form. Cf., however, the exercise at the end of § 73.

64. A Special Transformation of a Polynomial. Suppose that $f(x_1, x_2, x_3, x_4)$ is a homogeneous polynomial of the kth degree in the homogeneous coördinates $(x_1,\ x_2,\ x_3,\ x_4)$, so that the equation $f = 0$ represents a surface of the kth degree. If, in f, the term in x_4^k has the coefficient zero, the surface passes through the origin; and if the term in x_1^k (or x_2^k, or x_3^k) has the coefficient zero, the surface passes through the point at infinity on the axis of x_1 (or x_2, or x_3). It is clear that these peculiarities of the surface can be avoided, and that, too, in an infinite variety of ways, by subjecting the surface to a non-singular collineation which carries over any four non-complanar points, no one of which lies on the surface, into the origin and the three points at infinity on the coördinate axes. It is this fact, generalized to the case of n variables, which we now proceed to prove.

LEMMA. *If $f(x_1, \cdots x_n)$ is a homogeneous polynomial of the kth degree in which the term x_m^k is wanting, there exists a non-singular linear transformation of the variables $(x_1, \cdots x_n)$ which carries f into a new form f_1, in which the term in $x_m'^k$ has a coefficient different from zero, while the coefficients of the kth powers of the other variables have not been changed.*

In proving this theorem there is obviously no real loss of generality in taking as the variable x_m the last of the variables x_n.

Let us then consider the non-singular transformation

$$x_i = x_i' + a_i x_n' \qquad (i = 1, \cdots n-1).$$
$$x_n = x_n'$$

This transformation carries f into

$$f_1(x_1', \cdots x_n') \equiv f(x_1' + a_1 x_n', \cdots x_{n-1}' + a_{n-1} x_n', x_n'),$$

and evidently does not change the coefficients of the terms in $x_1^k, \cdots x_{n-1}^k$.

Now, since every term in f_1, except the term in $x_n'^k$, contains at least one of the variables $x_1', \cdots x_{n-1}'$, the coefficient of the term in $x_n'^k$ will be

$$f_1(0, \cdots 0, 1) = f(a_1, \cdots a_{n-1}, 1).$$

Our lemma will therefore be proved if we can show that the constants $a_1, \cdots a_{n-1}$ can be so chosen that this quantity is not zero.

Let us take any point $(b_1, \cdots b_n)$ for which $b_n \neq 0$; and consider a neighborhood of this point sufficiently small so that x_n does not vanish at any point in this neighborhood. Then, since f does not vanish identically, we can find a point $(c_1, \cdots c_n)$ in this neighborhood (so that $c_n \neq 0$) such that

$$f(c_1, \cdots c_n) \neq 0.$$

If now we take for $a_1, \cdots a_{n-1}$ the values $c_1/c_n, \cdots c_{n-1}/c_n$, we shall have, since f is homogeneous,

$$f(a_1, \cdots a_{n-1}, 1) \neq 0.$$

Thus our lemma is proved.

THEOREM 1. *If $f(x_1, \cdots x_n)$ is a homogeneous polynomial of the kth degree, there exists a non-singular linear transformation which carries f into a new form f_1 in which the terms in $x_1'^k, \cdots x_n'^k$ all have coefficients different from zero.*

The proof of this theorem follows at once from the preceding lemma. For we need merely to perform in succession the transformations which cause the coefficients first of x_1^k, then of x_2^k, etc., to become different from zero, and which our lemma assures us will exist and be non-singular, to obtain the transformation we want. To make sure of this it is necessary merely to notice that the coefficient of x_1^k obtained by the first transformation will not be changed by the subsequent transformations; that the same will be true of the coefficient of x_2^k obtained by the second trans formation; etc.

THEOREM 2. *If $f(x_1, \cdots x_n)$ is a polynomial of the kth degree which is not necessarily homogeneous, there exists a non-singular homogeneous linear transformation of $(x_1, \cdots x_n)$ which makes this polynomial of the kth degree in each of the variables $x_1', \cdots x_n'$ taken separately.*

If f is homogeneous, this is equivalent to Theorem 1. If f is non-homogeneous, we may write it in the form

$$f(x_1, \cdots x_n) \equiv \phi_k(x_1, \cdots x_n) + \phi_{k-1}(x_1, \cdots x_n) + \cdots + \phi_1(x_1, \cdots x_n) + \phi_0,$$

where each ϕ is a homogeneous polynomial of the degree indicated by its subscript or else is identically zero. We need now merely to apply Theorem 1 to the homogeneous polynomial ϕ_k, which is, of course, not identically zero.

This theorem, and therefore also Theorem 1, which is merely a special case of it, admits the following generalization to the case of a system of functions:

THEOREM 3. *If we have a system of polynomials*

$$f_1(x_1, \cdots x_n), \quad f_2(x_1, \cdots x_n), \quad \cdots\cdots f_m(x_1, \cdots x_n),$$

of degrees $k_1, k_2, \cdots k_m$ respectively, there exists a non-singular homogeneous linear transformation which makes these polynomials of degrees $k_1, \cdots k_m$ in each of the variables $x_1', \cdots x_n'$ taken separately.

This theorem may be proved either by the same method used in proving Theorems 1, 2; or by applying Theorem 2 to the product $f_1 f_2 \cdots f_m$.

CHAPTER XV

FACTORS AND COMMON FACTORS OF POLYNOMIALS IN ONE VARIABLE AND OF BINARY FORMS

65. Fundamental Theorems on the Factoring of Polynomials in One Variable and of Binary Forms. Theorem 2, § 6 may be stated in the following form:

THEOREM 1. *A polynomial of the nth degree in one variable is always reducible when $n > 1$. It can be resolved into the product of n linear factors in one, and essentially in only one, way.*

By means of § 62 we can deduce from this a similar theorem in the case of the binary form

$$(1) \qquad a_0 x_1^n + a_1 x_1^{n-1} x_2 + \cdots + a_n x_2^n.$$

Let us first assume that $a_0 \neq 0$. Then the non-homogeneous polynomial

$$(2) \qquad a_0 x_1^n + a_1 x_1^{n-1} + \cdots + a_n$$

corresponds to (1), according to the definition of § 62. Factoring (2), we get

$$a_0(x_1 - \alpha_1)(x_1 - \alpha_2) \cdots (x_1 - \alpha_n),$$

or, if we take n constants $\alpha_1'', \alpha_2'', \cdots \alpha_n''$ whose product is a_0,

$$(3) \qquad (\alpha_1'' x_1 - \alpha_1')(\alpha_2'' x_1 - \alpha_2') \cdots (\alpha_n'' x_1 - \alpha_n'),$$

where for brevity we have written

$$\alpha_i'' \alpha_i = \alpha_i' \qquad\qquad (i = 1, 2, \cdots n).$$

By Theorem 1, § 62, we now infer that the binary form (1) is identically equal to

$$(4) \qquad (\alpha_1'' x_1 - \alpha_1' x_2)(\alpha_2'' x_1 - \alpha_2' x_2) \cdots (\alpha_n'' x_1 - \alpha_n' x_2).$$

Moreover, there cannot be any way essentially different from this of factoring (1) into linear factors, for if there were we should, by setting $x_2 = 1$, have a way of factoring (2) into linear factors essentially different from (3). Thus our theorem is proved on the supposition that $a_0 \neq 0$.

Turning now to the case $a_0 = 0$, let us suppose that

$$a_0 = \cdots = a_{k-1} = 0, \; a_k \neq 0,$$

where $k \leqq n$. The form (1) then has the form

(5) $$a_k x_1^{n-k} x_2^k + \cdots + a_n x_2^n,$$

which is equal to the product of k factors x_2 and the binary form

$$a_k x_1^{n-k} + \cdots + a_n x_2^{n-k}$$

of degree $n - k$. Since the first coefficient in this form is not zero, it can, as we have just seen, be factored into $n - k$ linear factors. Thus, here also, we see that the binary form can be written in the form (4), the only peculiarity being that in this case k of the constants a'' are zero. We leave it to the reader to show that this factoring can be performed in essentially only one way. This being done, we have the result:

THEOREM 2. *A binary form of the nth degree is always reducible when $n > 1$. It can be resolved into the product of n linear factors in one, and essentially only one, way.*

EXERCISES

1. Prove that every real polynomial in one variable of degree higher than two is reducible in the domain of reals, and can be resolved into irreducible factors in one, and essentially only one, way.

2. Prove the corresponding theorem for real binary forms.

66. The Greatest Common Divisor of Positive Integers.* We will consider in this section the problem of finding the greatest common divisor of two positive integers a and b, which has the closest

---

* In this section we use the term *divisor* in the arithmetical sense, not in the algebraic sense defined in § 60. An integer b is said to be a divisor of an integer a if an integer c exists such that $a = bc$.

analogy with the algebraic problem of the next section. The solution of this problem, which was given by Euclid, is as follows:

If we divide a by b* and get a quotient q_0 and a remainder r_1, we may write
$$a = q_0 b + r_1,$$

where, if the division is carried out as far as possible, we have $r_1 < b$.

Then divide b by r getting a quotient q_1 and a remainder r_2 which, if the division is carried out as far as possible, is less than r_1. Proceeding in this way, we get the following system of equations, in which, since the remainders r_1, r_2, $\cdots$ are positive integers which continually decrease, we ultimately come to a point where the division leaves no remainder :

$$(1) \quad \begin{cases} a = q_0 b + r_1 & r_1 < b, \\ b = q_1 r_1 + r_2 & r_2 < r_1, \\ r_1 = q_2 r_2 + r_3 & r_3 < r_2, \\ \quad \cdot \quad \cdot \quad \cdot \quad \cdot \quad \cdot \\ \quad \cdot \quad \cdot \quad \cdot \quad \cdot \quad \cdot \\ r_{\rho-2} = q_{\rho-1} r_{\rho-1} + r_\rho & r_\rho < r_{\rho-1}, \\ r_{\rho-1} = q_\rho r_\rho & 0 < r_\rho. \end{cases}$$

From the first of these equations we see that every common factor of a and b is a factor of r_1; from the second, that every common factor of b and r_1 is a factor of r_2; etc.; finally, that every common factor of $r_{\rho-2}$ and $r_{\rho-1}$ is a factor of r_ρ. Hence *every common factor of a and b is a factor of r_ρ.*

On the other hand, we see from the last equation (1) that every factor of r_ρ is a factor of $r_{\rho-1}$; from the next to the last equation, that every common factor of r_ρ and $r_{\rho-1}$ is a factor of $r_{\rho-2}$; etc.; finally, that every common factor of r_2 and r_1 is a factor of b, and that every common factor of r_1 and b is a factor of a. Hence *every factor of r_ρ is a common factor of a and b.*

Since the largest factor of r_ρ is r_ρ itself, we have the result :

THEOREM 1. *In Euclid's algorithm* (1), *the greatest common divisor of a and b is r_ρ.*

In particular, a necessary and sufficient condition that a and b be relatively prime is that $r_\rho = 1$.

---

* This is possible even if $a < b$, the only peculiarity in this case being that the quotient is zero and the remainder equal to a.

We will next deduce from the equations (1) an important formula by means of which r_ρ is expressed in terms of a, b, and the q's.

From the first equation (1) we have

$$r_1 = a - q_0 b.$$

Substituting this value into the second equation, we get for r_2 the value

$$r_2 = - q_1 a + (q_0 q_1 + 1) b.$$

Substituting the values for r_1 and r_2 just found in the third equation, we get

$$r_3 = (q_1 q_2 + 1) a - (q_0 q_1 q_2 + q_2 + q_0) b.$$

Proceeding in this way, we can express each of the r's, and therefore ultimately r_ρ, in terms of a and b. In order to express conveniently the general formula here, we introduce the following notation :

$$(2) \quad \begin{cases} [\ ] = 1 \\ [\alpha_1] = \alpha_1, \\ [\alpha_1, \alpha_2] = \alpha_1 \alpha_2 + 1, \\ [\alpha_1, \alpha_2, \alpha_3] = \alpha_1 \alpha_2 \alpha_3 + \alpha_3 + \alpha_1, \\ \quad \cdots \cdots \cdots \cdots \cdots \\ \quad \cdots \cdots \cdots \cdots \cdots \\ [\alpha_1, \cdots \alpha_n] = [\alpha_1, \cdots \alpha_{n-1}] \alpha_n + [\alpha_1, \cdots \alpha_{n-2}]. \end{cases}$$

It will be seen that the values of r_1, r_2, r_3 found above are included in the formula

$$(3) \qquad r_k = (-1)^{k-1} [q_1, q_2, \cdots q_{k-1}] a + (-1)^k [q_0, q_1, q_2, \cdots q_{k-1}] b.$$

By the method of mathematical induction this formula will therefore be established for all values of $k \leqq \rho$ if, assuming that it holds when $k \leqq k_1$, we can show that it holds when $k = k_1 + 1$. This follows at once when, in the formula

$$r_{k_1+1} = r_{k_1-1} - q_{k_1} r_{k_1},$$

we substitute for r_{k_1} and r_{k_1-1} their values from (3) and reduce the resulting expression by means of the definitions (2).

We have therefore established the formula

$$(4) \qquad\qquad r_\rho = Aa + Bb,$$

where $\quad A = (-1)^{\rho-1} [q_1, q_2, \cdots q_{\rho-1}], \quad B = (-1)^\rho [q_0, q_1, \cdots q_{\rho-1}].$

Since the q's are integers, it is clear that A and B will be integers.

The most important application of the result just obtained is to the case in which a and b are relatively prime. Here $r_\rho = 1$, and we have

$$(5) \qquad\qquad Aa + Bb = 1.$$

Conversely, if two integers A and B exist which satisfy (5), a and b are relatively prime, as otherwise the left-hand side of (5) would have a factor greater than 1.

THEOREM 2. *A necessary and sufficient condition for a and b to be relatively prime is that there exist two integers A and B such that $Aa + Bb = 1$.*

EXERCISES

1. Prove that $[\alpha_1, \alpha_2, \cdots \alpha_n] = [\alpha_n, \alpha_{n-1}, \cdots \alpha_1]$.

[SUGGESTION. Use the method of mathematical induction.]

2. Prove that the numerical values of the integers A and B found above are respectively less than $\frac{1}{2}b$ and $\frac{1}{2}a$.

[SUGGESTION. Show that $a/b = [q_0, \cdots q_\rho]/[q_1, \cdots q_\rho]$, and that this second fraction is expressed in its lowest terms.]

3. Prove that there can exist only one pair of integers A and B satisfying the relation $Aa + Bb = 1$ and such that A and B are numerically less than $\frac{1}{2}b$ and $\frac{1}{2}a$ respectively.

67. The Greatest Common Divisor of Two Polynomials in One Variable. In place of the integers a and b of the last section, we consider here the two polynomials:

$$(1) \qquad \begin{cases} f(x) \equiv a_0 x^n + a_1 x^{n-1} + \cdots + a_n, \\ \phi(x) \equiv b_0 x^m + b_1 x^{m-1} + \cdots + b_m. \end{cases}$$

By the greatest common divisor of these two polynomials is meant their common factor of greatest degree.* It will turn out in the course of our work that (except in the case in which f and ϕ are both identically zero) this greatest common divisor is completely determinate except for an arbitrary constant factor which may be introduced into it.

---

* Many English and American text-books use the term *highest* common factor ; but as there is not the slightest possibility that the word *greatest*, here, should refer to the *value* of the polynomial, since the polynomial has an infinite number of values for different values of the argument, it seems better to retain the traditional term.

We will assume that neither f nor ϕ is a mere constant, and that the notation has been so introduced that f is of at least as high degree as ϕ; that is, we assume

$$a_0 \neq 0, \quad b_0 \neq 0, \quad n \geqq m > 0.$$

Let us apply Euclid's algorithm to f and ϕ precisely as in § 66 we applied it to a and b. We thus get the system of identities

$$(2) \quad \begin{cases} f(x) \equiv Q_0(x)\phi(x) + R_1(x), \\ \phi(x) \equiv Q_1(x)R_1(x) + R_2(x), \\ R_1(x) \equiv Q_2(x)R_2(x) + R_3(x), \\ \cdot \quad \cdot \quad \cdot \quad \cdot \quad \cdot \quad \cdot \quad \cdot \quad \cdot \quad \cdot \\ \cdot \quad \cdot \quad \cdot \quad \cdot \quad \cdot \quad \cdot \quad \cdot \quad \cdot \\ R_{\rho-1}(x) \equiv Q_\rho(x)R_\rho(x) + R_{\rho+1}. \end{cases}$$

For the sake of uniformity we will write

$$\phi(x) \equiv R_0(x).$$

Then R_0, R_1, R_2, $\cdots$ are polynomials of decreasing degrees, so that after a finite number of steps a remainder is reached which is a constant. This remainder we have indicated by $R_{\rho+1}$.

From this algorithm we infer, as in § 66, that every common factor of f and ϕ is a factor of all the R's, and, on the other hand, that every common factor of two successive R's is a factor of all the preceding R's and therefore of f and ϕ. Accordingly, if f and ϕ have a common factor which is not a constant, this common factor must be a factor of the constant $R_{\rho+1}$, and therefore $R_{\rho+1} = 0$. Conversely, if $R_{\rho+1} = 0$, the polynomial $R_\rho(x)$ is itself a common factor of R_ρ and $R_{\rho+1}$, and therefore of f and ϕ. Hence the two theorems:

THEOREM 1. *A necessary and sufficient condition that two polynomials in one variable f and ϕ, neither of which is a constant, be relatively prime is that in Euclid's algorithm, (2), $R_{\rho+1} \neq 0$.*

THEOREM 2. *If in Euclid's algorithm, (2), $R_{\rho+1} = 0$, then $R_\rho(x)$ is the greatest common divisor of f and ϕ.*

By means of this theorem we are in a position to compute the greatest common divisor, not only of two, but of any finite number, of polynomials in one variable. Thus if we want the greatest common divisor of $f(x)$, $\phi(x)$, $\psi(x)$, we should first compute, as above, the greatest common divisor $R_\rho(x)$ of f and ϕ, and then, by the same method, compute the greatest common divisor of $R_\rho(x)$ and $\psi(x)$.

From the identities (2) we can compute the value of each of the remainders in terms of f, ϕ, and the quotients Q. The formulæ here are precisely like those of § 66, and give for $R_{\rho+1}$ the value

$$(3) \qquad R_{\rho+1} \equiv (-1)^\rho [Q_1(x),\, Q_2(x),\, \cdots\, Q_\rho(x)]f(x)$$
$$+ (-1)^{\rho+1} [Q_0(x),\, Q_1(x),\, \cdots\, Q_\rho(x)]\phi(x).$$

Suppose, now, that f and ϕ are relatively prime. We may then divide (3) by $R_{\rho+1}$ and get

$$(4) \qquad\qquad F(x)f(x) + \Phi(x)\phi(x) \equiv 1,$$

where

$$(5) \qquad \begin{cases} F(x) \equiv \dfrac{(-1)^\rho}{R_{\rho+1}} \Big[Q_1(x),\, Q_2(x),\, \cdots\, Q_\rho(x) \Big], \\[3mm] \Phi(x) \equiv \dfrac{(-1)^{\rho+1}}{R_{\rho+1}} \Big[Q_0(x),\, Q_1(x),\, \cdots\, Q_\rho(x) \Big]. \end{cases}$$

From the definitions (2), § 66, we see that F and Φ are polynomials in x. The existence of two polynomials F and Φ which satisfy (4) is therefore a necessary condition that f and ϕ be relatively prime. It is also a sufficient condition ; for from (4) we see that every common factor of f and ϕ must be a factor of 1, that is, must be a constant. Thus we have proved the theorem:

THEOREM 3. *A necessary and sufficient condition that the polynomials $f(x)$ and $\phi(x)$ be relatively prime is that two polynomials $F(x)$ and $\Phi(x)$ exist which satisfy* (4).*

We can make this statement a little more precise by noting the degrees of F and Φ as given by (5). For this purpose let us first notice that if $\alpha_1, \cdots \alpha_n$ are polynomials of degrees $k_1, \cdots k_n$ respectively, $[\alpha_1, \cdots \alpha_n]$ will, by (2), § 66, not be of degree greater than $k_1 + \cdots + k_n$. Now let the degree of $R_i(x)$ be m_i, and, as above, the degrees of f and ϕ, n and m respectively. Then (cf. (2)) the degrees of $Q_0, Q_1, Q_2, \cdots$ will be $n - m$, $m - m_1$, $m_1 - m_2$, $\cdots$ respectively. Accordingly, by (5), the degrees of F and Φ are respectively not greater than

$$(m - m_1) + (m_1 - m_2) + \cdots + (m_{\rho-1} - m_\rho) = m - m_\rho,$$

and $(n - m) + (m - m_1) + (m_1 - m_2) + \cdots + (m_{\rho-1} - m_\rho) = n - m_\rho$.

Hence, since $m_\rho > 0$, F is of degree less than m, and Φ of degree less than n.

---

* The proof we have given of this theorem applies only when neither f nor ϕ is a constant. The truth of the theorem is at once obvious if f or ϕ is a constant.

Conversely, we will now show that if F_1 is a polynomial of degree less than m, and Φ_1 a polynomial of degree less than n, and if

$$(6) \qquad\qquad F_1(x)f(x) + \Phi_1(x)\phi(x) \equiv 1,$$

then $\qquad\qquad F_1(x) \equiv F(x), \quad \Phi_1(x) \equiv \Phi(x).$

To prove this, subtract (6) from (4), getting
$$(F - F_1)f \equiv (\Phi_1 - \Phi)\phi.$$

If we resolve the two sides of this identity into their linear factors, we see that, since f and ϕ are relatively prime, f must be a factor of $\Phi_1 - \Phi$ and ϕ a factor of $F - F_1$. This, however, is possible only if $\Phi_1 - \Phi$ and $F - F_1$ vanish identically, as otherwise they would be of lower degree than f and ϕ respectively. We have thus proved the theorem :

THEOREM 4. *If $f(x)$ and $\phi(x)$ are relatively prime, and neither is a constant, there exists one, and only one, pair of polynomials $F(x)$ $\Phi(x)$, whose degrees are respectively less than the degrees of ϕ and f, and which satisfy the identity* (4).

Before proceeding to the general applications of the principles here developed which will be found in the next section, the reader will do well to familiarize himself somewhat with the ideas involved by considering the special case of two polynomials of the second degree :

$$f(x) \equiv a_0 x^2 + a_1 x + a_2 \qquad\qquad a_0 \neq 0,$$
$$\phi(x) \equiv b_0 x^2 + b_1 x + b_2 \qquad\qquad b_0 \neq 0.$$

If the condition that these two polynomials be relatively prime be worked out by a direct application of Euclid's algorithm, it will be found necessary to consider separately the cases in which $a_1 b_0 - a_0 b_1$ is or is not zero. By collating these results it will be found that in all cases the desired condition is:

$$(a_2 b_0 - a_0 b_2)^2 + (a_1 b_0 - a_0 b_1)(a_1 b_2 - a_2 b_1) \neq 0.$$

This condition may be found more neatly and quickly by obtaining the condition that two polynomials of the form

$$F(x) \equiv p_0 x + p_1,$$
$$\Phi(x) \equiv q_0 x + q_1$$

exist which satisfy the identity (4).

It is this last method which we shall apply to the general case in the next section.

68. The Resultant of Two Polynomials in One Variable. Let

$$f(x) \equiv a_0 x^n + a_1 x^{n-1} + \cdots + a_{n-1} x + a_n \qquad a_0 \neq 0,\ n > 0,$$
$$\phi(x) \equiv b_0 x^m + b_1 x^{m-1} + \cdots + b_{m-1} x + b_m \qquad b_0 \neq 0,\ m > 0.$$

The condition that these polynomials be relatively prime consists, as we see from Theorem 4, § 67, in the existence of constants p_0, p_1, $\cdots$ p_{m-1}, q_0, q_1, $\cdots$ q_{n-1} such that

$$(p_0 x^{m-1} + p_1 x^{m-2} + \cdots + p_{m-1})(a_0 x^n + a_1 x^{n-1} + \cdots + a_n)$$
$$+ (q_0 x^{n-1} + q_1 x^{n-2} + \cdots + q_{n-1})(b_0 x^m + b_1 x^{m-1} + \cdots + b_m) \equiv 1.$$

Equating coefficients of like powers of x, we see that the following system of equations is equivalent to the last written identity:

$$
\begin{cases}
a_0 p_0 & + b_0 q_0 & = 0, \\
a_1 p_0 + a_0 p_1 & + b_1 q_0 + b_0 q_1 & = 0, \\
\cdots & \cdots & \\
a_m p_0 + a_{m-1} p_1 + \cdots + a_1 p_{m-1} + b_m q_0 + b_{m-1} q_1 + \cdots + b_0 q_m & = 0, \\
a_{m+1} p_0 + a_m p_1 + \cdots + a_2 p_{m-1} \qquad + b_m q_1 + \cdots\cdots + b_0 q_{m+1} & = 0, \\
\cdots & \cdots & \\
a_n p_0 + a_{n-1} p_1 + \cdots + a_{n-m+1} p_{m-1} \qquad\quad + b_m q_{n-m} + \cdots\cdots + b_1 q_{n-1} & = 0, \\
a_n p_1 + \cdots + a_{n-m+2} p_{m-1} \qquad\qquad + b_m q_{n-m+1} + \cdots + b_2 q_{n-1} & = 0, \\
\cdots & \cdots & \\
a_n p_{m-1} \qquad\qquad\qquad + b_m q_{n-1} & = 1.
\end{cases}
$$

In writing these equations we have assumed for the sake of definiteness that $n \geq m$, though the change would be immaterial if this were not the case. This is a system of $m + n$ linear equations in the $m + n$ unknowns $p_0, \cdots p_{m-1}, q_0, \cdots q_{n-1}$, whose determinant, after an interchange of rows and columns and a shifting of the rows, is

$$
R\!\begin{pmatrix} a_0, \cdots a_n \\ b_0, \cdots b_m \end{pmatrix} =
\begin{vmatrix}
a_0 & \cdot & \cdot & \cdot & \cdot & \cdot & \cdot & a_n & 0 & \cdot & \cdot & \cdot & 0 \\
0 & a_0 & \cdot & \cdot & \cdot & \cdot & \cdot & & a_n & 0 & \cdot & \cdot & 0 \\
\cdot & \cdot & \cdot & \cdot & \cdot & \cdot & \cdot & \cdot & \cdot & \cdot & \cdot & \cdot & \cdot \\
\cdot & \cdot & \cdot & \cdot & \cdot & \cdot & \cdot & \cdot & \cdot & \cdot & \cdot & \cdot & \cdot \\
\cdot & \cdot & \cdot & \cdot & \cdot & \cdot & \cdot & \cdot & \cdot & \cdot & \cdot & \cdot & \cdot \\
0 & \cdot & \cdot & \cdot & 0 & a_0 & \cdot & \cdot & \cdot & \cdot & \cdot & \cdot & a_n \\
0 & \cdot & \cdot & \cdot & \cdot & 0 & b_0 & \cdot & \cdot & \cdot & \cdot & \cdot & b_m \\
0 & \cdot & \cdot & \cdot & \cdot & 0 & b_0 & \cdot & \cdot & \cdot & \cdot & b_m & 0 \\
\cdot & \cdot & \cdot & \cdot & \cdot & \cdot & \cdot & \cdot & \cdot & \cdot & \cdot & \cdot & \cdot \\
\cdot & \cdot & \cdot & \cdot & \cdot & \cdot & \cdot & \cdot & \cdot & \cdot & \cdot & \cdot & \cdot \\
b_0 & \cdot & \cdot & \cdot & \cdot & b_m & 0 & \cdot & \cdot & \cdot & \cdot & \cdot & 0
\end{vmatrix},
$$

a determinant which, it should be noticed, has $m + n$ rows and columns.

If $R \neq 0$, the set of equations (1) has one and only one solution, and f and ϕ are relatively prime. If $R = 0$, two cases seem at first sight possible (cf. § 16): either the system of equations has no solution, or it has an infinite number of solutions. This latter alternative cannot, however, really arise, for we have seen in Theorem 4, § 67, that not more than one pair of polynomials F and Φ can exist which satisfy formula (4) of that section and whose degrees do not exceed $m-1$ and $n-1$ respectively. Accordingly, if $R = 0$, the set of equations has *no* solution and f and ϕ have a common factor.

R is called the *resultant* of f and ϕ.*

The term *resultant* has thus been defined only on the supposition that f and ϕ are both of at least the first degree. It is desirable to extend this definition to the case in which one or both of these polynomials is a constant. Except in the extreme case $m = n = 0$, we will continue to use the determinant R as the definition of the resultant. Thus when $m = 0$, $n > 0$ we have

$$R \begin{pmatrix} a_0, \cdots a_n \\ b_0 \end{pmatrix} = (-1)^{\frac{n(n-1)}{2}} b_0^n.$$

If $b_0 \neq 0$ we have $R \neq 0$, and moreover in this case f and ϕ are relatively prime since the constant ϕ has no factors other than constants. If, however, $b_0 = 0$, we have $R = 0$, and every factor of f is a factor of ϕ, since ϕ is now identically zero.

Similarly when $n = 0$, $m > 0$, we have

$$R \begin{pmatrix} a_0 \\ b_0, \cdots b_m \end{pmatrix} = a_0^m,$$

and we see that a necessary and sufficient condition that f and ϕ be relatively prime is that $R \neq 0$.

Finally, when $n = m = 0$, we define the symbol $R \begin{pmatrix} a_0 \\ b_0 \end{pmatrix}$, which we still use to denote the resultant, by the formula

$$R \begin{pmatrix} a_0 \\ b_0 \end{pmatrix} = \begin{cases} 1 & \text{when } a_0 \text{ and } b_0 \text{ are not both zero,} \\ 0 & \text{when } a_0 = b_0 = 0. \end{cases}$$

We may now say with entire generality:

THEOREM. *A necessary and sufficient condition for two polynomials in one variable to be relatively prime is that their resultant do not vanish.*

For another method of approach to the resultant, cf. Exercise 4 at the end of § 76.

---

* It should be noticed that the resultant of ϕ and f may be the negative of the resultant of f and ϕ. This change of sign is of no consequence for most purposes.

69. The Greatest Common Divisor in Determinant Form.

DEFINITION. *By the ith subresultant R_i of two polynomials in one variable is understood the determinant obtained by striking out the first i and the last i rows and also the first i and the last i columns from the resultant of these polynomials.*

Thus if the polynomials are of degrees 5 and 3 respectively, the resultant R is a determinant of the eighth order, R_1 of the sixth, R_2 of the fourth, and R_3 of the second, as indicated below :

$$
R=\;R_1=\;R_2=\;R_3=\;
\begin{vmatrix}
a_0 & a_1 & a_2 & a_3 & a_4 & a_5 & 0 & 0 \\
0 & a_0 & a_1 & a_2 & a_3 & a_4 & a_5 & 0 \\
0 & 0 & a_0 & a_1 & a_2 & a_3 & a_4 & a_5 \\
0 & 0 & 0 & 0 & b_0 & b_1 & b_2 & b_3 \\
0 & 0 & 0 & b_0 & b_1 & b_2 & b_3 & 0 \\
0 & 0 & b_0 & b_1 & b_2 & b_3 & 0 & 0 \\
0 & b_0 & b_1 & b_2 & b_3 & 0 & 0 & 0 \\
b_0 & b_1 & b_2 & b_3 & 0 & 0 & 0 & 0
\end{vmatrix}
$$

We now state the following results, leaving their proof to the reader:

LEMMA. *If $f_1(x)$ and $\phi_1(x)$ are polynomials, and*

$$f(x) \equiv (x - \alpha)f_1(x), \qquad \phi(x) \equiv (x - \alpha)\phi_1(x),$$

the resultant of f_1 and ϕ_1 and their successive subresultants are equal respectively to the successive subresultants of f and ϕ.

THEOREM 1. *The degree of the greatest common divisor of $f(x)$ and $\phi(x)$ is equal to the subscript of the first of the subresultants $R_0 = R,\ R_1,\ R_2,\ \cdots$ which does not vanish.*

THEOREM 2. *If i is the degree of the greatest common divisor of two polynomials $f(x)$ and $\phi(x)$, then this greatest common divisor may be obtained from the ith subresultant of f and ϕ by replacing the last* *element in the last row of coefficients of f by $f(x)$, the element just above this by $xf(x)$, the element above this by $x^2 f(x)$, etc.; and replacing the last element in the first row of coefficients of ϕ by $\phi(x)$, the element below this by $x\phi(x)$, etc.*

If, for instance, the degrees of f and ϕ are 5 and 3 respectively, and $i = 1$, the greatest common divisor is

$$\begin{vmatrix} a_0 & a_1 & a_2 & a_3 & a_4 & xf(x) \\ 0 & a_0 & a_1 & a_2 & a_3 & f(x) \\ 0 & 0 & 0 & b_0 & b_1 & \phi(x) \\ 0 & 0 & b_0 & b_1 & b_2 & x\phi(x) \\ 0 & b_0 & b_1 & b_2 & b_3 & x^2\phi(x) \\ b_0 & b_1 & b_2 & b_3 & 0 & x^3\phi(x) \end{vmatrix}.$$

70. Common Roots of Equations. Elimination. Consider the equations

$$f(x) \equiv a_0 x^n + a_1 x^{n-1} + \cdots + a_n = 0 \qquad a_0 \neq 0,$$
$$\phi(x) \equiv b_0 x^m + b_1 x^{m-1} + \cdots + b_m = 0 \qquad b_0 \neq 0,$$

whose roots are $\alpha_1, \alpha_2, \cdots \alpha_n$ and $\beta_1, \beta_2, \cdots \beta_m$, respectively; and suppose $f(x)$ and $\phi(x)$ resolved into their linear factors:

$$f(x) \equiv a_0 (x - \alpha_1)(x - \alpha_2) \cdots (x - \alpha_n),$$
$$\phi(x) \equiv b_0 (x - \beta_1)(x - \beta_2) \cdots (x - \beta_m).$$

Since, by Theorem 1, § 65, these sets of factors are unique, it is evident that the equations $f(x) = 0$ and $\phi(x) = 0$ will have a common root when, and only when, $f(x)$ and $\phi(x)$ have a common factor, that is, when, and only when, the resultant R of f and ϕ is zero.

To eliminate x between two equations $f(x) = 0$ and $\phi(x) = 0$, is often taken in elementary algebra to mean: to find a relation between the coefficients of f and ϕ which must hold if the two equations are both satisfied; that is, to find a *necessary* condition for the two equations to have a common root. For most purposes, however, when we eliminate we want a relation between the coefficients which not only holds when the two equations have a common root, but such that, conversely, when it holds the equations will have a common root. From this broader point of view, to eliminate x between two equations $f(x) = 0$ and $\phi(x) = 0$ means simply to find a *necessary and sufficient* condition that these equations have a common root. Hence the result of this elimination is $R = 0$. Let us, however, look at this question from a little different point of view.

In the equations

$$(1) \qquad a_0x^5 + a_1x^4 + a_2x^3 + a_3x^2 + a_4x + a_5 = 0 \qquad a_0 \neq 0,$$

$$(2) \qquad b_0x^3 + b_1x^2 + b_2x + b_3 = 0 \qquad b_0 \neq 0,$$

let us consider the different powers of x as so many distinct unknowns. We have, then, two non-homogeneous, linear equations in the five unknowns x, x^2, x^3, x^4, x^5. Multiplying (1) through by x and then by x^2, and (2) by x, x^2, x^3, x^4, in turn, we have

$$a_0x^7 + a_1x^6 + a_2x^5 + a_3x^4 + a_4x^3 + a_5x^2 = 0,$$
$$a_0x^6 + a_1x^5 + a_2x^4 + a_3x^3 + a_4x^2 + a_5x = 0,$$
$$a_0x^5 + a_1x^4 + a_2x^3 + a_3x^2 + a_4x + a_5 = 0,$$
$$b_0x^3 + b_1x^2 + b_2x + b_3 = 0,$$
$$b_0x^4 + b_1x^3 + b_2x^2 + b_3x = 0,$$
$$b_0x^5 + b_1x^4 + b_2x^3 + b_3x^2 = 0,$$
$$b_0x^6 + b_1x^5 + b_2x^4 + b_3x^3 = 0,$$
$$b_0x^7 + b_1x^6 + b_2x^5 + b_3x^4 = 0,$$

a system of eight non-homogeneous, linear equations in seven unknowns.

If a value of x satisfies both (1) and (2), it will evidently satisfy all the above equations. These equations are therefore consistent, so that by Theorem 1, § 16, we have.

$$\begin{vmatrix} a_0 & a_1 & a_2 & a_3 & a_4 & a_5 & 0 & 0 \\ 0 & a_0 & a_1 & a_2 & a_3 & a_4 & a_5 & 0 \\ 0 & 0 & a_0 & a_1 & a_2 & a_3 & a_4 & a_5 \\ 0 & 0 & 0 & 0 & b_0 & b_1 & b_2 & b_3 \\ 0 & 0 & 0 & b_0 & b_1 & b_2 & b_3 & 0 \\ 0 & 0 & b_0 & b_1 & b_2 & b_3 & 0 & 0 \\ 0 & b_0 & b_1 & b_2 & b_3 & 0 & 0 & 0 \\ b_0 & b_1 & b_2 & b_3 & 0 & 0 & 0 & 0 \end{vmatrix} = 0.$$

Hence the vanishing of this determinant is a *necessary* condition for (1) and (2) to have a common root.

This device is known as *Sylvester's Dialytic Method of Elimination.*

* For the sake of simplicity we have taken the special case where $n = 5$ and $m = 3$. The method, however, is perfectly general.

The above determinant is seen to be exactly the same as the resultant R of (1) and (2), so that Sylvester's method leads to the same condition for two equations to have a common root as that found above, namely $R = 0$. It does not prove, however, that this condition is sufficient, but merely that it is necessary. Thus Sylvester's method, while brief, is very imperfect.

The number of roots common to two equations, $f(x) = 0$ and $\phi(x) = 0$, and also an equation for computing the common roots, may be obtained at once from § 69.

71. The Cases $a_0 = 0$ and $b_0 = 0$. It is important for us to note that according to the definitions we have given, the determinant $R\begin{pmatrix} a_0, \cdots a_n \\ b_0, \cdots b_m \end{pmatrix}$ will be the resultant of the two polynomials

$$f(x) \equiv a_0 x^n + a_1 x^{n-1} + \cdots + a_n,$$
$$\phi(x) \equiv b_0 x^m + b_1 x^{m-1} + \cdots + b_m,$$

only when f and ϕ are precisely of degrees n and m respectively, that is, only when $a_0 \neq 0$, $b_0 \neq 0$. Thus, for instance, the resultant of the polynomials

$$f(x) \equiv a_1 x^{n-1} + a_2 x^{n-2} + \cdots + a_n,$$
$$\phi(x) \equiv b_0 x^m + b_1 x^{m-1} + \cdots + b_m,$$

is not the $(m+n)$-rowed determinant $R\begin{pmatrix} 0, a_1, \cdots a_n \\ b_0, \cdots b_m \end{pmatrix}$ but, if $a_1 \neq 0$, $b_0 \neq 0$, the $(m+n-1)$-rowed determinant $R\begin{pmatrix} a_1, \cdots a_n \\ b_0, \cdots b_m \end{pmatrix}$ or, if a_1 or b_0 is zero, a determinant of still lower order.*

Let us indicate by R the $(m+n)$-rowed determinant $R\begin{pmatrix} a_0, \cdots a_n \\ b_0, \cdots b_m \end{pmatrix}$ and by r the resultant of f and ϕ, and consider the case $a_0 = 0$, $a_1 \neq 0$, $b_0 \neq 0$. Since every element of the first column of R except the last is zero, we may write

$$R = (-1)^{m+n-1} b_0 r.$$

In a similar way we see that if the degree of f is $n - i$, and $b_0 \neq 0$, we may write

$$R = \pm b_0^i r,$$

and if the degree of ϕ is $m - i$, and $a_0 \neq 0$, we have

$$R = a_0^i r.$$

Accordingly, except when $a_0 = b_0 = 0$, R differs from r only by a non-vanishing factor.

* As an illustration take the two polynomials $f(x) \equiv (\alpha + \beta) x^2 + x - \beta$ and $\phi(x) \equiv \alpha x + 1$. If $\alpha + \beta \neq 0$ and $\alpha \neq 0$, the resultant here is $(\alpha^2 - 1)\beta$. But if $\alpha = -\beta \neq 0$, the resultant is $1 - \alpha^2$.

THEOREM. *Although* $R\begin{pmatrix} a_0, & \cdots & a_n \\ b_0, & \cdots & b_m \end{pmatrix}$ *is the resultant of f and ϕ only when $a_0 \neq 0$ and $b_0 \neq 0$ (or when $m = 0$ or $n = 0$), nevertheless its vanishing still forms a necessary and sufficient condition that f and ϕ have a common factor even when $a_0 = 0$ or $b_0 = 0$, provided merely that both a_0 and b_0 are not zero.*

That this last restriction can not be removed is at once evident; for, if $a_0 = b_0 = 0$, every element in the first column of the determinant is zero, and hence the determinant vanishes irrespective of whether f and ϕ have a common factor or not.* All that we can say, if we do not wish to make this exception, is, therefore, that in all cases the vanishing of R forms a *necessary* condition that f and ϕ have a common factor.

72. The Resultant of Two Binary Forms. Let us now consider the binary forms

$$f(x_1,\ x_2) \equiv a_0 x_1^n + a_1 x_1^{n-1} x_2 + \cdots + a_n x_2^n \qquad (n \geqq 1)$$
$$\phi(x_1,\ x_2) \equiv b_0 x_1^m + b_1 x_1^{m-1} x_2 + \cdots + b_m x_2^m \qquad (m \geqq 1).$$

By the side of these forms we write the polynomials in one variable

$$F(x) \equiv a_0 x^n + a_1 x^{n-1} + \cdots + a_n,$$
$$\Phi(x) \equiv b_0 x^m + b_1 x^{m-1} + \cdots + b_m.$$

The determinant $\qquad R\begin{pmatrix} a_0, & \cdots & a_n \\ b_0, & \cdots & b_m \end{pmatrix}$

will be the resultant of F and Φ only when neither a_0 nor b_0 is zero. We will, however, call it the resultant of the binary forms f and ϕ in all cases.

---

*By looking at the question from the side of the theory of common roots of two equations (cf. § 70), and by introducing the conception of *infinite roots*, we may avoid even this last exception. An equation

$$a_0 x^n + a_1 x^{n-1} + \cdots + a_n = 0$$

has n roots, distinct or coincident, provided $a_0 \neq 0$. If we allow a_0 to approach the value zero, one or more of these roots becomes in absolute value larger and larger, as is seen by the transformation $x' = 1/x$. Hence it is natural to say that if $a_0 = 0$ the equation has an infinite root. If then we consider two equations each of which has an infinite root as having a common root, we may say:

A necessary and sufficient condition that the equations

$$a_0 x^n + a_1 x^{n-1} + \cdots + a_n = 0 \qquad\qquad n > 0,$$
$$b_0 x^m + b_1 x^{m-1} + \cdots + b_m = 0 \qquad\qquad m > 0,$$

have a common root is in all cases the vanishing of $R\begin{pmatrix} a_0, & \cdots & a_n \\ b_0, & \cdots & b_m \end{pmatrix}$.

THEOREM. *A necessary and sufficient condition for two binary forms to have a common factor other than a constant is that their resultant be zero.*

If a_0 and b_0 are both different from zero, the non-homogeneous polynomials F and Φ correspond to the forms f and ϕ according to the definition of § 62. Accordingly, by Theorem 2 of that section, a necessary and sufficient condition that f and ϕ have a common factor other than a constant is, in this case, the vanishing of their resultant.

On the other hand, if $a_0 = b_0 = 0$, f and ϕ have the common factor x_2, and the resultant of f and ϕ obviously vanishes.

A similar remark applies to the case in which all the a's or all the b's are zero.

There remain then only the following two cases to be considered,

$$(1) \qquad a_0 \neq 0 \; ; \; b_0 = b_1 = \cdots = b_k = 0, \; b_{k+1} \neq 0 \qquad (k < m),$$

$$(2) \qquad b_0 \neq 0 \; ; \; a_0 = a_1 = \cdots = a_k = 0, \; a_{k+1} \neq 0 \qquad (k < n).$$

In Case (1), F corresponds to f, and, if we write

$$\phi(x_1, x_2) \equiv x_2^{k+1} \phi_1(x_1, x_2),$$

Φ corresponds to ϕ_1. Now we know in this case (cf. § 71) that $R \neq 0$ is a necessary and sufficient condition that F and Φ be relatively prime. Accordingly, by Theorem 2, § 62, it is also a necessary and sufficient condition that f and ϕ_1 be relatively prime. But since x_2 is not a factor of f, the two forms f and ϕ will be relatively prime when and only when f and ϕ_1 are relatively prime. Thus our theorem is proved in this case.

The proof in Case (2) is precisely similar to that just given.

CHAPTER XVI

FACTORS OF POLYNOMIALS IN TWO OR MORE VARIABLES

73. Factors Involving only One Variable of Polynomials in Two Variables. We have seen in the last chapter that polynomials in one variable are always reducible when they are of degree higher than the first. Polynomials in two, or more, variables are, in general, not reducible, as we have already noticed in the special case of quadratic forms.

Let $f(x, y)$ be any polynomial in two variables, and suppose it arranged according to powers of x,

$$f(x, y) \equiv a_0(y)x^n + a_1(y)x^{n-1} + \cdots + a_{n-1}(y)x + a_n(y),$$

the a's being polynomials in y.

THEOREM 1. *A necessary and sufficient condition that a polynomial in y alone, $\psi(y)$, be a factor of $f(x, y)$ is that it be a factor of all the a's.*

The condition is clearly sufficient. To prove that it is necessary, let us suppose that $\psi(y)$ is a factor of $f(x, y)$. Then

$$(1) \qquad a_0(y)x^n + \cdots + a_n(y) \equiv \psi(y)[b_0(y)x^n + \cdots + b_n(y)],$$

where the b's are polynomials in y. For any particular value of y we deduce from (1), which is then an identity in x, the following equations :

$$\begin{cases} a_0(y) = \psi(y)b_0(y), \\ a_1(y) = \psi(y)b_1(y), \\ \quad \cdot \quad \cdot \quad \cdot \quad \cdot \quad \cdot \\ a_n(y) = \psi(y)b_n(y). \end{cases}$$

Since these equations hold for every value of y, they are identities, and $\psi(y)$ is a factor of all the a's.

THEOREM 2. *If $f(x, y)$ and $\phi(x, y)$ are any two polynomials in x and y, and $\psi(y)$ is an irreducible polynomial in y alone* which is a factor of the product $f\phi$, then ψ is a factor of f or of ϕ.*

Let
$$f(x, y) \equiv a_0(y)x^n + a_1(y)x^{n-1} + \cdots + a_n(y),$$
and
$$\phi(x, y) \equiv b_0(y)x^m + b_1(y)x^{m-1} + \cdots + b_m(y);$$
then
$$f(x, y)\phi(x, y) \equiv a_0 b_0 x^{n+m} + (a_0 b_1 + a_1 b_0)x^{n+m-1}$$
$$+ (a_0 b_2 + a_1 b_1 + a_2 b_0)x^{n+m-2} + \cdots + a_n b_m.$$

In order to prove that ψ is a factor either of f or of ϕ we must prove that it is either a factor of all the a's or of all the b's. If this were not the case, we could find a first a in the sequence $a_0, a_1, \cdots a_n$ of which ψ is not a factor. Call this function a_i. There would also be a first b in the sequence of $b_0, b_1, \cdots b_m$ which is not divisible by ψ. Call this function b_j. Our theorem will be proved if we can show that this assumption, that a_i and b_j are not divisible by ψ while all the functions $a_0, \cdots a_{i-1}, b_0, \cdots b_{j-1}$, are divisible by ψ, leads to a contradiction. For this purpose let us consider in the product $f\phi$ the coefficient of $x^{(n-i)+(m-j)}$, which may be written

$$a_0 b_{i+j} + \cdots + a_{i-1} b_{j+1} + a_i b_j + a_{i+1} b_{j-1} + \cdots + a_{i+j} b_0,$$

provided we agree that the a's and b's with subscripts greater than n and m respectively shall be identically zero. Since $f\phi$ is by hypothesis divisible by ψ, it follows from Theorem 1 that the last written expression must be divisible by ψ. This being obviously the case for all the terms which preceed and for all which succeed the term $a_i b_j$, it follows that this term must also be divisible by ψ, so that among the linear factors of the function $a_i b_j$ must be found ψ. But by Theorem 1, § 65, the function $a_i b_j$ can be resolved into its linear factors in essentially only one way, and one way of so resolving it is to resolve a_i and b_j into their linear factors. Since ψ is *not* one of these factors, we are led to a contradiction, and our theorem is proved.

An important corollary of our theorem is:

COROLLARY. *Let $f(x, y)$ and $\phi(x, y)$ be polynomials in (x, y), and let $\psi(y)$ be a polynomial in y alone. If ψ is a factor of the product of $f\phi$ but is relatively prime to ϕ, then ψ is a factor of f.*

* That is, a linear polynomial.

If ψ is irreducible, this corollary is identical with the theorem. Let us suppose ψ resolved into its irreducible factors none of which are constants, that is, into its factors of the first degree:

$$\psi(y) \equiv \psi_1(y)\,\psi_2(y) \cdots \psi_k(y).$$

Now consider the identity which expresses the fact that ψ is a factor of $f\phi$:

$$(2) \qquad f(x,\, y)\phi(x,\, y) \equiv \psi_1(y)\psi_2(y) \cdots \psi_k(y)G(x,\, y).$$

This shows that $\psi_1(y)$ is a factor of $f\phi$ and hence, by Theorem 2, it is a factor either of f or of ϕ. Since ψ and ϕ are relatively prime, ψ_1 cannot be a factor of ϕ. It must then be a factor of f:

$$f(x,\, y) \equiv \psi_1(y)f_1(x,\, y).$$

Substituting this in (2), and cancelling out ψ_1, as we have a right to do since it is not identically zero, we get

$$(3) \qquad f_1(x,\, y)\phi(x,\, y) \equiv \psi_2(y) \cdots \psi_k(y)G(x,\, y).$$

From this we infer that ψ_2, being a factor of $f_1\phi$, must be a factor of f_1:

$$f_1(x,\, y) \equiv \psi_2(y)f_2(x,\, y).$$

We substitute this in (3) and cancel out ψ_2. Proceeding in this way we get

$$f(x,\, y) \equiv \psi_1(y)\psi_2(y) \cdots \psi_k(y)f_k(x,\, y) \equiv \psi(y)f_k(x,\, y),$$

an identity which proves our corollary.

EXERCISE

If $f(x,\, y)$ and $\phi(x,\, y)$ are polynomials, then any two sets of polynomials

$$P_1(y),\ Q_1(x,\, y),\ R_1(x,\, y),$$
$$P_2(y),\ Q_2(x,\, y),\ R_2(x,\, y),$$

will be proportional to each other provided,

    (a) they satisfy the identities

$$P_1(y)f(x,\, y) \equiv Q_1(x,\, y)\phi(x,\, y) + R_1(x,\, y),$$
$$P_2(y)f(x,\, y) \equiv Q_2(x,\, y)\phi(x,\, y) + R_2(x,\, y);$$

    (b) there is no factor other than a constant common to P_1, Q_1, and also no factor other than a constant common to P_2, Q_2;

    (c) R_1 and R_2 are both of lower degree in x than ϕ.

    (Cf. Theorem 2, § 63.)

74. The Algorithm of the Greatest Common Divisor for Polynomials in Two Variables. We will consider the two polynomials in x and y,

$$f(x,\ y) \equiv a_0(y)x^n + a_1(y)x^{n-1} + \cdots + a_n(y),$$

$$\phi(x,\ y) \equiv b_0(y)x^m + b_1(y)x^{m-1} + \cdots + b_m(y),$$

and assume $a_0 \not\equiv 0,\ b_0 \not\equiv 0,\ n \geqq m > 0$.

Theorem 1 of the last section in combination with the results of § 67 enables us to get all the common factors of f and ϕ which involve y only; for such factors must be common factors of all the a's and all the b's.

It remains, then, merely to devise a method of obtaining the common factors of f and ϕ which do not themselves contain factors in y alone. We will show how this can be done by means of the algorithm of the greatest common divisor.

Dividing f by ϕ (cf. § 63, Theorem 2), we get the identity

$$P_0(y)f(x,\ y) \equiv Q_0(x,\ y)\phi(x,\ y) + R_1(x,\ y),$$

when R_1 is either identically zero, or is of lower degree in x than ϕ. If $R_1 \not\equiv 0$, divide ϕ by R_1, getting the identity

$$P_1(y)\phi(x,\ y) \equiv Q_1(x,\ y)R_1(x,\ y) + R_2(x,\ y),$$

where R_2 is either identically zero, or is of lower degree in x than R_1. If $R_2 \not\equiv 0$, divide R_2 by R_1. Proceeding in this way, we get the following system of identities in which the degrees in x of $R_1, R_2, \cdots$ continually decrease, so that after a certain number of steps we reach an R, say $R_{\rho+1}$, which is independent of x :

$$(1) \quad \begin{cases} P_0(y)f(x,\ y) \equiv Q_0(x,\ y)\phi(x,\ y) + R_1(x,\ y), \\ P_1(y)\phi(x,\ y) \equiv Q_1(x,\ y)R_1(x,\ y) + R_2(x,\ y), \\ P_2(y)R_1(x,\ y) \equiv Q_2(x,\ y)R_2(x,\ y) + R_3(x,\ y), \\ \cdot \quad \cdot \quad \cdot \quad \cdot \quad \cdot \quad \cdot \quad \cdot \quad \cdot \quad \cdot \quad \cdot \quad \cdot \quad \cdot \\ \cdot \quad \cdot \quad \cdot \quad \cdot \quad \cdot \quad \cdot \quad \cdot \quad \cdot \quad \cdot \quad \cdot \quad \cdot \quad \cdot \\ P_{\rho-1}(y)R_{\rho-2}(x,\ y) \equiv Q_{\rho-1}(x,\ y)R_{\rho-1}(x,\ y) + R_\rho(x,\ y), \\ P_\rho(y)R_{\rho-1}(x,\ y) \equiv Q_\rho(x,\ y)R_\rho(x,\ y) + R_{\rho+1}(y). \end{cases}$$

THEOREM 1. *A necessary and sufficient condition that f and ϕ have a common factor which involves x is*

$$R_{\rho+1}(y) \equiv 0.$$

In order to prove this theorem we first note that, by the first of the identities (1), any common factor of f and ϕ is a factor of R_1, hence, by the second of the identities, it is a factor of R_2, etc. Finally we see that *every common factor of f and ϕ is a factor of all the R's.* But $R_{\rho+1}$ does not contain x. Hence if f and ϕ have a common factor which contains x, $R_{\rho+1}(y) \equiv 0$.

Now suppose conversely that $R_{\rho+1}(y) \equiv 0$, and let

$$(2) \qquad R_\rho(x, y) \equiv S(y)G(x, y),$$

where G has no factor in y alone.* The last identity (1) then tells us that $P_\rho(y)$ is a factor of

$$Q_\rho(x, y)S(y)G(x, y),$$

and since by hypothesis G has no factor in y alone, $P_\rho(y)$ must, by the Corollary of Theorem 2, § 73, be a factor of $Q_\rho S$, that is

$$(3) \qquad Q_\rho(x, y)S(y) \equiv P_\rho(y)H(x, y).$$

Substituting first (2) and then (3) in the last identity (1), and cancelling out the factor $P_\rho(y)$ from the resulting identity, as we have a right to do since $P_\rho(y) \not\equiv 0$, we get the result

$$R_{\rho-1}(x, y) \equiv H(x, y)G(x, y).$$

That is, G is a factor not only of R_ρ but also of $R_{\rho-1}$. Accordingly we may write the next to the last identity (1) in the form

$$P_{\rho-1}(y)R_{\rho-2}(x, y) \equiv J(x, y)G(x, y).$$

By the corollary of Theorem 2, § 73, we see that $P_{\rho-1}(y)$ is a factor of J, so that $P_{\rho-1}(y)$ can be cancelled out of this last written identity, and we see that G is a factor of $R_{\rho-2}$.

Proceeding in this way, we see that G is a factor of all the R's, and therefore, finally, of f and ϕ. Moreover, we see from (2) that G is of at least the first degree in x, as otherwise R_ρ would not contain x, while $R_{\rho+1}$ was assumed to be the first of the R's which did not involve x.

Thus our theorem is proved.

Since, as we saw above, every common factor of f and ϕ is also a factor of all the R's, it follows from (2) that, if ψ is a common factor of f and ϕ,

$$G(x, y)S(y) \equiv \psi(x, y)K(x, y).$$

---

* If R_ρ has no factor in y alone, S reduces to a constant.

If then ψ contains no factor in y alone, S must, by the Corollary of Theorem 2, § 73, be a factor of K. Consequently by cancelling out S from the last written identity, we see that ψ is a factor of G. That is,

Theorem 2. *If in Euclid's algorithm $R_{\rho+1} \equiv 0$, the greatest common divisor of f and ϕ which contains no factor in y alone is the polynomial $G(x, y)$ obtained by striking from $R_{\rho}(x, y)$ all factors in y alone.*

We note that if $R_{\rho+1}$ is a constant different from zero, f and ϕ are relatively prime; but that the converse of this is not true as the simple example

$$f \equiv 2x^2 + 3y^2, \qquad \phi \equiv x$$

shows.

Going back to the identities (1), we get from the first of these identities, by mere transposition, the value of R_1 in terms of f and ϕ (and P_0, Q_0). Substituting this value in the second identity, we get a value for R_2 in terms of f, ϕ, and certain P's and Q's. Proceeding in this way, we finally get the formula

$$(4) \qquad R_{\rho+1}(y) \equiv F(x, y)f(x, y) + \Phi(x, y)\phi(x, y)$$

where F and Φ are polynomials in (x, y).

75. Factors of Polynomials in Two Variables.

Theorem 1. *If $f(x, y)$ and $\phi(x, y)$ are any two polynomials in x and y, and $\psi(x, y)$ is an irreducible polynomial which is a factor of the product $f\phi$, then ψ is a factor of f or of ϕ.*

If ψ does not contain both x and y, this theorem reduces to Theorem 2, § 73. It remains, then, only to consider the case that ψ involves both variables. In this case, at least one of the polynomials f, ϕ must be of at least the first degree in x. Without loss of generality we may assume this to be f. If ψ is a factor of f, our theorem is true. Suppose ψ is not a factor of f; then, since ψ is irreducible, f and ψ are relatively prime, and if we apply the algorithm of the greatest common divisor to f and ψ (as we did in the last section to f and ϕ) the first remainder $R_{\rho+1}(y)$ which does not involve x is not identically zero. The identity (4) of the last section now becomes

$$(1) \qquad R_{\rho+1}(y) \equiv F(x, y)f(x, y) + \Psi(x, y)\psi(x, y).$$

If we multiply this by $\phi(x, y)$, the second member becomes a polynomial which has ψ as a factor, since, by hypothesis, $f\phi$ has ψ as a factor. We may therefore write

$$(2) \qquad R_{\rho+1}(y)\phi(x, y) \equiv \psi(x, y)\chi(x, y).$$

Now no factor other than a constant of $R_{\rho+1}$ can be a factor of ψ, since ψ is irreducible. Consequently, by the Corollary of Theorem 2, § 73, $R_{\rho+1}$ is a factor of $\chi(x, y)$. Cancelling out $R_{\rho+1}$ from (2), as we have a right to do since it does not vanish identically, we get an identity of the form

$$\phi(x, y) \equiv \psi(x, y)\chi_1(x, y);$$

that is, ψ is a factor of ϕ, and our theorem is proved.

By applying this theorem a number of times, we get the

COROLLARY. *If the product of any number of polynomials in two variables,*

$$f_1(x, y)f_2(x, y) \cdots f_k(x, y),$$

is divisible by an irreducible polynomial $\psi(x, y)$, then ψ is a factor of at least one of the f's.

We come now to the fundamental theorem of the whole subject of divisibility of polynomials in two variables.

THEOREM 2. *A polynomial in two variables which is not identically zero can be resolved into the product of irreducible factors no one of which is a constant in one, and essentially in only one, way.*

That a polynomial $f(x, y)$ can be resolved into the product of irreducible factors no one of which is a constant in at least one way may be seen as follows. If f is irreducible, no factoring is possible or necessary. If f is reducible, we have

$$f(x, y) \equiv f_1(x, y)f_2(x, y),$$

where neither f_1 nor f_2 is a constant. If f_1 and f_2 are both irreducible, we have a resolution of f of the form demanded. If not, resolve such of these polynomials f_1 and f_2 as are reducible into the product of two factors neither of which is a constant. We thus get f expressed as the product of three or four factors. This is the resolution of f demanded if all the factors are irreducible. If not, resolve such as are reducible into the product of two factors, etc. This process must stop after a finite number of steps, for each time we factor a polynomial into two factors, the degrees of the factors are lower than the degree of the original polynomial. We shall thus ultimately resolve f by this process into the product of irreducible factors, no one of which is a constant.

Suppose now that f can be resolved in two ways into the product of irreducible factors none of which are constants,

$$f(x, y) \equiv f_1(x, y)f_2(x, y) \cdots f_k(x, y)$$
$$\equiv \phi_1(x, y)\phi_2(x, y) \cdots \phi_l(x, y).$$

Since ϕ_1 is a factor of f, it must, by the Corollary of Theorem 1, be a factor of one of the polynomials $f_1, f_2, \cdots f_k$. Suppose the f's so arranged that it is a factor of f_1. Then, since f_1 is irreducible, f_1 and ϕ_1 can differ only by a constant factor, and since $\phi_1 \not\equiv 0$, we may cancel it from the identity above, getting

$$c_1 f_2 f_3 \cdots f_k \equiv \phi_2 \phi_3 \cdots \phi_l.$$

In the same way we see from this identity that f_2 and one of the ϕ's, say ϕ_2, differ only by a constant factor. Cancelling ϕ_2, we get

$$c_1 c_2 f_3 \cdots f_k \equiv \phi_3 \cdots \phi_l.$$

Proceeding in this way, we should use up the ϕ's before the f's if $l < k$, the f's before the ϕ's if $l > k$. Neither of these cases is possible, for we should then have ultimately a constant on one side of the identity, and a polynomial different from a constant on the other. Thus we must have $k = l$. Moreover we see that the f's can be arranged in such an order that each f is proportional to the corresponding ϕ, and this is what we mean (cf. Definition 7, § 60) by saying that the two methods of factoring are not essentially different.

Thus our theorem is proved.

THEOREM 3. *If two polynomials f and ϕ in (x, y) are relatively prime, there are only a finite number of pairs of values of (x, y) for which f and ϕ both vanish.**

For if f and ϕ both vanished at the points

$$(3) \qquad\qquad (x_1, y_1), (x_2, y_2), \cdots,$$

and if these points were infinite in number, there would be among them either an infinite number of distinct x's or an infinite number of

---

* Stated geometrically, this theorem tells us that two algebraic plane curves $f(x, y) = 0$, $\phi(x, y) = 0$ can intersect in an infinite number of points only when they have an entire algebraic curve in common.

distinct y's. By a suitable choice of notation we may suppose that there are an infinite number of distinct y's. Then it is clear that f and ϕ must be of at least the first degree in x, since a polynomial in y alone which does not vanish identically cannot vanish for an infinite number of values of y. We may then apply to f and ϕ the algorithm of the greatest common divisor as in § 74, thus getting (cf. (4), § 74) an identity of the form

$$(4) \qquad F(x,\, y)f(x,\, y) + \Phi(x,\, y)\phi(x,\, y) \equiv R_{\rho+1}(y) \not\equiv 0.$$

Since the first member of (4) vanishes at all the points (3), $R_{\rho+1}(y)$ would vanish for an infinite number of distinct values of y, and this is impossible.

An important corollary of the theorem just proved is that if f and ϕ are two irreducible polynomials in $(x,\, y)$, and if the equations $f = 0$ and $\phi = 0$ have the same locus, then f and ϕ differ merely by a constant factor. This would, however, no longer be necessarily true if f and ϕ were not irreducible, as the example,

$$f \equiv xy^2, \quad \phi \equiv x^2 y,$$

shows; for the two curves $f = 0$ and $\phi = 0$ are here identical, since the curve in each case consists of the two coördinate axes, and yet f and ϕ are not proportional. By means of the following convention, however, the statement made above becomes true in all cases:

Let f be resolved into its irreducible factors,

$$f \equiv f_1^{\alpha_1} f_2^{\alpha_2} \cdots f_k^{\alpha_k},$$

where $f_1,\, \cdots f_k$ are irreducible polynomials in $(x,\, y)$, no two of which are proportional to each other. The curve $f = 0$ then consists of the k pieces, $\qquad f_1 = 0,\, f_2 = 0,\, \cdots f_k = 0.$

To each of these pieces we attach the corresponding positive integer α_i which we call the *multiplicity* of this piece; and we then regard two curves given by algebraic equations as identical only when they consist of the same irreducible pieces, and each of these pieces has the same multiplicity in both cases. With this convention we may say:

COROLLARY. *If f and ϕ are polynomials in $(x,\, y)$ neither of which is identically zero, a necessary and sufficient condition that the two curves $f = 0$, $\phi = 0$ be identical is that the polynomials f and ϕ differ only by a constant factor.*

EXERCISES

1. Let $f(x)$, $\phi(x)$, $\psi(x)$ be polynomials in x whose coefficients lie in a certain domain of rationality. Then if ψ is irreducible in this domain and is a factor of the product $f\phi$, prove that ψ is a factor of f or of ϕ.

2. Let $f(x)$ be a polynomial in x, which is not identically zero, and whose coefficients lie in a certain domain of rationality. Prove that f can be resolved into a product of polynomials whose coefficients lie in this domain, which are irreducible in this domain, and no one of which is a constant, in one and essentially in only one way.

3. Extend the results of this section to polynomials in two variables whose coefficients lie in a certain domain of rationality.

76. Factors of Polynomials in Three or More Variables. The results so far obtained in this chapter may be extended to polynomials in three variables without, in the main, essentially modifying the methods already used. We proceed therefore to state the theorems in the order in which they should be proved, leaving the proofs of most of them to the reader. The extension to the case of n variables then presents no difficulty, and is left entirely to the reader (cf. Exercise 1).

Let $f(x, y, z)$ be any polynomial in three variables, and suppose it arranged according to powers of x,

$$f(x, y, z) \equiv a_0(y, z)x^n + a_1(y, z)x^{n-1} + \cdots + a_n(y, z),$$

the a's being polynomials in (y, z).

Corresponding to Theorems 1, 2 of § 73 we have

THEOREM 1. *A necessary and sufficient condition that a polynomial in (y, z) be a factor of f is that it be a factor of all the a's.*

THEOREM 2. *If $f(x, y, z)$ and $\phi(x, y, z)$ are any two polynomials in (x, y, z) and $\psi(y, z)$ is an irreducible polynomial in (y, z) only which is a factor of the product $f\phi$, then ψ is a factor of f or of ϕ.*

COROLLARY. *Let $f(x, y, z)$ and $\phi(x, y, z)$ be polynomials in (x, y, z), and let $\psi(y, z)$ be a polynomial in (y, z) alone. If ψ is a factor of the product of $f\phi$, but is relatively prime to ϕ, then it is a factor of f.*

To find the greatest common divisor of two polynomials in three variables we proceed exactly as in the case of two variables, getting a set of identities similar to (1), § 74, the P's and $R_{\rho+1}$ being now functions of (y, z), while the other R's and the Q's are functions of (x, y, z).   Corresponding to Theorems 1, 2 of § 74 we now have

THEOREM 3.   *A necessary and sufficient condition that $f(x, y, z)$, and $\phi(x, y, z)$ have a common factor which involves x is that $R_{\rho+1}(y, z) \equiv 0$.*

THEOREM 4.   *If $R_{\rho+1}(y, z) \equiv 0$, the greatest common divisor of $f(x, y, z)$ and $\phi(x, y, z)$ which contains no factor in (y, z) alone is the polynomial $G(x, y, z)$ obtained by striking out from $R_\rho(x, y, z)$ all factors in (y, z) alone.*

From the algorithm of the greatest common divisor for the two polynomials $f(x, y, z)$, $\phi(x, y, z)$ we also deduce the identity

$$(1) \qquad R_{\rho+1}(y, z) \equiv F(x, y, z)f(x, y, z) + \Phi(x, y, z)\phi(x, y, z),$$

similar to (4), § 74.

Corresponding to Theorems 1, 2 of § 75 we have

THEOREM 5.   *If $f(x, y, z)$ and $\phi(x, y, z)$ are any two polynomials and $\psi(x, y, z)$ is an irreducible polynomial which is a factor of the product $f\phi$, then ψ is a factor of f or of ϕ.*

COROLLARY.   *If the product of any number of polynomials*

$$f_1(x, y, z)f_2(x, y, z) \cdots f_k(x, y, z),$$

is divisible by an irreducible polynomial $\psi(x, y, z)$, then ψ is a factor of at least one of the f's.

THEOREM 6.   *A polynomial in three variables which is not identically zero can be resolved into the product of irreducible factors no one of which is a constant in one, and essentially in only one, way.*

When we come to Theorem 3, § 75, however, we find that it does not admit of immediate extension to the case of three variables ; for $R_{\rho+1}(y)$, which came into the proof of that theorem, becomes now $R_{\rho+1}(y, z)$, and we can no longer say that this does not vanish at an infinite number of points (y, z).   Not only is the proof thus seen to fail, but the obvious extension of the theorem itself is seen to be false when we recall that two surfaces intersect, in general, in a curve.

This theorem may, however, be replaced by the following one :

THEOREM 7.   *If $f(x, y, z)$ and $\phi(x, y, z)$ are any two polynomials in three variables of which ϕ is irreducible, and if f vanishes at all points (x, y, z) at which ϕ vanishes, then ϕ is a factor of f.*

In proving this theorem we may, without loss of generality, assume that ϕ actually contains one of the variables, say x; for if ϕ contains none of the variables x, y, z, the theorem is trivial and obviously true.

Suppose ϕ were not a factor of f. Then, since ϕ is irreducible, f and ϕ are relatively prime. Hence, in the identity (1) above, $R_{\rho+1}(y, z) \not\equiv 0$. Let us write

$$(2) \qquad \phi(x, y, z) \equiv b_0(y, z)x^m + b_1(y, z)x^{m-1} + \cdots + b_m(y, z) \qquad (m \geqq 1),$$

where, without loss of generality, we may assume $b_0(y, z) \not\equiv 0$. Then

$$(3) \qquad R_{\rho+1}(y, z)b_0(y, z) \not\equiv 0.$$

Accordingly we can find a point (y_1, z_1) such that

$$(4) \qquad R_{\rho+1}(y_1, z_1) \neq 0, \qquad b_0(y_1, z_1) \neq 0.$$

Consequently $\phi(x, y_1, z_1)$ is a polynomial in x alone which is of at least the first degree, and which therefore (Theorem 1, § 6) vanishes for some value x_1 of x. That is

$$\phi(x_1, y_1, z_1) = 0.$$

Accordingly, by hypothesis,

$$f(x_1, y_1, z_1) = 0.$$

Referring now to the identity (1), we see that

$$R_{\rho+1}(y_1, z_1) = 0.$$

This, however, is in contradiction with (4). Thus our theorem is proved.

If to each part of a reducible algebraic surface we attach a *multiplicity* in precisely the same way as was explained in the last section for plane curves, we infer at once the

COROLLARY. *If f and ϕ are polynomials in (x, y, z) neither of which is identically zero, a necessary and sufficient condition that the two surfaces*

$$f = 0, \phi = 0$$

be identical is that the polynomials f and ϕ differ only by a constant factor.

Theorem 7 admits also the following generalization :

THEOREM 8. *If $f(x, y, z)$ and $\phi(x, y, z)$ are any two polynomials in three variables which both vanish at the point (x_0, y_0, z_0) and of which ϕ is irreducible, and if in the neighborhood N of (x_0, y_0, z_0) f vanishes at all points at which ϕ vanishes, then ϕ is a factor of f.*

We assume, as before, that ϕ contains x and can therefore be written in the form (2). Let us first consider the case in which $b_0(y_0, z_0) \neq 0$. Here the proof is very similar to the proof of Theorem 7.

We obtain relation (3) precisely as above, and from it we infer that a point (y_1, z_1) *in as small a neighborhood M of* (y_0, z_0) *as we please* can be found at which the relations (4) are true.

Now consider the equation

$$(5) \qquad\qquad \phi(x, y_1, z_1) = 0.$$

By writing ϕ in the form (2), we see that by taking the neighborhood M of (y_0, z_0) sufficiently small, we can make the coefficients of (5) differ from the coefficients of

$$(6) \qquad\qquad \phi(x, y_0, z_0) = 0$$

by as little as we please (cf. Theorem 3, § 5). Now x_0 is by hypothesis a root of equation (6). Consequently by taking M sufficiently small, we can cause (5) to have at least one root x_1 which differs from x_0 by as little as we please (cf. Theorem 4, § 6). Thus we see that a point (x_1, y_1, z_1) in the given neighborhood N of (x_0, y_0, z_0) can be found at which

$$\phi(x_1, y_1, z_1) = 0.$$

Accordingly, by hypothesis,

$$f(x_1, y_1, z_1) = 0.$$

From the identity (1) we have then

$$R_{\rho+1}(y_1, z_1) = 0,$$

which is in contradiction with (4). Thus our theorem is proved on the supposition that $b_0(y_0, z_0) \neq 0$.*

---

* The proof just given will, in fact, apply to the case in which not all the b's in (2) vanish at the point (y_0, z_0), if we use the extension of Theorem 4, § 6, which is there mentioned in a footnote. It is only the extreme case in which all the b's vanish at this point which requires the special treatment which we now proceed to give. The reader is advised to consider the geometrical meaning of this extreme case.

In order to treat the case in which $b_0(y_0, z_0) = 0$, let us denote by k the degree of $\phi(x, y, z)$, and let us subject this polynomial to a non-singular linear transformation

$$(7) \qquad \begin{cases} x = \alpha_1 x' + \beta_1 y' + \gamma_1 z' \\ y = \alpha_2 x' + \beta_2 y' + \gamma_2 z' \\ z = \alpha_3 x' + \beta_3 y' + \gamma_3 z' \end{cases}$$

which makes the degree of ϕ in x equal to the total degree k of ϕ (cf. Theorem 2, §64).

Suppose that this transformation carries over the point (x_0, y_0, z_0) into the point (x_0', y_0', z_0'). Then it is possible, since (7) is non-singular, to take such a small neighborhood N' of (x_0', y_0', z_0') that all points in this neighborhood correspond to points in the given neighborhood N of (x_0, y_0, z_0).

Moreover, by means of (7), ϕ has gone over into

$$(8) \qquad \phi'(x', y', z') \equiv b_0' x'^k + b_1'(y', z')x'^{k-1} + \cdots + b_k'(y', z'),$$

where b_0' is a constant different from zero. Let us denote by $f'(x', y', z')$ the polynomial into which f is transformed. Then it is clear that, since, in the neighborhood N, f vanishes whenever ϕ does, in the neighborhood N' (which corresponds to a *part* of N), f' vanishes whenever ϕ' does. Accordingly we can apply the part of the theorem already proved to the two polynomials f' and ϕ', since the first coefficient of ϕ' in the form (8), being a constant different from zero, does not vanish at (y_0', z_0'). We infer that ϕ' is a factor of f',

$$f'(x', y', z') \equiv \phi'(x', y', z')\psi'(x', y', z').$$

If here we replace x', y', z' by their values in terms of x, y, z from (7), we see that ϕ is a factor of f; and our theorem is proved.

EXERCISES

1. State and prove the eight theorems of this section for the case of polynomials in n variables.

2. Extend the result of the exercise at the the end of §73 to the case of polynomials in n variables.

3. Extend the results of the two preceding exercises to the case in which we consider only polynomials whose coefficients lie in a certain domain of rationality.

4. The resultant of two polynomials in one variable

$$f(x) \equiv a_0 x^n + a_1 x^{n-1} + \cdots + a_n,$$

$$\phi(x) \equiv b_0 x^m + b_1 x^{m-1} + \cdots + b_m,$$

is sometimes defined as the polynomial R in the a's and b's of lowest degree which satisfies an identity of the form

$$Ff + \Phi\phi \equiv R,$$

where F and Φ are polynomials in $(a_0, \cdots a_n; b_0, \cdots b_m; x)$, and the identity is an identity in all these arguments. Prove that the resultant as thus defined differs only by a constant factor different from zero from the resultant as we defined it in § 68.

CHAPTER XVII

GENERAL THEOREMS ON INTEGRAL RATIONAL INVARIANTS

77. The Invariance of the Factors of Invariants. Let us consider the general n-ary form of the kth degree which we will represent by $f(x_1, \cdots x_n ; a_1, a_2, \cdots)$, the x's being the variables and the a's the coefficients. By suitably changing the a's, this symbol may be used to represent *any* such form. Hence, if we subject such a form to a linear transformation, the new form, being n-ary and of the same degree as the old, may be represented by the same functional letter : $f(x_1', \cdots x_n' ; a_1', a_2', \cdots)$. This new form will evidently be homogeneous and linear in the a's ; that is, each of the a''s is a homogeneous linear polynomial in the a's. It is also clear that each of the a''s is a homogeneous polynomial of the kth degree in the coefficients of the transformation.

It follows from the very definition of invariants that if we have a number of integral rational relative invariants of a form or system of forms, their product will also be an integral rational relative invariant. It is the converse of this that we wish to prove in this section. We begin by stating this converse in the simple case of a single form.

THEOREM 1. *If $I(a_1, a_2, \cdots)$ is an integral rational invariant of the n-ary form*
$$f(x_1, \cdots x_n ; a_1, a_2, \cdots),$$
and is reducible, then all its factors are invariants.

It will evidently be sufficient to prove that the irreducible factors of I are invariants. Let $f_1, f_2, \cdots f_l$ be the irreducible factors of I.
Subjecting f to the linear transformation
$$\begin{cases} x_1 = c_{11}x_1' + \cdots + c_{1n}x_n', \\ \quad \cdot \quad \cdot \quad \cdot \quad \cdot \quad \cdot \quad \cdot \\ \quad \cdot \quad \cdot \quad \cdot \quad \cdot \quad \cdot \quad \cdot \\ x_n = c_{n1}x_1' + \cdots + c_{nn}x_n', \end{cases}$$

whose determinant we call c, and denoting the coefficients of the transformed form by a_1', a_2', $\cdots$, we have

$$(1) \qquad\qquad I(a_1',\, a_2',\, \cdots) \equiv c^\lambda I(a_1,\, a_2,\, \cdots);$$

an identity which may also be written

$$f_1(a_1',\, a_2',\, \cdots) \cdots f_l(a_1',\, a_2',\, \cdots) \equiv c^\lambda f_1(a_1,\, a_2,\, \cdots) \cdots f_l(a_1,\, a_2,\, \cdots).$$

We have here a polynomial in the c's and a's which, on the second side of the identity, is resolved into its irreducible factors, since by Theorem 1, §61, the determinant c is irreducible. Hence each factor on the first side is equal to the product of some of the factors on the second. That is

$$(2) \qquad f_i(a_1',\, a_2',\, \cdots) \equiv c^{\lambda_i}\phi_i(a_1,\, a_2,\, \cdots) \qquad\qquad (i = 1,\, 2,\, \cdots l)$$

where the ϕ's are polynomials.

Now let

$$c_{11} = c_{22} = \cdots = c_{nn} = 1,$$

and let all the other c's be zero. Our transformation becomes the identical transformation, the determinant $c = 1$, and each a' is equal to the corresponding a. The identities (2) therefore reduce to

$$f_i(a_1,\, a_2,\, \cdots) \equiv \phi_i(a_1,\, a_2,\, \cdots) \qquad\qquad (i = 1,\, 2,\, \cdots l).$$

Substituting this value of ϕ_i in (2), we see that f_i is an invariant, and our theorem is proved.

The general theorem, now, is the following :

THEOREM 2. *If* $I\,(a_1,\, a_2,\, \cdots;\, b_1,\, b_2,\, \cdots;\, \cdots)$ *is an integral rational invariant of the system of forms*

$$f_1(x_1,\, \cdots x_n;\ a_1,\, a_2,\, \cdots)$$

$$f_2(x_1,\, \cdots x_n;\ b_1,\, b_2,\, \cdots)$$

$$\cdot \quad \cdot \quad \cdot \quad \cdot \quad \cdot \quad \cdot \quad \cdot$$

$$\cdot \quad \cdot \quad \cdot \quad \cdot \quad \cdot \quad \cdot \quad \cdot$$

and is reducible, then all its factors are invariants.

The proof of this theorem is practically identical with that of Theorem 1.

EXERCISE

If $I(a_1, a_2, \cdots; b_1, b_2, \cdots; \cdots; y_1, \cdots y_n; z_1, \cdots z_n; \cdots)$ is an integral rational covariant of the system of forms

$$f_1(x_1, \cdots x_n; a_1, a_2, \cdots),$$
$$f_2(x_1, \cdots x_n; b_1, b_2, \cdots),$$
$$\cdot \quad \cdot \quad \cdot \quad \cdot \quad \cdot \quad \cdot \quad \cdot \quad \cdot$$
$$\cdot \quad \cdot \quad \cdot \quad \cdot \quad \cdot \quad \cdot \quad \cdot \quad \cdot$$

and the system of points $(y_1, \cdots y_n)$, $(z_1, \cdots z_n)$, $\cdots$, and is reducible, then all its factors are covariants (or invariants).

78. A More General Method of Approach to the Subject of Relative Invariants. We have called a polynomial I in the coefficients of an n-ary form f a relative invariant of this form if it has the property of being merely multiplied by a power of the determinant of the transformation when f is subjected to a linear transformation. It is natural to inquire what class of functions I we should obtain if we make the less specific demand that I be multiplied by a polynomial in the coefficients of the transformation. We should expect to get in this way a class of functions more general than the invariants we have so far considered. As a matter of fact, we get precisely the same class of functions, as is shown by the following theorem:

THEOREM. *Let I be a polynomial not identically zero in the coefficients $(a_1, a_2, \cdots)$ of an n-ary form f, and let $(a_1', a_2', \cdots)$ be the coefficients of the form obtained by subjecting f to the linear transformation*

$$\begin{cases} x_1 = c_{11}x_1' + \cdots + c_{1n}x_n', \\ \cdot \quad \cdot \quad \cdot \quad \cdot \quad \cdot \quad \cdot \quad \cdot \\ \cdot \quad \cdot \quad \cdot \quad \cdot \quad \cdot \quad \cdot \quad \cdot \\ x_n = c_{n1}x_1' + \cdots + c_{nn}x_n'. \end{cases}$$

If $$I(a_1', a_2', \cdots) \equiv \psi(c_{11}, \cdots c_{nn}) \, I(a_1, a_2, \cdots),$$

where ψ is a polynomial in the c's, and this is an identity in the a's and c's, then ψ is a power of the determinant of the transformation.

We will first show that $\psi \neq 0$ when $c \neq 0$. If possible let $d_{11}, \cdots d_{nn}$ be a particular set of values of the c_{ij}'s such that

$$\psi(d_{11}, \cdots d_{nn}) = 0,$$

while

$$\begin{vmatrix} d_{11} \cdots d_{1n} \\ \cdot \ \cdot \ \cdot \ \cdot \ \cdot \\ \cdot \ \cdot \ \cdot \ \cdot \ \cdot \\ d_{n1} \cdots d_{nn} \end{vmatrix} \neq 0.$$

Then the transformation

$$\begin{cases} x_1 = d_{11}x_1' + \cdots + d_{1n}x_n', \\ \ \cdot \quad \cdot \quad \cdot \quad \cdot \quad \cdot \quad \cdot \quad \cdot \quad \cdot \\ \ \cdot \quad \cdot \quad \cdot \quad \cdot \quad \cdot \quad \cdot \quad \cdot \quad \cdot \\ x_n = d_{n1}x_1' + \cdots + d_{nn}x_n', \end{cases}$$

has an inverse

$$\begin{cases} x_1' = \delta_{11}x_1 + \cdots + \delta_{1n}x_n, \\ \ \cdot \quad \cdot \quad \cdot \quad \cdot \quad \cdot \quad \cdot \quad \cdot \quad \cdot \\ \ \cdot \quad \cdot \quad \cdot \quad \cdot \quad \cdot \quad \cdot \quad \cdot \quad \cdot \\ x_n' = \delta_{n1}x_1 + \cdots + \delta_{nn}x_n. \end{cases}$$

Let us consider a special set of a's such that $I(a_1, a_2, \cdots) \neq 0$. Then

$$I(a_1', a_2', \cdots) = \psi(d_{11}, \cdots d_{nn})\, I(a_1, a_2, \cdots) = 0.$$

Now apply the inverse transformation, and we have

$$I(a_1, a_2, \cdots) = \psi(\delta_{11}, \cdots \delta_{nn})\, I(a_1', a_2', \cdots) = 0,$$

which is contrary to our hypothesis.

Having thus proved that ψ can vanish only when $c = 0$, let us break up ψ into its irreducible factors,

$$\psi(c_{11}, \cdots c_{nn}) \equiv \psi_1(c_{11}, \cdots c_{nn})\, \psi_2(c_{11}, \cdots c_{nn}) \cdots \psi_\lambda(c_{11}, \cdots c_{nn}).$$

Since ψ vanishes whenever $\psi_i = 0$, ψ_i can vanish only when $c = 0$. Hence by the theorem for n variables which corresponds to Theorem 7, § 76, ψ_i must be a factor of c. But c is irreducible. Hence ψ_i can differ from c only by a constant factor, and we may write

$$\psi \equiv Kc^\lambda.$$

It remains then merely to prove that the constant K has the value 1. For this purpose consider the identity

$$I(a_1', a_2', \cdots) \equiv Kc^\lambda I(a_1, a_2 \cdots),$$

and give to the c_{ij}'s the values which they have in the identical transformation. Then $c = 1$, and the a''s are equal to the corresponding a's. The last written identity therefore becomes

$$I(a_1, a_2, \cdots) \equiv KI(a_1, a_2, \cdots);$$

from which we infer that $K = 1$.

EXERCISES

1. Prove that if a polynomial I in the coefficients a_1, a_2, $\cdots$ of an n-ary form and the coördinates $(y_1, \cdots y_n)$ of a point has the property of being merely multiplied by a certain rational function ψ of the coefficients of the transformation when the form and the point are subjected to a linear transformation, then ψ is a positive or negative power of the determinant of the transformation, and I is a covariant.

2. Generalize the theorem of this section to the case of invariants of a system of forms.

3. Generalize the theorem of Exercise 1 to the case of a system of forms and a system of points.

4. Prove that every rational invariant of a form or system of forms is the ratio of two integral rational invariants.

5. Generalize the theorem of Exercise 4 to the case of covariants.

79. The Isobaric Character of Invariants and Covariants. In many investigations, and in particular in the study of invariants and covariants, it is desirable to attach a definite *weight* to each of the variables with which we have to deal. To a product of two or more such variables we then attach a weight equal to the sum of the weights of the factors, and this weight is supposed to remain unchanged if the product is multiplied by a constant coefficient. Thus if z_1, z_2, z_3 are regarded as having weights w_1, w_2, w_3 respectively, the term

$$5\, z_1 z_2 z_3^2$$

would have the weight $\quad w_1 + w_2 + 2w_3.$

If, then, having thus attached a definite weight to each of the variables, we consider a polynomial, each term of this polynomial will be of a definite weight, and *by the weight of a polynomial we understand the greatest weight of any of its terms whose coefficient is not zero.* If moreover all the terms of a polynomial are of the same weight, the polynomial is said to be *isobaric*.

It may be noticed that, according to this definition, a polynomial which vanishes identically is the only one which has no weight, while a polynomial which reduces to a constant different from zero is of weight zero. Moreover if two polynomials are of weights w_1 and w_2, their product is of weight $w_1 + w_2$.*

---

* The conception of degree of a polynomial is merely the special case of the conception of weight in which all the variables are supposed to have weight 1. The conception of being isobaric then reduces to the conception of homogeneity.

We will apply this conception of weight first to the case in which the variables of which we have been speaking are the coefficients $a_1, a_2, \cdots$ of the n-ary form

$$f(x_1, \cdots x_n; a_1, a_2, \cdots).$$

We shall find it desirable to admit n different determinations of the weights of these a's; one determination corresponding to each of the variables $x_1, \cdots x_n$.

DEFINITION 1.   *If a_i is the coefficient of the term*

$$a_i x_1^{p_1} x_2^{p_2} \cdots x_n^{p_n}$$

in an n-ary form, we assign to a_i the weights $p_1, p_2, \cdots p_n$ respectively with regard to the variables $x_1, x_2, \cdots x_n$.

In the case of a binary form,

$$a_0 x_1^k + a_1 x_1^{k-1} x_2 + \cdots + a_k x_2^k,$$

the subscripts of the coefficients indicate their weights with regard to x_2, while their weights with regard to x_1 are equal to the differences between the degree of the form and these subscripts.

As a second example, we mention the quadratic form

$$\sum_1^n a_{ij} x_i x_j.$$

Here the weight of any coefficient with regard to one of the variables, say x_j, is equal to the number of times j occurs as a subscript to this coefficient.*

In connection with this subject of weight, the special linear transformation

$$(1) \qquad \begin{cases} x_i = x_i' \\ x_j = k x_j' \end{cases} \qquad (i \neq j)$$

is useful.   If a_i is a coefficient which is of weight λ with regard to x_j, the term in which this coefficient occurs contains x_j^λ, and therefore

$$a_i' = k^\lambda a_i.$$

---

* For forms of higher degree, a similar notation for the coefficients by means of multiple subscripts might be used. The weight of each coefficient could then be at once read off from the subscripts.

That is

THEOREM 1. *The weight with regard to x_j of a coefficient of an n-ary form is the exponent of the power of k by which this coefficient is multiplied after the special transformation* (1).

From this it follows at once that an isobaric polynomial of weight λ with regard to x_j in the coefficients $(a_1, a_2, \cdots)$ of an n-ary form is simply multiplied by k^λ if the form is subjected to the linear transformation (1).

Moreover, the converse of this is also true. For if $a_1', a_2', \cdots$ are the coefficients of the n-ary form after the transformation (1), and if $\phi(a_1, a_2, \cdots)$ is a polynomial which has the property that

$$\phi(a_1', a_2', \cdots) \equiv k^\lambda \phi(a_1, a_2, \cdots),$$

this being an identity in the a's and also in k, we can infer, as follows, that ϕ is isobaric of weight λ. Let us group the terms of ϕ together according to their weights, thus writing ϕ in the form

$$\phi(a_1, a_2, \cdots) \equiv \phi_1(a_1, a_2, \cdots) + \phi_2(a_1, a_2, \cdots) + \cdots$$

where $\phi_1, \phi_2, \cdots$ are isobaric of weights $\lambda_1, \lambda_2, \cdots$. We have then

$$\phi(a_1', a_2', \cdots) \equiv k^{\lambda_1} \phi_1(a_1, a_2, \cdots) + k^{\lambda_2} \phi_2(a_1, a_2, \cdots) + \cdots.$$

But on the other hand

$$\phi(a_1', a_2', \cdots) \equiv k^\lambda \phi(a_1, a_2, \cdots) \equiv k^\lambda \phi_1(a_1, a_2, \cdots) + k^\lambda \phi_2(a_1, a_2, \cdots) + \cdots.$$

Comparing the last members of these two identities, we see that

$$\lambda = \lambda_1 = \lambda_2 = \cdots$$

as was to be proved. We have thus established the theorem:

THEOREM 2. *A necessary and sufficient condition that a polynomial ϕ in the coefficients of an n-ary form be simply multiplied by k^λ when the form is subjected to transformations of the form* (1) *is that ϕ be isobaric of weight λ with regard to x_j.*

By means of this theorem we can show that the use of the word *weight* introduced in §31 is in accord with the definition given in the present section. For an integral rational invariant of an n-ary form which, according to the definition of §31, is of weight λ will, if the form is subjected to the transformation (1), be merely multiplied by k^λ and must therefore, according to Theorem 2, be isobaric of weight λ with regard to x_j.   That is:

THEOREM 3. *If I is an integral rational invariant of a form f which according to the definition of § 31 is of weight λ, it will also be of weight λ with regard to each of the variables x_j of f according to the definitions of this section, and it will be isobaric with regard to each of these variables.*

As an illustration of this theorem we may mention the discriminant
$$a_0 a_2 - a_1^2$$
of the binary quadratic form
$$a_0 x_1^2 + 2a_1 x_1 x_2 + a_2 x_2^2$$
which is isobaric of weight 2 both with regard to x_1 and with regard to x_2.

The reader should consider in the same way the discriminant of the general quadratic form.

All of the considerations of the present section may be extended immediately to the case in which we have to deal, not with a single form, but with a system of forms.   We state here merely the theorem which corresponds to Theorem 3.

THEOREM 4. *If I is an integral rational invariant of a system of forms which according to the definition of § 31 is of weight λ, it will also be of weight λ with regard to each of the variables x_j of the system, and it will be isobaric with regard to each of these variables.*

The reader may consider as an illustration of this theorem the resultant of a system of linear forms, and also the invariants obtained in Chapters XII and XIII.

We saw in Theorem 5, § 31, that the weight of an integral rational invariant cannot be negative.   This fact now becomes still more evident, since the weight of no coefficient is negative.   Moreover, we can now add the following further fact:

THEOREM 5. *An integral rational invariant of a form or system of forms cannot be of weight zero.*

For consider any term of the invariant whose coefficient is not zero.   This term involves the product of a number of coefficients of the system of forms.   Since none of these coefficients can be of negative weight, the weight of the term will be at least as great as the weight of any one of them. But any one of them is at least of weight 1 with regard to some one of the variables. Hence the invariant is at least of weight 1 with regard to some one of the variables, and hence with regard to any of the variables.

In order, finally, to be able to extend the considerations of this section to the case of covariants, we must lay down the following additional definition:

DEFINITION 2. *If the sets of variables $(y_1, \cdots y_n)$, $(z_1, \cdots z_n)$, $\cdots$ are cogredient with the variables $(x_1, \cdots x_n)$ of a system of n-ary forms, we will assign to y_j, z_j, $\cdots$ the weight -1 with regard to x_j, to all the other y's, z's, etc. the weight 0.*

It will be noticed that here too, when we perform the transformation (1), each of the variables is multiplied by a power of k whose exponent is the weight of the variable. It is therefore easy* to extend the considerations of this section to this case, and we thus get the theorem:

THEOREM 6. *If I is an integral rational covariant of a system of forms and a system of points which is of weight λ according to the definition of § 31, it will also be of weight λ with regard to each of the variables of the system, and it will be isobaric with regard to each of these variables.*

As an example of this theorem we note that the polar

$$a_0 y_1 z_1 + a_1 (y_1 z_2 + y_2 z_1) + a_2 y_2 z_2$$

of a binary quadratic form is isobaric of weight zero. The reader may satisfy himself that the same is true of the polar of the general quadratic form.

30. Geometric Properties and the Principle of Homogeneity. It is a familiar fact that many geometric properties of plane curves or surfaces are expressed by the vanishing of an integral rational function of the coefficients of their equations. Take, for instance, the surface

(1) $$f(x,\ y,\ z;\ a_1,\ a_2,\ \cdots) = 0,$$

---

* Slight additional care must be taken here on account of the possible presence of terms of negative weight.

where f is a polynomial of the kth degree in the non-homogeneous coördinates x, y, z, and a_1, a_2, $\cdots$ are the coefficients of this polynomial; and consider the relation

$$(2) \qquad\qquad \phi(a_1, a_2, \cdots) = 0,$$

where ϕ is a polynomial, which we will assume to be of at least the first degree, in the coefficients a_1, a_2, $\cdots$. By Theorem 3, § 6, there are an infinite number of polynomials of the kth degree in (x, y, z) whose coefficients satisfy the relation (2) and also an infinite number whose coefficients do not satisfy this relation. In other words, all polynomials of the kth degree in (x, y, z) may be divided into two classes, A and B, of which the first is characterized by condition (2) being fulfilled, while the second is characterized by this condition not being fulfilled. We may therefore say that (2) is a necessary and sufficient condition that f have a certain *property*, namely, the property of belonging in class A.

The simplest examples, however, show that this property of f need not correspond to a geometric property of the surface (1). To illustrate this, let $k = 1$, so that we have

$$f \equiv a_1 x + a_2 y + a_3 z + a_4,$$

and consider first the polynomial in the a's :

$$\phi \equiv a_4.$$

The vanishing of ϕ gives a necessary and sufficient condition that f belong to the class of homogeneous polynomials of the first degree in (x, y, z), and thus expresses a property of the polynomial. This same condition, $a_4 = 0$, also expresses a property of the plane $f = 0$, namely, the property that it pass through the origin.

Suppose, however, that instead of the function ϕ we take the polynomial

$$\phi_1 \equiv a_4 - 1.$$

The vanishing of this polynomial also gives a necessary and suffi cient condition that the polynomial f have a certain property, namely, that its constant term have the value 1. It does *not* serve to distinguish planes into two classes, since we may write the equation of any plane (except those through the origin) either with the constant term 1 or with the constant term different from 1 by merely multiplying the equation through by a constant.

From the foregoing it will be seen that saying that a surface has a certain *property* amounts to the same thing as saying that it belongs to a certain class of surfaces.*

THEOREM 1. *The equation* (2) *expresses a necessary and sufficient condition for a geometric property of the surface* (1) *when, and only when, the polynomial ϕ is homogeneous.*

For if ϕ is non-homogeneous, let us write it in the form

$$\phi \equiv \phi_n + \phi_{n-1} + \cdots + \phi_1 + \phi_0,$$

where ϕ_n is a homogeneous polynomial of the nth degree and each of the other ϕ's which is not identically zero is a homogeneous polynomial of the degree indicated by its subscript. Let $a_1', a_2', \cdots$ be a set of values of the a's for which ϕ_n and at least one of the other ϕ_i's is not zero, and consider the surface

$$(3) \qquad f(x, y, z; \; ca_1', ca_2', \cdots) = 0.$$

The condition (2) for this surface is

$$c^n \phi_n(a_1', a_2', \cdots) + c^{n-1} \phi_{n-1}(a_1', a_2', \cdots) + \cdots + c\phi_1(a_1', a_2', \cdots)$$
$$+ \phi_0(a_1', a_2', \cdots) = 0.$$

This is an equation of the nth degree in c, and since at least one of the coefficients after the first is different from zero, it will have at least one root $c_1 \neq 0$. On the other hand, we can find a value $c_2 \neq 0$ which is not a root of this equation. Hence the surface (3) satisfies condition (2) if we let $c = c_1$ and does not satisfy it if $c = c_2$. But a change in the value of c merely multiplies the equation (3) by a constant and does not change the surface represented by it. Thus we see that one and the same surface can be regarded both as satisfying and as not satisfying condition (2). In other words, if ϕ is non-homogeneous, (2) does not express a property of the surface (1).

Assume now that ϕ is homogeneous of the nth degree, and consider the class A of polynomials f whose coefficients satisfy equation (2) and the class B whose coefficients do not satisfy this equation. Our theorem will be proved if we can show that we have hereby divided the surfaces (1) into two classes, that is, that if

---

* This brief explanation must not be regarded as an attempt to define the conception *property*, for no specific class can be defined without the use of some property.

a_1', a_2', $\cdots$ are the coefficients of a polynomial of class A and a_1'', a_2'', $\cdots$ the coefficients of a polynomial of class B, then the two surfaces

$$f(x,\ y,\ z;\ a_1',\ a_2',\ \cdots) = 0,$$
$$f(x,\ y,\ z;\ a_1'',\ a_2'',\ \cdots) = 0,$$

cannot be the same. If they were the same, the coefficients a_1', a_2', $\cdots$ would be proportional to a_1'', a_2'', $\cdots$ (cf. Theorem 7, Corollary, § 76),

$$a_1'' = ca_1',\ a_2'' = ca_2',\ \cdots$$

and therefore $\qquad \phi(a_1'',\ a_2'',\ \cdots) = c^n \phi(a_1',\ a_2',\ \cdots).$

But this is impossible since by hypothesis

$$\phi(a_1',\ a_2',\ \cdots) = 0,\ \ \phi(a_1'',\ a_2'',\ \cdots) \neq 0.$$

Thus our theorem is proved.

This theorem admits of generalization in various directions. Suppose first that instead of a single surface (1) we have a system of algebraic surfaces, and that ϕ is a polynomial in the coefficients of all these surfaces. Then precisely the reasoning just used shows that the equation $\phi = 0$ gives a necessary and sufficient condition for a geometric property of this system of surfaces when and only when ϕ is homogeneous in the coefficients of each surface taken separately.

On the other hand, we may use homogeneous coördinates in writing the equations of the surfaces, and the results so far stated will obviously hold without change:

THEOREM 2. *Let*

$$f_1(x,\ y,\ z,\ t;\ a_1,\ a_2,\ \cdots),\ f_2(x,\ y,\ z,\ t;\ b_1,\ b_2,\ \cdots),\ \cdots$$

be a system of homogeneous polynomials in $(x,\ y,\ z,\ t)$ *whose coefficients are* $a_1,\ a_2,\ \cdots;\ b_1,\ b_2,\ \cdots;\ etc.;\ and\ let*

$$\phi(a_1,\ a_2,\ \cdots;\ b_1,\ b_2,\ \cdots;\ \cdots)$$

be a polynomial in the a's, b's, etc. Then the equation $\phi = 0$ *expresses a necessary and sufficient condition that the system of surfaces*

$$f_1 = 0,\ f_2 = 0,\ \cdots$$

have a geometric property when, and only when, the polynomial ϕ *is homogeneous in the a's alone, also in the b's alone, etc.*

In conclusion we note that all the results of this section can be extended at once to algebraic curves in the plane ; or, indeed, to the case of space of any number of dimensions.

EXERCISE

If, in Theorem 2, besides the surfaces $f_1 = 0$, $f_2 = 0$, $\cdots$ we also have a system of points

$$(x_1,\ y_1,\ z_1,\ t_1),\quad (x_2,\ y_2,\ z_2,\ t_2),\ \cdots$$

and if ϕ is a polynomial not merely of the a's, b's, etc., but also of the coördinates of these points, prove that $\phi = 0$ expresses a necessary and sufficient condition that this system of surfaces and points have a geometric property when and only when ϕ is homogeneous in the a's alone, in the b's alone, etc., and also in $(x_1,\ y_1,\ z_1,\ t_1)$ alone, in $(x_2,\ y_2,\ z_2,\ t_2)$ alone, etc.

81. Homogeneous Invariants. From the developments of the last section it is clear that the only integral rational invariants which will be of importance in geometrical applications are those which are homogeneous in the coefficients of each of the ground-forms taken separately.* Such invariants we will speak of as *homogeneous invariants*. It will be found that all the invariants which we have met so far are of this kind.

An important relation between the weight and the various degrees connected with a homogeneous invariant is given by the following theorem :

THEOREM 1. *If we have a system of n-ary forms,*

$$(1)\qquad \begin{cases} f_1(x_1,\ \cdots\ x_n\ ;\ a_1,\ a_2,\ \cdots), \\ f_2(x_1,\ \cdots\ x_n\ ;\ b_1,\ b_2,\ \cdots), \\ \cdot\quad\cdot\quad\cdot\quad\cdot\quad\cdot\quad\cdot\quad\cdot \\ \cdot\quad\cdot\quad\cdot\quad\cdot\quad\cdot\quad\cdot\quad\cdot \end{cases}$$

of degrees m_1, m_2, $\cdots$ respectively, and if

$$I\,(a_1,\ a_2,\ \cdots\ ;\ b_1,\ b_2,\ \cdots\ ;\ \cdots)$$

---

* This statement must not be taken too literally. It is true if in the geometrical application in question we consider the variables as homogeneous coördinates and if we have to deal with the loci obtained by equating the ground-forms to zero. While this is the ordinary way in which we interpret invariants geometrically, other interpretations are possible. For instance, instead of interpreting the variables (x, y) as homogeneous coördinates on a line and equating the binary quadratic forms

$$f_1 \equiv a_1 x^2 + 2\, a_2 xy + a_3 y^2,$$
$$f_2 \equiv b_1 x^2 + 2\, b_2 xy + b_3 y^2,$$

to zero, thus getting two pairs of points on a line, we may interpret (x, y) as non-homogeneous coördinates in the plane, and consider the two conics $f_1 = 1$, $f_2 = 1$. With this interpretation, the vanishing of the invariant

$$a_1 a_3 - a_2^2 + b_1 b_3 - b_2^2,$$

which is not homogeneous in the a's alone or in the b's alone, has a geometric meaning.

is a homogeneous invariant of this system, of weight λ, *and of degree* α *in the a's, β in the b's, etc., then*

$$(2) \qquad m_1\alpha + m_2\beta + \cdots = n\lambda.$$

Subjecting the forms (1) to the linear transformation

$$(3) \qquad \begin{cases} x_1 = c_{11}x_1' + \cdots + c_{1n}x_n', \\ \cdot \quad \cdot \quad \cdot \quad \cdot \quad \cdot \quad \cdot \quad \cdot \quad \cdot \\ \cdot \quad \cdot \quad \cdot \quad \cdot \quad \cdot \quad \cdot \quad \cdot \quad \cdot \\ x_n = c_{n1}x_1' + \cdots + c_{nn}x_n', \end{cases}$$

whose determinant we will denote by c, we get

$$f_1(x_1', \cdots x_n' ; a_1', a_2', \cdots),$$
$$f_2(x_1', \cdots x_n' ; b_1', b_2', \cdots),$$
$$\cdot \quad \cdot \quad \cdot \quad \cdot \quad \cdot \quad \cdot \quad \cdot$$
$$\cdot \quad \cdot \quad \cdot \quad \cdot \quad \cdot \quad \cdot \quad \cdot$$

and, since by hypothesis I is an invariant of weight λ,

$$(4) \quad I(a_1', a_2', \cdots ; b_1', b_2', \cdots ; \cdots) \equiv c^\lambda I(a_1, a_2, \cdots ; b_1, b_2, \cdots ; \cdots).$$

Every a' is a homogeneous polynomial in the c_{ij}'s of degree m_1, every b' of degree m_2, etc. ; and since I is itself homogeneous of degree α in the a's, β in the b's, etc., we see that the left-hand side of (4) is a homogeneous polynomial of degree $m_1\alpha + m_2\beta + \cdots$ in the c_{ij}'s. Equating this to the degree of the right-hand side of (4) in the c_{ij}'s, which is evidently $n\lambda$, our theorem is proved.

An additional reason for the importance of these homogeneous invariants is that the non-homogeneous integral rational invariants can be built up from them, as is stated in the following theorem:

THEOREM 2. *If an integral rational invariant I of the system* (1) *be written in the form*

$$I \equiv I_1 + I_2 + \cdots + I_k$$

where each of the I_i's is a polynomial in the a's, b's, etc., which is homogeneous in the a's alone, and also in the b's alone, etc., and such that the sum of no two I_i's has this property, then each of the functions

$$I_1, I_2, \cdots I_k$$

is a homogeneous invariant of the system (1).

This theorem follows immediately from the definition of **an** invariant.   For from the identity,

$$I_1(a_1', a_2', \cdots ;\ b_1', b_2', \cdots ;\ \cdots) + \cdots + I_k(a_1', a_2', \cdots ;\ b_1', b_2', \cdots ;\ \cdots)$$
$$\equiv c^\lambda [\, I_1(a_1, a_2, \cdots ;\ b_1, b_2, \cdots ;\ \cdots)$$
$$+ \cdots + I_k(a_1, a_2, \cdots ;\ b_1, b_2, \cdots ;\ \cdots)\,],$$

we infer at once the identities,

$$I_1(a_1', a_2' \cdots ;\ b_1', b_2', \cdots ;\ \cdots) \equiv c^\lambda I_1(a_1, a_2, \cdots ;\ b_1, b_2, \cdots ;\ \cdots),$$
$$\cdots \cdots \cdots \cdots \cdots \cdots \cdots \cdots \cdots$$
$$I_k(a_1', a_2', \cdots ;\ b_1', b_2', \cdots ;\ \cdots) \equiv c^\lambda I_k(a_1, a_2, \cdots ;\ b_1, b_2, \cdots ;\ \cdots).$$

In the case of a single n-ary form, but in that case only, we have the theorem :

Theorem 3. *An integral rational invariant of a single n-ary form is always homogeneous.*

Let
$$f(x_1, \cdots x_n;\ a_1, a_2, \cdots)$$

be the ground-form, and let I be the invariant.   By Theorem 2 we may write
$$I \equiv I_1 + I_2 + \cdots + I_k$$

where $I_1, \cdots I_k$ are homogeneous invariants.   Let the degrees of these homogeneous invariants in the a's be $\alpha_1, \cdots \alpha_k$ respectively.   Their weights are all the same as the weight of I, which we will call λ. If, then, we call the degree of f, m, we have, by Theorem 1,

$$m\alpha_1 = n\lambda,\ m\alpha_2 = n\lambda,\ \cdots m\alpha_k = n\lambda,$$

from which, since $m > 0$, we infer

$$\alpha_1 = \alpha_2 = \cdots = \alpha_k.$$

That is, $I_1, \cdots I_k$ are of the same degree, and I is homogeneous.

Theorem 4. *If we have a system of n-ary forms $f_1, f_2, \cdots$ and a polynomial ϕ in their coefficients, the equation $\phi = 0$ gives a necessary and sufficient condition for a projective property of the system of loci in space of $n-1$ dimensions,*

$$f_1 = 0,\ \ f_2 = 0,\ \cdots\cdots,$$

when, and only when, ϕ is a homogeneous invariant of the system of forms f.

If ϕ is a homogeneous invariant, its vanishing gives a necessary and sufficient condition for a geometric property (cf. § 80), and this property must be a projective property since when we subject the loci to a non-singular collineation, ϕ is merely multiplied by a non-vanishing constant.

On the other hand let $\phi = 0$ be a necessary and sufficient condition for a projective property. In order to prove that ϕ is an invariant (it must be homogeneous by § 80) let $a_1, a_2, \cdots$ be the coefficients of f_1; $b_1, b_2, \cdots$ the coefficients of f_2, etc.; and suppose that the linear transformation,

$$(5) \qquad \begin{cases} x_1 = c_{11}x_1' + \cdots + c_{1n}x_n', \\ \quad \cdot \quad \cdot \quad \cdot \quad \cdot \quad \cdot \quad \cdot \quad \cdot \\ \quad \cdot \quad \cdot \quad \cdot \quad \cdot \quad \cdot \quad \cdot \quad \cdot \\ x_n = c_{n1}x_1' + \cdots + c_{nn}x_n', \end{cases}$$

carries over f_1 into f_1' with coefficients $a_1', a_2' \cdots$; f_2 into f_2' with coefficients $b_1', b_2', \cdots$; etc. The polynomial ϕ formed for the transformed forms is

$$\phi(a_1', a_2', \cdots; \ b_1', b_2', \cdots; \ \cdots),$$

and may, since the a''s, b''s, $\cdots$ are polynomials in the a's, b's, $\cdots$ and the c's, be itself regarded as a polynomial in the a's, b's, $\cdots$ and the c's. Looking at it from this point of view, let us resolve it into its irreducible factors,

$$(6) \quad \phi(a_1', a_2', \cdots; \ b_1', b_2', \cdots; \ \cdots) \equiv \phi_1(a_1, a_2, \cdots; \ b_1, b_2, \cdots; \ \cdots c_{11}, \cdots c_{nn})$$
$$\cdots\cdots \phi_k(a_1, a_2, \cdots; \ b_1, b_2, \cdots; \ \cdots c_{11}, \cdots c_{nn}).$$

It is clear that at least one of the factors on the right must contain the c's. Let ϕ_1 be such a factor, and let us arrange it as a polynomial in the c's whose coefficients are polynomials in the a's, b's, etc. Let

$$\psi(a_1, a_2, \cdots; \ b_1, b_2, \cdots; \ \cdots)$$

be one of these coefficients which is not identically zero and which is the coefficient of a term in which at least one of the c's has an exponent greater than zero. We can, now, give to the a's, b's, $\cdots$ values which we will denote by A's, B's, $\cdots$ such that neither ϕ nor ψ vanish; and consider a neighborhood N of the point

$$(A_1, A_2, \cdots; \ B_1, B_2, \cdots; \ \cdots)$$

throughout which

$$(7) \qquad \phi(a_1, a_2, \cdots; \ b_1, b_2, \cdots; \ \cdots) \neq 0,$$
$$(8) \qquad \psi(a_1, a_2, \cdots; \ b_1, b_2, \cdots; \ \cdots) \neq 0.$$

Let us now restrict the a's, b's, $\cdots$ to the neighborhood N and ask ourselves under what circumstances we can have $\phi_1 = 0$. If this equation is fulfilled, we see from (6) that ϕ vanishes for the transformed loci, while, by (7), it does not vanish for the original loci. Since, by hypothesis, the vanishing of ϕ gives a necessary and sufficient condition for a projective property, a transformation (5) which causes ϕ to vanish when it did not vanish before must be a singular transformation. That is, *if the a's, b's, $\cdots$ are in the neighborhood N, whenever ϕ_1 vanishes the determinant c of (5) vanishes.* Moreover, ϕ_1 does vanish for values of the a's, b's, $\cdots$ in N, for if we assign to the a's, b's, $\cdots$ any such values, ϕ_1 becomes a polynomial in the c_{ij}'s, which, by (8), is of at least the first degree, and therefore vanishes for suitably chosen values of the c_{ij}'s. We can therefore apply the theorem for more than three variables analogous to Theorem 8, § 76, and infer that ϕ_1 is a factor of the determinant c ; and consequently, since this determinant is irreducible (Theorem 1, § 61), that ϕ_1 is merely a constant multiple of c.

The reasoning we have just applied to ϕ_1 applies equally to any of the factors on the right of (6) which are of at least the first degree in the c_{ij}'s. Accordingly (6) reduces to the form

$$(9) \qquad \phi(a_1', a_2', \cdots ; b_1', b_2' \cdots ; \cdots) \equiv c^\lambda \chi(a_1, a_2, \cdots ; b_1, b_2, \cdots ; \cdots),$$

where χ no longer involves the c_{ij}'s. To determine this polynomial χ, let us assign to the c_{ij}'s the values 0, 1 which reduce (5) to the identical transformation. Then the a'''s, b'''s, $\cdots$ reduce to the a's, b's $\cdots$, while $c = 1$; so that from (9) we see that

$$\phi(a_1, a_2, \cdots ; b_1, b_2 \cdots ; \cdots) \equiv \chi(a_1, a_2, \cdots ; b_1, b_2, \cdots ; \cdots).$$

Substituting this value of χ in (9), we see that ϕ is really an invariant.

In order to avoid all misunderstanding, we state here explicitly that if we have two or more polynomials, ϕ_1, ϕ_2, $\cdots$ in the coefficients of the forms f_i, the equations $\phi_1 = \phi_2 = \cdots = 0$ may be a necessary and sufficient condition for a projective property of the loci $f_i = 0$, even though ϕ_1, ϕ_2, $\cdots$ are not invariants. For instance, a necessary and sufficient condition that the two lines

$$a_1 x_1 + a_2 x_2 + a_3 x_3 = 0,$$
$$b_1 x_1 + b_2 x_2 + b_3 x_3 = 0$$

coincide is the vanishing of the three two-rowed determinants of the matrix

$$\left\| \begin{array}{ccc} a_1 & a_2 & a_3 \\ b_1 & b_2 & b_3 \end{array} \right\|,$$

none of which is an invariant. Or, again, a necessary and sufficient condition that a quadric surface break up into two planes, distinct or coincident, is the vanishing of all the three-rowed determinants of its matrix, and these are not invariants. In this case we can also express the condition in question by the *identical* vanishing of a certain contravariant, namely, the adjoint of the quadratic form; and this — a projective relation expressed by the identical vanishing of a covariant or contravariant — is typical of what we shall usually have when a single equation $\phi = 0$ is not sufficient to express the condition. There are, however, cases where the condition is given by the vanishing of two or more invariants; cf. Exercise 6, § 90.

EXERCISES

1. Prove that if in Theorem 1 our system consists not merely of the ground-forms (1) but also of certain points

$$(y_1, \cdots y_n), \ (z_1, \cdots z_n), \cdots,$$

and we have not an invariant I but a covariant of weight λ, and of degree α in the a's, β in the b's, etc., η in the y's, ζ in the z's, etc., then

$$m_1\alpha + m_2\beta + \cdots = n\lambda + \eta + \zeta + \cdots.$$

2. Extend Theorem 2 to the case of covariants. Does Theorem 3 admit of such extension?

3. Extend Theorem 4 to the case of covariants.

4. Show that an integral rational invariant of a single binary form of odd degree must be of even degree.

5. Show that the weight of an integral rational invariant of a single binary form can never be smaller than the degree of the form.

6. Express the condition that (a) two lines, and (b) two planes coincide, in the form of the identical vanishing of a covariant or contravariant.

7. Prove that a polynomial in the coefficients of a system of n-ary forms which is homogeneous in the coefficients of each form taken by themselves, and which is unchanged when the forms are subjected to any linear transformation of determinant $+1$, is an invariant of the system of forms.

8. Generalize Exercise 7 to the case of covariants.

82. Resultants and Discriminants of Binary Forms.   If we interpret (x_1, x_2) as homogeneous coördinates in one dimension, the equations obtained by equating the two binary forms

$$f(x_1, x_2) \equiv a_0 x_1^n + a_1 x_1^{n-1} x_2 + \cdots + a_n x_2^n,$$
$$\phi(x_1, x_2) \equiv b_0 x_1^m + b_1 x_1^{m-1} x_2 + \cdots + b_m x_2^m$$

to zero represent sets of points on a line.   The points given by the equation $f = 0$ are the points at which the linear factors of f vanish, and the points corresponding to $\phi = 0$ are the points at which the linear factors of ϕ vanish.   Since two binary linear forms obviously vanish at the same point when, and only when, these linear forms are proportional, it follows that the loci of the two equations $f = 0$, $\phi = 0$ have a point in common when, and only when, f and ϕ have a common factor other than a constant.   Hence, by § 72, *a necessary and sufficient condition that the two loci $f = 0$, $\phi = 0$ have a point in common is that the resultant R of the binary forms f, ϕ vanish.*

The property of these two loci having a point in common is, however, a projective property.   Thus, by Theorem 4, § 81,

THEOREM 1.   *The resultant of two binary forms is a homogeneous invariant of this pair of forms.*

From the determinant form of R given in § 68 it is clear that R is of degree m in the a's and of degree n in the b's.   Hence by formula (2), § 81,
$$\lambda = mn.$$

THEOREM 2.   *The weight of the resultant of two binary forms of degrees m and n is mn.*

The following geometrical problem will lead us to an important invariant of a single binary form.

Let us resolve the form f, which we assume not to be identically zero, into its linear factors (cf. formula (4), § 65),

$$f(x_1, x_2) \equiv (\alpha_1'' x_1 - \alpha_1' x_2)(\alpha_2'' x_1 - \alpha_2' x_2) \cdots (\alpha_n'' x_1 - \alpha_n' x_2).$$

The equation $f = 0$ represents n distinct points provided no two of these linear factors are proportional to each other.   If, however, two of these factors are proportional, we say that f has a multiple linear factor, and in this case two or more of the n points represented by the equation $f = 0$ coincide.   Let us inquire under what conditions this will occur.

Form the partial derivatives :

$$(1)\begin{cases}\begin{aligned}\frac{\partial f}{\partial x_1} &\equiv \alpha_1''(\alpha_2'' x_1 - \alpha_2' x_2) \cdots (\alpha_n'' x_1 - \alpha_n' x_2)\\ &\quad + \alpha_2''(\alpha_1'' x_1 - \alpha_1' x_2)(\alpha_3'' x_1 - \alpha_3' x_2) \cdots (\alpha_n'' x_1 - \alpha_n' x_2)\\ &\quad + \cdots + \alpha_n''(\alpha_1'' x_1 - \alpha_1' x_2) \cdots (\alpha_{n-1}'' x_1 - \alpha_{n-1}' x_2),\\ \frac{\partial f}{\partial x_2} &\equiv -\alpha_1'(\alpha_2'' x_1 - \alpha_2' x_2) \cdots (\alpha_n'' x_1 - \alpha_n' x_2)\\ &\quad - \alpha_2'(\alpha_1'' x_1 - \alpha_1' x_2)(\alpha_3'' x_1 - \alpha_3' x_2) \cdots (\alpha_n'' x_1 - \alpha_n' x_2)\\ &\quad - \cdots - \alpha_n'(\alpha_1'' x_1 - \alpha_1' x_2) \cdots (\alpha_{n-1}'' x_1 - \alpha_{n-1}' x_2).\end{aligned}\end{cases}$$

From these formulæ we see that any multiple linear factor of f is a factor of both of these partial derivatives.

Conversely, if these partial derivatives have a common linear factor, it must be a factor of f on account of the formula,

$$x_1 \frac{\partial f}{\partial x_1} + x_2 \frac{\partial f}{\partial x_2} \equiv nf,\text{*}$$

a formula which follows immediately from the expressions,

$$(2)\begin{cases}\dfrac{\partial f}{\partial x_1} \equiv na_0 x_1^{n-1} + (n-1)a_1 x_1^{n-2} x_2 + \cdots + a_{n-1} x_2^{n-1},\\[2ex] \dfrac{\partial f}{\partial x_2} \equiv a_1 x_1^{n-1} + 2a_2 x_1^{n-2} x_2 + \cdots + na_n x_2^{n-1}.\end{cases}$$

But, by (1), no linear factor of f can be a factor of $\partial f/\partial x_1$ unless it is a multiple factor of f. Thus we have proved

THEOREM 3. *A necessary and sufficient condition that f have a multiple linear factor is that the resultant of $\partial f/\partial x_1$ and $\partial f/\partial x_2$ vanish.*

DEFINITION. *The resultant of $\partial f/\partial x_1$ and $\partial f/\partial x_2$ is called the discriminant of f.*

From (2) we see that the discriminant of f may be written as a determinant of order $2n - 2$ whose elements, so far as they are not zero, are numerical multiples of the coefficients $a_0, a_1, \cdots a_n$ of f. That is, this discriminant is a polynomial in the a's. Moreover, its vanishing gives a necessary and sufficient condition that the locus $f = 0$ have a projective property (namely, that two points of this locus coincide). Hence, by Theorem 4, § 81, this discriminant is a

---

* This is merely Euler's Theorem for Homogeneous Functions.

homogeneous invariant, whose degree and weight are readily determined.   Thus we get the theorem :

THEOREM 4.   *The discriminant of a binary form of the nth
degree is a homogeneous invariant of this form of degree $2(n-1)$ and
of weight $n(n-1)$.*

A slight modification in the definition of the discriminant is often
desirable.   Let us write the binary form f, not in the above form
where the coefficients are $a_0,\ a_1, \cdots a_n$, but, by the introduction of
binomial coefficients, in the form

$$f(x_1, x_2) \equiv a_0 x_1^n + n a_1 x_1^{n-1} x_2 + \frac{n(n-1)}{2\,!} a_2 x_1^{n-2} x_2^2 + \cdots + n a_{n-1} x_1 x_2^{n-1} + a_n x_2^n.$$

Then we may write

$$\frac{1}{n}\frac{\partial f}{\partial x_1} \equiv a_0 x_1^{n-1} + (n-1) a_1 x_1^{n-2} x_2 + \frac{(n-1)(n-2)}{2\,!} a_2 x_1^{n-3} x_2^2 + \cdots + a_{n-1} x_2^{n-1},$$

$$\frac{1}{n}\frac{\partial f}{\partial x_2} \equiv a_1 x_1^{n-1} + (n-1) a_2 x_1^{n-2} x_2 + \frac{(n-1)(n-2)}{2\,!} a_3 x_1^{n-3} x_2^2 + \cdots + a_n x_2^{n-1}.$$

We may then define the discriminant of f as the resultant of the two
binary forms just written.   We thus get for the discriminant a
polynomial in the a's which differs from the discriminant as above
defined only by a numerical factor, and for which Theorems 3 and 4
obviously still hold.   If this last definition be applied to the case of
a binary quadratic form, it will be seen that it leads us precisely to
what we called the discriminant of this quadratic form in the earlier
chapters of this book.

EXERCISES

1. Prove that the resultant of two binary forms of degrees n and m respectively is irreducible.

[SUGGESTION.   When $b_0 = 0$, R is equal to a_0 times the resultant of two binary
forms of degrees n and $m-1$ respectively.   Show that if this last resultant is irreducible, R is also irreducible, and use the method of induction, starting with the case
$n=1,\ m=1$.]

2. Prove by the methods of this chapter that the bordered determinants of
Chapter XII are invariants of weight 2.

3. The following account of Bézout's method of elimination is sometimes
given :

If f and ϕ are polynomials in x which are both of degree n, the expression

$$f(x)\phi(y) - \phi(x)f(y)$$

vanishes, independently of y, for a value of x for which both f and ϕ vanish, and is divisible by $x - y$, since it is zero for $x = y$. Hence

$$F(x, y) = \frac{f(x)\phi(y) - \phi(x)f(y)}{x - y}$$

is a polynomial of degree $(n-1)$ in x which vanishes for all values of y when x is a common root of f and ϕ. Arranging F according to the powers of y, we have the expression

$$
\begin{aligned}
F = \ & c_{00} + c_{01}x + c_{02}x^2 + \cdots + c_{0,\,n-1}x^{n-1} \\
& + y(c_{10} + c_{11}x + c_{12}x^2 + \cdots + c_{1,\,n-1}x^{n-1}) \\
& + y^2(c_{20} + c_{21}x + c_{22}x^2 + \cdots + c_{2,\,n-1}x^{n-1}) \\
& + \ \cdots\ \cdots\ \cdots\ \cdots\ \cdots \\
& + y^{n-1}(c_{n-1,\,0} + c_{n-1,\,1}x + c_{n-1,\,2}x^2 + \cdots + c_{n-1,\,n-1}x^{n-1}).
\end{aligned}
$$

If this function is to vanish independently of y, the coefficient of each power of y must be zero. This gives n equations between which we can eliminate the n quantities, $1\ x, x^2, \cdots x^{n-1}$, obtaining the resultant in the form of the determinant,

$$
R = \begin{vmatrix}
c_{00} & c_{01} \cdots & c_{0,\,n-1} \\
c_{10} & c_{11} \cdots & c_{1,\,n-1} \\
\cdot & \cdot \quad \cdot \quad \cdot & \cdot \\
\cdot & \cdot \quad \cdot \quad \cdot & \cdot \\
c_{n-1,\,0} & c_{n-1,\,1} \cdots & c_{n-1,\,n-1}
\end{vmatrix} = 0 .
$$

With the help of the auxiliary function F we have, in this case, reduced the resultant to a determinant of the nth order, while that obtained by the method of Sylvester was of order $2n$.

Criticise this treatment and make it rigorous, applying it, in particular, to the case of homogeneous variables.

4. If f and ϕ are polynomials in (x, y) of degrees n and m respectively and are relatively prime, prove that the curves $f = 0$, $\phi = 0$ cannot have more than mn points of intersection.

[SUGGESTION. Show first that the coördinate axes can be turned in such a way that no two points of intersection have the same abscissa, and that the equations of the two curves are of degrees n and m respectively, after the transformation, *in y alone*. Then eliminate y between the two equations by Sylvester's dyalitic method.]

5. Prove that every integral rational invariant of the binary cubic is a constant multiple of a power of the discriminant.

[SUGGESTION. Show that if the discriminant is not zero, every binary cubic can be reduced by a non-singular linear transformation to the normal form $x_1^3 - x_2^3$. Then as in §48.]

CHAPTER XVIII

SYMMETRIC POLYNOMIALS

83. Fundamental Conceptions. Σ and S Functions.

DEFINITION 1. *A polynomial $F(x_1, \cdots x_n)$ is said to be* **symmetric** *if it is unchanged by any interchange of the variables $(x_1, \cdots x_n)$.*

It is not necessary, however, to consider all the possible permutations of the variables in order to show that a polynomial is symmetric. If we can show that it is unchanged by the interchange of *every pair* of the variables, this is sufficient, for any arrangement $(x_a, x_b, \cdots x_k)$ may be obtained from $(x_1, x_2, \cdots x_n)$ by interchanging the x's in pairs. Thus, if $a \neq 1$, interchange x_a with x_1; then interchange the second letter in the arrangement thus obtained with x_2; and so on. Hence we have the following theorem:

THEOREM 1. *A necessary and sufficient condition for a polynomial to be symmetric is that it be unchanged by every interchange of two variables.*

A special class of symmetric polynomials of much importance are the Σ-functions, defined as follows:

DEFINITION 2. Σ *before any term means the sum of this term and of all the similar ones obtained from it by interchanging the subscripts.*

Thus, for example,

$$\Sigma x_1^a \equiv x_1^a + x_2^a + \cdots + x_n^a,$$

$$\begin{aligned}
\Sigma x_1^a x_2^\beta \equiv\ & x_1^a x_2^\beta + x_1^a x_3^\beta + \cdots + x_1^a x_n^\beta \\
& + x_2^a x_1^\beta + x_2^a x_3^\beta + \cdots + x_2^a x_n^\beta \qquad (a \neq \beta) \\
& + \quad \cdot \quad \cdot \quad \cdot \quad \cdot \quad \cdot \quad \cdot \\
& + x_n^a x_1^\beta + x_n^a x_2^\beta + \cdots + x_n^a x_{n-1}^\beta,
\end{aligned}$$

$$\begin{aligned}
\Sigma x_1^a x_2^a \equiv\ & x_1^a x_2^a + x_1^a x_3^a + \cdots + x_1^a x_n^a \\
& + x_2^a x_3^a + \cdots + x_2^a x_n^a \\
& + \quad \cdot \quad \cdot \quad \cdot \\
& + x_{n-1}^a x_n^a.
\end{aligned}$$

It is clearly immaterial in what order the exponents α, β, $\cdots$ are written. Thus, $\Sigma x_1^a x_2^\beta x_3^\gamma \equiv \Sigma x_1^\beta x_2^\gamma x_3^a$.

If we consider any term of a symmetric polynomial, it is evident that the polynomial must contain all the terms obtained from this one by interchanging the x's. This aggregate of terms is merely a constant multiple of one of the Σ's just defined. In the same way it is clear that all the other terms of the symmetric polynomial must arrange themselves in groups each of which is a constant multiple of a Σ. That is,

THEOREM 2. *Every symmetric polynomial is a linear combination with constant coefficients of a certain number of Σ's.*

Among these Σ's the simplest are the sums of powers of the x's. For the sake of brevity the notation is used:

$$S_k \equiv \Sigma x_1^k \equiv x_1^k + x_2^k + \cdots + x_n^k \qquad (k = 1, 2, \cdots).$$

It is sometimes convenient to write $S_0 = n$.

THEOREM 3. *Any symmetric polynomial in the x's can be expressed as a polynomial in a certain number of the S's.*

Since every symmetric polynomial is a linear combination of a certain number of Σ's, in order to prove our theorem we have only to show that every Σ can be expressed as a polynomial in the S's. Now

$$S_\alpha \equiv x_1^\alpha + x_2^\alpha + \cdots + x_n^\alpha,$$
$$S_\beta \equiv x_1^\beta + x_2^\beta + \cdots + x_n^\beta.$$

Hence, if $\alpha \neq \beta$,

$$S_\alpha S_\beta \equiv x_1^{\alpha+\beta} + x_2^{\alpha+\beta} + \cdots + x_n^{\alpha+\beta} + x_1^\alpha x_2^\beta + x_1^\alpha x_3^\beta + \cdots$$
$$\equiv S_{\alpha+\beta} + \Sigma x_1^\alpha x_2^\beta.$$

From this we get the formula:

$$(1) \qquad \Sigma x_1^\alpha x_2^\beta = S_\alpha S_\beta - S_{\alpha+\beta} \qquad (\alpha \neq \beta).$$

If $\alpha = \beta$, we have

$$S_\alpha^2 \equiv x_1^{2\alpha} + x_2^{2\alpha} + \cdots + x_n^{2\alpha} + 2x_1^\alpha x_2^\alpha + 2x_1^\alpha x_3^\alpha + \cdots$$
$$\equiv S_{2\alpha} + 2\Sigma x_1^\alpha x_2^\alpha.$$

Hence

$$(2) \qquad \Sigma x_1^\alpha x_2^\alpha \equiv \tfrac{1}{2}(S_\alpha^2 - S_{2\alpha}).$$

Similarly, by multiplying $\Sigma x_1^\alpha x_2^\beta$ by S_γ, we get the following formulæ where the three integers α, β, γ are supposed to be distinct:

$$(3) \qquad \Sigma x_1^\alpha x_2^\beta x_3^\gamma \equiv S_\alpha S_\beta S_\gamma - S_{\alpha+\beta} S_\gamma - S_{\alpha+\gamma} S_\beta - S_{\beta+\gamma} S_\alpha + 2 S_{\alpha+\beta+\gamma}$$

$$(4) \qquad \Sigma x_1^\alpha x_2^\alpha x_3^\gamma \equiv \tfrac{1}{2}(S_\alpha^2 S_\gamma - S_{2\alpha} S_\gamma - 2 S_{\alpha+\gamma} S_\alpha + 2 S_{2\alpha+\gamma}),$$

$$(5) \qquad \Sigma x_1^\alpha x_2^\alpha x_3^\alpha \equiv \tfrac{1}{6}(S_\alpha^3 - 3 S_{2\alpha} S_\alpha + 2 S_{3\alpha}).$$

The proof indicated in these two special cases may be extended to the general case as follows:

If we multiply together the two symmetric polynomials

$$(6) \qquad \Sigma x_1^a x_2^\beta \cdots x_k^\kappa, \qquad S_\lambda \equiv \Sigma x_1^\lambda, \qquad (k < n)$$

we get terms of various sorts which are readily seen to be all contained in one or the other of the following polynomials, each of these polynomials being actually represented:

$$(7) \quad \Sigma x_1^{a+\lambda} x_2^\beta \cdots x_k^\kappa, \; \Sigma x_1^a x_2^{\beta+\lambda} \cdots x_k^\kappa, \; \cdots\cdots \; \Sigma x_1^a x_2^\beta \cdots x_k^{\kappa+\lambda}, \; \Sigma x_1^a x_2^\beta \cdots x_k^\kappa x_{k+1}^\lambda.$$

Consequently, since the product of the two polynomials (6) is symmetric, it must have the form

$$c_1 \Sigma x_1^{a+\lambda} x_2^\kappa \cdots x_k^\kappa + c_2 \Sigma x_1^a x_2^{\beta+\lambda} \cdots x_k^\kappa + \cdots\cdots + c_k \Sigma x_1^a x_2^\beta \cdots x_k^{\kappa+\lambda}$$
$$+ c_{k+1} \Sigma x_1^a x_2^\beta \cdots x_k^\kappa x_{k+1}^\lambda,$$

where $c_1, \cdots c_{k+1}$ are positive integers.

Transposing, we may write

$$\Sigma x_1^a x_2^\beta \cdots x_{k+1}^\lambda \equiv \frac{1}{c_{k+1}}\Big[\Sigma x_1^a x_2^\beta \cdots x_k^\kappa \cdot \Sigma x_1^\lambda - c_1 \Sigma x_1^{a+\lambda} x_2^\beta \cdots x_k^\kappa$$
$$- c_2 \Sigma x_1^a x_2^{\beta+\lambda} \cdots x_k^\kappa - \cdots\cdots - c_k \Sigma x_1^a x_2^\beta \cdots x_k^{\kappa+\lambda} \Big].$$

Hence, if our theorem is true for $\Sigma x_1^a \cdots x_k^\kappa$, it is also true for $\Sigma x_1^a \cdots x_{k+1}^\lambda$. But we know it is true for $k = 1$ (by definition of the S's), hence it is true for $k = 2$, hence for $k = 3$, and so on. Thus our theorem is completely proved.

84. Elementary Symmetric Functions. The notation $\Sigma x_1^a x_2^\beta \cdots x_n^\nu$ may be used to represent *any* Σ in n variables. If $\beta = \gamma = \cdots = \nu = 0$, this becomes Σx_1^a or S_a; if $\gamma = \cdots = \nu = 0$, it becomes $\Sigma x_1^a x_2^\beta$; and so on.

Let us now consider $\Sigma x_1^a x_2^\beta \cdots x_n^\nu$ where $\alpha, \beta, \cdots \nu$, are all 0 or 1. The following n cases arise:

$$\begin{array}{lll} \alpha = 1, & \beta = \gamma = \cdots = \nu = 0, & \Sigma x_1, \\ \alpha = \beta = 1, & \gamma = \cdots = \nu = 0, & \Sigma x_1 x_2, \\ \cdots & \cdots & \cdots \\ \alpha = \beta = \cdots = \mu = 1, & \nu = 0, & \Sigma x_1 x_2 \cdots x_{n-1}, \\ \alpha = \beta = \cdots = \nu = 1, & & x_1 x_2 \cdots x_n. \end{array}$$

The extreme case $\alpha = \beta = \cdots = \nu = 0$ is of no interest. We will represent these n symmetric polynomials by $p_1, p_2, \cdots p_n$, respectively. They are called the *elementary symmetric functions.*

THEOREM 1. *Any symmetric polynomial in the x's may be expressed as a polynomial in the p's.*

Since any symmetric polynomial in the x's may be expressed as a polynomial in the S's, it is sufficient to show that every S may be expressed as a polynomial in the p's.

Let us introduce a new variable x and consider the polynomial

$$f(x;\, x_1, x_2, \cdots x_n) \equiv (x - x_1)(x - x_2) \cdots (x - x_n)$$
$$\equiv x^n - p_1 x^{n-1} + p_2 x^{n-2} - \cdots + (-1)^n p_n.$$

Using the factored form of f, we may write

$$\frac{\partial f}{\partial x} \equiv \frac{f}{x - x_1} + \frac{f}{x - x_2} + \cdots + \frac{f}{x - x_n}.$$

Since f vanishes identically when $x = x_i$, we may write

$$f \equiv (x^n - x_i^n) - p_1(x^{n-1} - x_i^{n-1}) + \cdots.$$

Accordingly,

$$\frac{f}{x - x_i} \equiv x^{n-1} + (x_i - p_1)x^{n-2} + (x_i^2 - p_1 x_i + p_2)x^{n-3} + \cdots,$$

$$\frac{\partial f}{\partial x} \equiv nx^{n-1} + (S_1 - np_1)x^{n-2} + (S_2 - p_1 S_1 + np_2)x^{n-3} + \cdots.$$

On the other hand, we have

$$\frac{\partial f}{\partial x} \equiv nx^{n-1} - (n-1)p_1 x^{n-2} + (n-2)p_2 x^{n-3} - \cdots.$$

Hence, equating the coefficients of like powers of x in these two expressions, we have

$$\begin{cases} S_1 - np_1 \equiv -(n-1)p_1, \\ S_2 - p_1 S_1 + np_2 \equiv (n-2)p_2, \\ \cdot \quad \cdot \quad \cdot \quad \cdot \quad \cdot \quad \cdot \quad \cdot \quad \cdot \quad \cdot \quad \cdot \\ S_{n-1} - p_1 S_{n-2} + p_2 S_{n-3} - \cdots + (-1)^{n-1} np_{n-1} \equiv (-1)^{n-1} p_{n-1}, \end{cases}$$

or

$$(1) \begin{cases} S_1 - p_1 \equiv 0, \\ S_2 - p_1 S_1 + 2p_2 \equiv 0, \\ \cdot \quad \cdot \quad \cdot \quad \cdot \quad \cdot \quad \cdot \quad \cdot \quad \cdot \\ S_{n-1} - p_1 S_{n-2} + p_2 S_{n-3} - \cdots + (-1)^{n-1}(n-1)p_{n-1} \equiv 0. \end{cases}$$

Now consider the identities

$$x_i^n - p_1 x_i^{n-1} + p_2 x_i^{n-2} - \cdots + (-1)^n p_n \equiv 0 \qquad (i = 1, 2, \cdots n).$$

Multiplying these identities by $x_1^{k-n}, \cdots x_n^{k-n}$ respectively and adding the results, we have

$$(2) \qquad S_k - p_1 S_{k-1} + p_2 S_{k-2} - \cdots + (-1)^n p_n S_{k-n} \equiv 0 \qquad (k = n,\, n+1,\, \cdots).$$

Formulæ (1) and (2) are known as *Newton's Formulæ*. By means of them we can compute in succession the values of $S_1,\, S_2,\, \cdots$ as polynomials in the p's:

$$(3) \qquad \begin{cases} S_1 \equiv p_1, \\ S_2 \equiv p_1^2 - 2p_2, \\ S_3 \equiv p_1^3 - 3p_1 p_2 + 3p_3, \\ \quad \cdot \quad \cdot \quad \cdot \quad \cdot \quad \cdot \quad \cdot \quad \cdot \\ \quad \cdot \quad \cdot \quad \cdot \quad \cdot \quad \cdot \quad \cdot \quad \cdot \end{cases}$$

Thus our theorem is proved.

It will be noted that Newton's formulæ (1) cannot be obtained from (2) by giving to k values less than n. The necessity for two different sets of formulæ may, however, be avoided by introducing the notation

$$p_{n+1} \equiv p_{n+2} \equiv \cdots \equiv 0.$$

Then all of Newton's formulæ may be included in the following form:

$$(4) \qquad S_k - p_1 S_{k-1} + \cdots + (-1)^{k-1} p_{k-1} S_1 + (-1)^k k p_k \equiv 0 \qquad (k = 1,\, 2,\, \cdots).$$

Using this notation, we see that the explicit formulæ (3) for expressing the S's in terms of the p's are wholly independent of the number n of the x's.

Since the formulæ referred to in the last section for expressing the Σ's in terms of the S's are also independent of n, we have established

THEOREM 2. *If we introduce the notation* $p_{n+1} \equiv p_{n+2} \equiv \cdots \equiv 0$ *and use Newton's Formulæ in the form* (4), *the formula for expressing any* Σ *as a polynomial in the* p's *is independent of the number* n *of the* x's.

When we have k polynomials in n variables

$$f_1(x_1, \cdots x_n),\, f_2(x_1, \cdots x_n),\, \cdots f_k(x_1, \cdots x_n),$$

we say that there exists a *rational relation* between them when, and only when, a polynomial in k variables

$$F(z_1, \cdots z_k)$$

exists which is not identically zero, but which becomes identically zero as a polynomial in the x's when each z is replaced by the corresponding f,

$$F(f_1, \cdots f_k) \equiv 0.$$

THEOREM 3. *There exists no rational relation between the elementary symmetric functions in n variables $p_1, \cdots p_n$.*

For let $F(z_1, \cdots z_n)$ be any polynomial in n variables which is not identically zero, and let $(a_1, \cdots a_n)$ be a point at which this polynomial does not vanish. Determine $(x_1, \cdots x_n)$ as the roots of the equation

$$x^n - a_1 x^{n-1} + a_2 x^{n-2} - \cdots + (-1)^n a_n = 0.$$

For these values of the x's, the p's have the values $a_1, \cdots a_n$, and therefore $F(p_1, \cdots p_n)$ does not vanish for these x's, and is consequently not identically zero as a polynomial in the x_i's. Thus our theorem is proved.

COROLLARY. *There is only one way in which a symmetric polynomial in $(x_1, \cdots x_n)$ can be expressed as a polynomial in the elementary symmetric functions $p_1, \cdots p_n$.*

For if f is a symmetric polynomial, and if we had two expressions for it,

$$f(x_1, \cdots x_n) \equiv \phi_1(p_1, \cdots p_n),$$
$$f(x_1, \cdots x_n) \equiv \phi_2(p_1, \cdots p_n),$$

then by subtracting these identities from one another we should have as an identity in the x's,

$$\phi_1(p_1, \cdots p_n) - \phi_2(p_1, \cdots p_n) \equiv 0.$$

This, however, would give us a rational relation between the p's, unless

$$\phi_1(z_1, \cdots z_n) \equiv \phi_2(z_1, \cdots z_n).$$

Thus we see that the two expressions for f are really the same.

EXERCISES

1. Obtain the expressions for the following symmetric polynomials in terms of the elementary symmetric functions:

$$\Sigma x_1^2 x_2, \qquad\qquad \Sigma x_1^2 x_2^2, \qquad\qquad \Sigma x_1^3 x_2^2 x_3.$$

2. Prove that every symmetric polynomial in $(x_1, \cdots x_n)$ can be expressed in one, and only one, way as a polynomial in $S_1, \cdots S_n$.

85. The Weights and Degrees of Symmetric Polynomials.

We will attach to each of the elementary symmetric functions p_i a weight equal to its subscript, cf. § 79.

THEOREM 1. *A homogeneous symmetric polynomial of degree m in the x's, when expressed in terms of the p's, is isobaric of weight m.*

Let

$$(1) \qquad f(x_1, x_2, \cdots x_n) \equiv \phi(p_1, p_2, \cdots p_n)$$

be such a polynomial. Since p_1 is a homogeneous polynomial of the first degree in the x's, p_2 of the second, etc., any term of ϕ, when written in the x's, must be a homogeneous polynomial of degree equal to the original weight of the term. Thus, for example, the term $6\,p_1^2 p_2 p_3^3$ whose weight is 13, when written in the x's will be a homogeneous polynomial of degree 13. Accordingly an isobaric group of terms when expressed in terms of the x's will, since by Theorem 3, § 84, it cannot reduce identically to zero, be homogeneous of the same degree as its original weight. If then ϕ were not isobaric, f would not be homogeneous, and our theorem is proved.

COROLLARY. *If f is non-homogeneous and of the mth degree, ϕ is non-isobaric and of weight m.*

THEOREM 2. *A symmetric polynomial in $(x_1, \cdots x_n)$, when written in terms of the elementary symmetric functions $p_1, \cdots p_n$, will be of the same degree in the p's as it was at first in **any one** of the x's.*

Let f be the symmetric polynomial, and write

$$f(x_1, x_2, \cdots x_n) \equiv \phi(p_1, p_2, \cdots p_n),$$

and suppose f is of degree m in x_1 (and therefore, on account of the symmetry, in any one of the x's), and that ϕ is of degree μ in the p's. We wish to prove that $m = \mu$. Since the p's are of the first degree in x_1, it is clear that $m \leqq \mu$.

If ϕ is non-homogeneous, we can break it up into the sum of a number of homogeneous polynomials by grouping together all the terms of like degree. Each of these homogeneous polynomials in the p's can be expressed (by substituting for the p's their values in terms of the x's) as a symmetric polynomial in the x's. If our theorem were established in the case in which the polynomial in the p's is homogeneous, its truth in the general case would then follow at once.

Let us then assume that ϕ is a homogeneous polynomial. The theorem is obviously true when $n = 1$, since then $p_1 \equiv -x_1$. It will therefore be completely proved by the method of mathematical induction if, assuming it to hold when the number of x's is $1, 2, \cdots n-1$, we can **prove** that it holds when the number of x's is n.

For this purpose let us first assume that p_n is not a factor of every term of ϕ. Then $\phi(p_1, \cdots p_{n-1}, 0)$ is not identically zero but is still a homogeneous polynomial of degree μ in $(p_1, \cdots p_{n-1})$. Now let $x_n = 0$. This makes $p_n = 0$, and gives the identity

$$(2) \qquad f(x_1, \cdots x_{n-1}, 0) \equiv \phi(p_1' \cdots p_{n-1}', 0),$$

where $p_1', \cdots p_{n-1}'$ are the elementary symmetric functions of $(x_1, \cdots x_{n-1})$, and $f(x_1, \cdots x_{n-1}, 0)$ is a symmetric polynomial of degree m_1 in x_1, where $m_1 \leq m$. From the assumption that our theorem holds when the number of x's is $n-1$, we infer from (2) that $\mu = m_1 \leq m$; and since we saw above that μ cannot be less than m, we infer that $\mu = m$, as was to be proved.

There remains merely the case to be considered in which p_n is a factor of every term of ϕ. Let p_n^k be the highest power of p_n which occurs as a factor in ϕ. Then

$$\phi(p_1, \cdots p_n) \equiv p_n^k \phi_1(p_1, \cdots p_n),$$

where ϕ_1 is a polynomial of degree $\mu - k$. Putting in for the p's their values in terms of the x's, we get

$$(3) \qquad f(x_1, \cdots x_n) \equiv x_1^k x_2^k \cdots x_n^k f_1(x_1, \cdots x_n),$$

where

$$(4) \qquad f_1(x_1, \cdots x_n) \equiv \phi_1(p_1, \cdots p_n).$$

From (3) we see that f_1 is of degree $m - k$ in x_1, and from (4), since ϕ_1 does not contain p_n as a factor, that the degrees of f_1 in x_1, and of ϕ_1 in the p's are equal,

$$m - k = \mu - k.$$

From which we see that $m = \mu$, as was to be proved.

The two theorems of this section are not only of theoretical importance, they may also be put to the direct practical use of facilitating the computation of the values of symmetric polynomials in terms of the p's.

In order to illustrate this, let us consider the symmetric function

$$f(x_1, \cdots x_n) \equiv \Sigma x_1^2 x_2 x_3.$$

Since f is homogeneous of the fourth degree in the x's, it will, by Theorem 1, be isobaric of weight 4 in the p's. Since it is of the second degree in x_1, it will, by Theorem 2, be of the second degree in the p's. Hence

$$(5) \qquad \Sigma x_1^2 x_2 x_3 \equiv A p_1 p_3 + B p_2^2 + C p_4,$$

where A, B, and C are independent of the number n (Theorem 2, § 84), and may be determined by the ordinary method of undetermined coefficients.

Take $n = 3$, so that $p_4 = 0$. Letting $x_1 = 0$, $x_2 = x_3 = 1$, we have $p_1 = 2$, $p_2 = 1$, $p_3 = 0$. Substituting these values in (5), we find $B = 0$.

Letting $x_1 = -1$, $x_2 = x_3 = 1$, we have $p_1 = 1$, $p_2 = -1$, $p_3 = -1$, which gives $A = 1$.

Now let $\qquad n = 4$, $x_1 = x_2 = x_3 = x_4 = 1$.

From this we find $\quad p_1 = 4$, $p_2 = 6$, $p_3 = 4$, $p_4 = 1$.

Substituting this in (5) gives $C = -4$. Hence

$$\Sigma x_1^2 x_2 x_3 \equiv p_1 p_3 - 4 p_4.$$

EXERCISES

1. The symmetric function

$$f(x_1, \cdots x_n) \equiv \Sigma x_1^2 x_2 x_3 + \Sigma x_1^2 x_2^2 + \Sigma x_1 x_2 x_3 x_4$$

is homogeneous of the fourth degree in the x's, and is of the second degree in x_1; hence, when written in terms of the p's, it will have the same form, $A p_1 p_3 + B p_2^2 + C p_4$, as the above example. Compute the values of A, B, and C.

2. If $f(x_1, x_2, x_3) \equiv (x_1 - x_2)^2 (x_1 - x_3)^2 (x_2 - x_3)^2$, show that

$$f(x_1, x_2, x_3) \equiv -27 p_3^2 - 4 p_2^3 + 18 p_1 p_2 p_3 - 4 p_1^3 p_3 + p_1^2 p_2^2.$$

86. The Resultant and the Discriminant of Two Polynomials in One Variable. Let

$$f(x) \equiv x^n + a_1 x^{n-1} + a_2 x^{n-2} + \cdots + a_n$$

$$\equiv (x - \alpha_1)(x - \alpha_2) \cdots (x - \alpha_n),$$

$$\phi(x) \equiv x^m + b_1 x^{m-1} + b_2 x^{m-2} + \cdots + b_m$$

$$\equiv (x - \beta_1)(x - \beta_2) \cdots (x - \beta_m),$$

be two polynomials in x, and consider the product of the mn factors

$$(1) \qquad \begin{cases} (\alpha_1 - \beta_1)(\alpha_1 - \beta_2) \cdots (\alpha_1 - \beta_m) \\ (\alpha_2 - \beta_1)(\alpha_2 - \beta_2) \cdots (\alpha_2 - \beta_m) \\ \qquad \cdot \quad \cdot \quad \cdot \quad \cdot \quad \cdot \quad \cdot \\ (\alpha_n - \beta_1)(\alpha_n - \beta_2) \cdots (\alpha_n - \beta_m). \end{cases}$$

This product vanishes when, and only when, at least one of the α's is equal to one of the β's. Its vanishing therefore gives a necessary and sufficient condition that f and ϕ have a common factor. Moreover, the product (1), being a symmetrical polynomial in the α's and also in the β's, can be expressed as a polynomial in the elementary symmetric functions of the α's and β's, and therefore as a polynomial in the a's and b's. This will be still more evident if we notice that the product (1) may be written

$$\phi(\alpha_1)\,\phi(\alpha_2)\,\cdots\,\phi(\alpha_n).$$

In this form it is a symmetric polynomial in the α's whose coefficients are polynomials in the b's, and it remains merely to bring in the a's in place of the α's.

We thus see that the product (1) may be expressed as a polynomial $F(a_1, \cdots a_n;\ b_1, \cdots b_m)$ in the a's and b's whose vanishing gives a necessary and sufficient condition that f and ϕ have a common factor. In §68 we also found a polynomial in the a's and b's whose vanishing gives a necessary and sufficient condition that f and ϕ have a common factor, namely the resultant R of f and ϕ.

We will now identify these two polynomials by means of the following theorem:

THEOREM 1. *The product* (1) *differs from the resultant R of f and ϕ only by a constant factor, and the resultant is an irreducible polynomial in the a's and b's.*

In order to prove this theorem we will first show that this product (1), which we will call $F(a_1, \cdots a_n;\ b_1, \cdots b_m)$, is irreducible. This may be done as follows: Suppose F is reducible, and let
$$F(a_1, \cdots a_n;\ b_1, \cdots b_m) \equiv F_1(a_1, \cdots a_n;\ b_1, \cdots b_m)\, F_2(a_1, \cdots a_n;\ b_1, \cdots b_m),$$
where F_1 and F_2 are polynomials neither of which is a constant. Then, since the a's and b's are symmetric polynomials in the α's and β's, F_1 and F_2 may be expressed as symmetric polynomials ϕ_1 and ϕ_2 in the α's and β's, and we may write

$$\phi_1(\alpha_1, \cdots \alpha_n;\ \beta_1, \cdots \beta_m)\,\phi_2(\alpha_1, \cdots \alpha_n;\ \beta_1, \cdots \beta_m)$$

$$\equiv \begin{cases} (\alpha_1 - \beta_1)\,(\alpha_1 - \beta_2)\cdots(\alpha_1 - \beta_m) \\ (\alpha_2 - \beta_1)\,(\alpha_2 - \beta_2)\cdots(\alpha_2 - \beta_m) \\ \quad\cdot\quad\cdot\quad\cdot\quad\cdot\quad\cdot\quad\cdot\quad\cdot\quad\cdot \\ (\alpha_n - \beta_1)\,(\alpha_n - \beta_2)\cdots(\alpha_n - \beta_m). \end{cases}$$

The factors on the right-hand side of this identity being irreducible, we see that ϕ_1 must be composed of some of these binomial factors and ϕ_2 of the others. This, however, is impossible, since neither ϕ_1 nor ϕ_2 would be symmetric. Hence F is irreducible.

Now, since $F = 0$ is a necessary and sufficient condition for $f(x)$ and $\phi(x)$ to have a common factor, and $R = 0$ is the same, any set of values of the a's and b's which make $F = 0$ will also make $R = 0$. Hence by the theorem for $n + m$ variables analogous to Theorem 7, § 76, F is a factor of R. Also, since F is a symmetric polynomial in the α's and β's of degree m in each of the α's and n in each of the β's, by Theorem 2, § 85, it must be of degree m in the a's and n in the b's. But R is of degree not greater than m in the a's and n in the b's, as is at once obvious from a glance at the determinant in § 68. Hence F, being a factor of R, and of degree not lower than R, can differ from it only by a constant factor. Thus our theorem is proved.

Let us turn now to the question: Under what conditions does the polynomial $f(x)$ have a multiple linear factor? Using the same notation as above, we see that the vanishing of the product

$$\left.\begin{array}{c} (\alpha_1 - \alpha_2)(\alpha_1 - \alpha_3) \cdots (\alpha_1 - \alpha_n) \\ (\alpha_2 - \alpha_3) \cdots (\alpha_2 - \alpha_n) \\ \cdot \quad \cdot \quad \cdot \quad \cdot \quad \cdot \quad \cdot \\ (\alpha_{n-1} - \alpha_n) \end{array}\right\} \equiv P(\alpha_1, \cdots \alpha_n)$$

is a necessary and sufficient condition for this. P is not symmetric in the α's, since an interchange of two subscripts changes P into $-P$. If, however, we consider P^2 in place of P, we have a symmetric polynominal, which can therefore be expressed as a polynomial in the a's,

$$[P(\alpha_1, \cdots \alpha_n)]^2 \equiv F(a_1, \cdots a_n).$$

Moreover, $F = 0$ is also a necessary and sufficient condition that $f(x)$ have a multiple linear factor.

On the other hand, it is easily seen that $f(x)$ has a multiple linear factor when and only when $f(x)$ and $f'(x)$ have a common linear factor. A necessary and sufficient condition for $f(x)$ to have a multiple linear factor is therefore the vanishing of the resultant of $f(x)$ and $f'(x)$. This resultant we will call the discriminant Δ of $f(x)$. It is obviously a polynomial in the coefficients of f.

Theorem 2. *The polynomials F and Δ differ only by a constant factor, and are irreducible.*

The proof of this theorem is similar to the proof of Theorem 1, and is left to the reader.

EXERCISES

1. Compute by the use of symmetric functions the product (1) for the two polynomials

$$x^2 + a_1 x + a_2,$$
$$x^2 + b_1 x + b_2,$$

and compare the result with the resultant obtained in determinant form.

2. Verify Theorem 2 by comparing the result of Exercise 2, § 85, with the discriminant in determinant form of the polynomial

$$x^3 + a_1 x^2 + a_2 x + a_3.$$

CHAPTER XIX

POLYNOMIALS SYMMETRIC IN PAIRS OF VARIABLES

87. Fundamental Conceptions. Σ and S Functions. The variables $(x_1, \cdots x_n)$ which we used in the last chapter may be regarded, if we wish, not as the coördinates of a point in space of n dimensions, but rather as the coördinates of n points on a line. In fact this is the interpretation which is naturally suggested to us by the ordinary applications of the theory of symmetric functions (cf. §86). Looked at from this point of view, it is natural to generalize the conception of symmetric functions by considering n points in a plane,

$$(1) \qquad (x_1,\, y_1),\, (x_2,\, y_2),\, \cdots (x_n,\, y_n).$$

DEFINITION. *A polynomial,*

$$F(x_1,\, y_1\,;\, x_2,\, y_2\,;\, \cdots x_n,\, y_n)$$

in the coördinates of the points (1) is said to be a symmetric polynomial in these pairs of variables if it is unchanged by every interchange of these pairs of variables.

As in the case of points on a line, we see that it is not necessary to consider all the possible permutations of the subscripts in order to show that a polynomial F is symmetric. It is sufficient to show that F is unchanged by the interchange of every pair of the points (1).

We will introduce the Σ notation here precisely as in the case of single variables. Thus, for example,

$$\Sigma x_1^{\alpha_1} y_1^{\beta_1} \equiv x_1^{\alpha_1} y_1^{\beta_1} + x_2^{\alpha_1} y_2^{\beta_1} + \cdots + x_n^{\alpha_1} y_n^{\beta_1},$$

$$\Sigma x_1^{\alpha_1} y_1^{\beta_1} x_2^{\alpha_2} y_2^{\beta_2} \equiv x_1^{\alpha_1} y_1^{\beta_1} x_2^{\alpha_2} y_2^{\beta_2} + x_1^{\alpha_1} y_1^{\beta_1} x_3^{\alpha_2} y_3^{\beta_2} + \cdots,$$

and so on.

As in the case of single variables, it is clear that the order in which the pairs of exponents $\alpha_1,\, \beta_1\,;\, \alpha_2,\, \beta_2\,;\, \ldots$ are written is immaterial; and also that *every symmetric polynomial in the pairs of variables* (1) *is a linear combination of a certain number of Σ's.*

We introduce the notation

$$S_{kl} \equiv \Sigma \, x_1^k y_1^l \equiv x_1^k y_1^l + x_2^k y_2^l + \cdots + x_n^k y_n^l \qquad \left(\begin{matrix} k = 0, \, 1, \, \cdots \\ l = 0, \, 1, \, \cdots \end{matrix} \right).$$

THEOREM. *Any symmetric polynomial $F(x_1, \, y_1 \, ; \, \cdots \, x_n, \, y_n)$ may be expressed as a polynomial in these S's.*

The proof of this theorem is exactly like that of Theorem 3, § 83, and is left to the reader.

88. Elementary Symmetric Functions of Pairs of Variables.
Every Σ function of n pairs of variables may, by giving to the α's and β's suitable values, be written in the form

$$(1) \qquad \Sigma \, x_1^{\alpha_1} y_1^{\beta_1} x_2^{\alpha_2} y_2^{\beta_2} \cdots x_n^{\alpha_n} y_n^{\beta_n}.$$

DEFINITION. *The function* (1) *is said to be an elementary symmetric function of the pairs of variables $(x_1, \, y_1), \, \cdots \, (x_n, \, y_n)$ when, and only when,*

$$\alpha_i + \beta_i = 0 \ or \ 1 \qquad (i = 1, \, 2, \, \cdots \, n),$$

but not all the α's and β's are zero.

We shall adopt the following notation for these elementary symmetric functions :

$$p_{10} \equiv \Sigma \, x_1, \qquad p_{01} \equiv \Sigma \, y_1,$$
$$p_{20} \equiv \Sigma \, x_1 x_2, \qquad p_{11} \equiv \Sigma \, x_1 y_2, \qquad p_{02} \equiv \Sigma \, y_1 y_2,$$
$$\cdots \cdots \cdots \cdots \cdots \cdots \cdots$$
$$p_{n0} \equiv x_1 x_2 \cdots x_n, \ \cdots \ p_{i,\,n-i} \equiv \Sigma x_1 \cdots x_i y_{i+1} \cdots y_n, \ \cdots \ p_{0n} \equiv y_1 y_2 \cdots y_n.$$

It is clear that there are a finite number, $\frac{1}{2} \, n \, (n + 3)$, of p_{ij}'s, but an infinite number of S_{ij}'s.

We will attach to each p a *weight with regard to the x's* equal to its first subscript and a *weight with regard to the y's* equal to its second subscript.   When we speak simply of the weight of p_{ij} we will mean its *total weight*, that is, the sum of its subscripts.

THEOREM. *Any symmetric polynomial $F(x_1, \, y_1 \, ; \, \cdots \, x_n, \, y_n)$ may be expressed as a polynomial in the p_{ij}'s.*

Since, by the theorem in § 87, any such polynomial may be expressed as a polynomial in the S_{ij}'s, it is sufficient to show that the S_{ij}'s may be expressed as polynomials in the p_{ij}'s.

Let $\quad \xi_1 \equiv \lambda x_1 + \mu y_1, \ \xi_2 \equiv \lambda x_2 + \mu y_2, \cdots \xi_n \equiv \lambda x_n + \mu y_n,$

and form the elementary symmetric functions of these ξ's :

$$\pi_1 \equiv \Sigma \xi_1 \equiv \lambda \Sigma x_1 + \mu \Sigma y_1 \equiv \lambda p_{10} + \mu p_{01},$$

$$\pi_2 \equiv \Sigma \xi_1 \xi_2 \equiv \Sigma (\lambda x_1 + \mu y_1)(\lambda x_2 + \mu y_2)$$

$$\equiv \lambda^2 \Sigma x_1 x_2 + \lambda \mu \Sigma x_1 y_2 + \mu^2 \Sigma y_1 y_2$$

$$\equiv \lambda^2 p_{20} + \lambda \mu p_{11} + \mu^2 p_{02},$$

$$\pi_3 \equiv \Sigma \xi_1 \xi_2 \xi_3 \equiv \lambda^3 p_{30} + \lambda^2 \mu p_{21} + \lambda \mu^2 p_{12} + \mu^3 p_{03},$$

$$\cdot \quad \cdot \quad \cdot \quad \cdot \quad \cdot \quad \cdot \quad \cdot \quad \cdot \quad \cdot \quad \cdot \quad \cdot \quad \cdot \quad \cdot \quad \cdot$$

$$\cdot \quad \cdot \quad \cdot \quad \cdot \quad \cdot \quad \cdot \quad \cdot \quad \cdot \quad \cdot \quad \cdot \quad \cdot \quad \cdot \quad \cdot \quad \cdot$$

$$\pi_n \equiv \xi_1 \xi_2 \cdots \xi_n \equiv \lambda^n p_{n0} + \lambda^{n-1} \mu p_{n-1,\,1} + \lambda^{n-2} \mu^2 p_{n-2,\,2} + \cdots + \mu^n p_{0n}.$$

Also let
$$\sigma_k \equiv \Sigma \xi_1^k \qquad\qquad (k = 1,\, 2,\, \cdots)$$

Let α and β be positive integers, or zero, but not both zero.

Then
$$\sigma_{\alpha+\beta} \equiv \Sigma \xi_1^{\alpha+\beta} \equiv \lambda^{\alpha+\beta} \Sigma x_1^{\alpha+\beta} + \lambda^{\alpha+\beta-1} \mu \Sigma x_1^{\alpha+\beta-1} y_1 + \cdots$$

$$\equiv \lambda^{\alpha+\beta} S_{\alpha+\beta,\,0} + \lambda^{\alpha+\beta-1} \mu S_{\alpha+\beta-1,\,1} + \cdots.$$

But by Theorem 1, § 84, we may write

$$\sigma_{\alpha+\beta} \equiv F(\pi_1,\, \pi_2,\, \cdots \pi_n),$$

where F is a polynomial.   Hence

$$\lambda^{\alpha+\beta} S_{\alpha+\beta,\,0} + \lambda^{\alpha+\beta-1} \mu S_{\alpha+\beta-1,\,1} + \cdots \equiv \Psi(p_{10},\, \cdots p_{0n},\, \lambda,\, \mu),$$

where Ψ is a polynomial.   Regarding this as an identity in $(\lambda,\, \mu)$ and equating the coefficients of the terms containing $\lambda^\alpha \mu^\beta$, we get an identity in the x's and y's,

$$S_{\alpha\beta} \equiv \Phi(p_{10},\, \cdots p_{0n}),$$

where Φ is a polynomial in the p's.   Thus our theorem is proved.

Theorem 3, § 84, does not hold in the case of pairs of variables, as relations between the $\frac{1}{2} n (n+3)$ p_{ij}'s do exist; for example, if $n = 2$, the polynomial

$$4 p_{20} p_{02} - p_{20} p_{01}^2 - p_{10}^2 p_{02} + p_{10} p_{11} p_{01} - p_{11}^2$$

vanishes identically when the p's are replaced by their values in terms of the x's.   It does not vanish identically when $n = 3$.

In view of the remark just made, it is clear that the representations of polynomials in pairs of variables in terms of the p_{ij}'s will not be unique.

For further information concerning the subjects treated in this section, the reader may consult Netto's *Algebra*, Vol. 2, p. 63.

EXERCISES

1. Prove that a polynomial symmetric in the pairs of variables (x_i, y_i) and which is homogeneous in the x's alone of degree n and in the y's alone of degree m can be expressed as a polynomial in the p_{ij}'s isobaric of weight n with regard to the x's, and m with regard to the y's.

2. Express the symmetric polynomial

$$\Sigma \, x_1^2 y_2 y_3$$

in terms of the p_{ij}'s by the method of undetermined coefficients, making use of Exercise 1.

3. A polynomial in $(x_1, y_1, z_1; \ x_2, y_2, z_2; \ \cdots x_n, y_n, z_n)$ which is unchanged by every interchange of the subscripts is called a symmetric polynomial in the n points (x_i, y_i, z_i).

Extend the results of this section and the last to polynomials of this sort.

89. Binary Symmetric Functions. The pairs of variables (x_1, y_1), $\cdots (x_n, y_n)$ may be regarded as the homogeneous coördinates of n points on a line as well as the non-homogeneous coördinates of n points in a plane. It will then be natural to consider only symmetric polynomials which are homogeneous in each pair of variables alone. Such polynomials we will call *binary symmetric functions*. Most of the p_{ij}'s of the last section are thus excluded. The last $n + 1$ of them $(p_{n0}, p_{n-1,1}, \cdots p_{0n})$, however, are homogeneous of the first degree in each pair of variables alone. We will call them the *elementary binary symmetric functions*.

THEOREM 1. *Any binary symmetric function in $(x_1, y_1; \ \cdots x_n, y_n)$ can be expressed as a polynomial in $(p_{n0}, p_{n-1,1}, \cdots p_{0n})$.*

If we break up our binary symmetric function into Σ's, it is clear that each of these Σ's will itself be a binary symmetric function, or, as we will say for brevity, a binary Σ. It is therefore sufficient to prove that our theorem is true for every binary Σ. The general binary Σ may be written

$$\Sigma \, x_1^{\alpha_1} y_1^{\beta_1} x_2^{\alpha_2} y_2^{\beta_2} \cdots x_n^{\alpha_n} y_n^{\beta_n} \qquad (\alpha_1 \geqq \alpha_2 \geqq \cdots \geqq \alpha_n),$$

where, if we denote by m the degree of this Σ in any one of the pairs of variables,

$$m = \alpha_1 + \beta_1 = \alpha_2 + \beta_2 = \cdots = \alpha_n + \beta_n.$$

Let us assume for the moment that none of the y's are zero, and let

$$X_1 = \frac{x_1}{y_1}, \ X_2 = \frac{x_2}{y_2}, \ \cdots X_n = \frac{x_n}{y_n}.$$

Now consider the elementary symmetric functions of these X's:

$$P_1 = \Sigma X_1 \qquad = \frac{p_{1,\,n-1}}{p_{0n}},$$

$$P_2 = \Sigma X_1 X_2 \qquad = \frac{p_{2,\,n-2}}{p_{0n}},$$

$$\cdot \quad \cdot \quad \cdot \quad \cdot \quad \cdot \quad \cdot \quad \cdot$$
$$\cdot \quad \cdot \quad \cdot \quad \cdot \quad \cdot \quad \cdot \quad \cdot$$

$$P_n = X_1 X_2 \cdots X_n = \frac{p_{n0}}{p_{0n}}.$$

We may write

$$(1) \qquad \frac{\Sigma x_1^{\alpha_1} y_1^{\beta_1} x_2^{\alpha_2} y_2^{\beta_2} \cdots x_n^{\alpha_n} y_n^{\beta_n}}{p_{0n}^m} = \Sigma X_1^{\alpha_1} X_2^{\alpha_2} \cdots X_n^{\alpha_n} = \Phi(P_1, \cdots P_n),$$

where, since we have assumed $\alpha_1 \geqq \alpha_2 \geqq \cdots \geqq \alpha_n$, Φ is a polynomial of degree α_1 in the P's (Theorem 2, § 85). Hence we may write

$$(2) \qquad \Phi(P_1, \cdots P_n) = \frac{\phi(p_{0n},\, p_{1,\,n-1},\, \cdots p_{n0})}{p_{0n}^{\alpha_1}},$$

where ϕ is a homogeneous polynomial of degree α_1.

We thus get from (1) and (2)

$$(3) \qquad \Sigma x_1^{\alpha_1} y_1^{\beta_1} \cdots x_n^{\alpha_n} y_n^{\beta_n} = p_{0n}^{\beta_1} \phi(p_{0n},\, p_{1,\,n-1},\, \cdots p_{n0}),$$

an equation which holds except when one of the y's is zero. Since each side of (3) can be regarded as a polynomial in the x's and y's, we infer, by Theorem 5, § 2, that this is an identity, and our theorem is proved.

By Theorem 1, § 85, Φ is isobaric of weight $\alpha_1 + \alpha_2 + \cdots + \alpha_n$ in the P's. Hence $\Sigma x_1^{\alpha_1} y_1^{\beta_1} \cdots x_n^{\alpha_n} y_n^{\beta_n}$, when expressed in terms of these $(n+1)$ p_{ij}'s, is isobaric of weight $\alpha_1 + \alpha_2 + \cdots + \alpha_n$, provided we count the weight of the p_{ij}'s with regard to the x's. Passing back now to an aggregate of a number of such Σ's, we get

THEOREM 2. *If a binary symmetric function is homogeneous in the n x's (or y's) of degree k, it will, when expressed in terms of $p_{n0},\, p_{n-1,1},\, \cdots p_{0n}$, be isobaric of weight k with regard to the x's (or y's)*

We have seen in the proof of Theorem 1 that the polynomial ϕ in (3) is a homogeneous polynomial of degree α_1 in the p's; so that $\Sigma x_1^{\alpha_1} y_1^{\beta_1} \cdots x_n^{\alpha_n} y_n^{\beta_n}$ is a homogeneous polynomial of degree $\alpha_1 + \beta_1 = m$ in the p's. Hence

THEOREM 3. *Any binary symmetric function of degree m in each pair of variables will, when written in terms of $p_{n0},\, p_{n-1,\,1},\, \cdots p_{0n}$ be a homogeneous polynomial of degree m in these p's.*

EXERCISES

1. Prove that no rational relation exists between $p_{n0}, \cdots p_{0n}$, and hence that a binary symmetric function can be expressed as a polynomial in them in only one way.

2. By a ternary symmetric function is meant a symmetric polynomial in n points (x_i, y_i, z_i) which is homogeneous in the coördinates of each point.

Extend the results of this section to ternary symmetric functions. Cf. Exercise 3, § 88.

90. Resultants and Discriminants of Binary Forms. It is the object of the present section to show how the subject of the resultants and discriminants of binary forms may be approached from the point of view of symmetric functions.

Let
$$f(x_1, x_2) \equiv a_0 x_1^n + a_1 x_1^{n-1} x_2 + \cdots + a_n x_2^n$$
$$\equiv (\alpha_1'' x_1 - \alpha_1' x_2)(\alpha_2'' x_1 - \alpha_2' x_2) \cdots (\alpha_n'' x_1 - \alpha_n' x_2),$$
$$\phi(x_1, x_2) \equiv b_0 x_1^m + b_1 x_1^{m-1} x_2 + \cdots + b_m x_2^m$$
$$\equiv (\beta_1'' x_1 - \beta_1' x_2)(\beta_2'' x_1 - \beta_2' x_2) \cdots (\beta_m'' x_1 - \beta_m' x_2),$$

be two binary forms. Each of these polynomials has here been written first in the unfactored and secondly in the factored form. By a comparison of these two forms we see at once that **the elementary binary symmetric fractions of the n points**

$$(\alpha_1', \alpha_1''), (\alpha_2', \alpha_2''), \cdots (\alpha_n', \alpha_n'')$$

are
$$a_0, -a_1, a_2, \cdots (-1)^n a_n ;$$

and of the m points $(\beta_1', \beta_1''),(\beta_2', \beta_2''), \cdots (\beta_m', \beta_m'')$

are
$$b_0, -b_1, b_2, \cdots (-1)^m b_m.$$

Let us now consider the two linear factors
$$\alpha_i'' x_1 - \alpha_i' x_2, \quad \beta_j'' x_1 - \beta_j' x_2.$$

A necessary and sufficient condition for these factors to be proportional is that the determinant
$$\alpha_i'' \beta_j' - \alpha_i' \beta_j''$$

vanish. Let us form the product of all such determinants :

$$P \equiv \left\{ \begin{matrix} (\alpha_1'' \beta_1' - \alpha_1' \beta_1'') \, (\alpha_2'' \beta_1' - \alpha_2' \beta_1'') \cdots (\alpha_n'' \beta_1' - \alpha_n' \beta_1'') \\ (\alpha_1'' \beta_2' - \alpha_1' \beta_2'') \, (\alpha_2'' \beta_2' - \alpha_2' \beta_2'') \cdots (\alpha_n'' \beta_2' - \alpha_n' \beta_2'') \\ \cdot \; \cdot \; \cdot \; \cdot \; \cdot \; \cdot \; \cdot \; \cdot \; \cdot \; \cdot \; \cdot \; \cdot \; \cdot \; \cdot \; \cdot \; \cdot \\ \cdot \; \cdot \; \cdot \; \cdot \; \cdot \; \cdot \; \cdot \; \cdot \; \cdot \; \cdot \; \cdot \; \cdot \; \cdot \; \cdot \; \cdot \; \cdot \\ (\alpha_1'' \beta_m' - \alpha_1' \beta_m'') \, (\alpha_2'' \beta_m' - \alpha_2' \beta_m'') \cdots (\alpha_n'' \beta_m' - \alpha_n' \beta_m'') \end{matrix} \right\}.$$

The vanishing of this product is a necessary and sufficient condition that at least one of the linear factors of f be proportional to one of the linear factors of ϕ, that is, that f and ϕ have a common factor which is not a constant.

We may obviously reduce P to the simple form

$$P \equiv f(\beta_1', \beta_1'') f(\beta_2', \beta_2'') \cdots f(\beta_m', \beta_m'').$$

In this form it appears as a homogeneous polynomial of the mth degree in the a's, and as a symmetric polynomial in the m points (β_i', β_i''). Moreover, it is obviously a binary symmetric function which is of the nth degree in the coördinates of each point. Consequently, by Theorem 3, § 89, it can be expressed as a homogeneous polynomial of the nth degree in the elementary binary symmetric functions of the points (β_i', β_i''), that is, in the b's. Thus we have shown that *the product P can be expressed as a polynomial in the a's and b's which is homogeneous in the a's of degree m and in the b's of degree n.*

In § 72 we found another polynomial in the a's and b's, whose vanishing also gives a necessary and sufficient condition for f and ϕ to have a common factor, namely, the resultant R. We will now identify these polynomials by means of the following theorem :

THEOREM 1. *The product P differs from the resultant R of f and ϕ only by a constant factor, and the resultant is an irreducible polynomial in the a's and b's.*

We may show, in exactly the same way as in the proof of Theorem 1, § 86, that P, when expressed as a polynomial in the a's and b's, is irreducible. Since $P = 0$ and $R = 0$ each give a necessary and sufficient condition for f and ϕ to have a common factor, any set of values of the a's and b's which make $P = 0$ will also make $R = 0$. Thus by Theorem 7, § 76, P is a factor of R. We have seen that P is of degree m in the a's and n in the b's. The same is also true of R, as may easily be seen by inspection of the determinant of § 68. Hence, P being a factor of R, and of the same degree, can differ from it only by a constant factor. Thus our theorem is proved.

Let us now inquire under what conditions the binary form $f(x_1, x_2)$ has a multiple linear factor. Using the same notation as above, we see that the vanishing of the product

$$\left.\begin{array}{l} (\alpha_1'' \alpha_2' - \alpha_1' \alpha_2'')(\alpha_1'' \alpha_3' - \alpha_1' \alpha_3'') \cdots (\alpha_1'' \alpha_n' - \alpha_1' \alpha_n'') \\ \qquad (\alpha_2'' \alpha_3' - \alpha_2' \alpha_3'') \cdots (\alpha_2'' \alpha_n' - \alpha_2' \alpha_n'') \\ \qquad\qquad \cdot \quad \cdot \quad \cdot \quad \cdot \quad \cdot \quad \cdot \quad \cdot \quad \cdot \\ \qquad\qquad\qquad (\alpha_{n-1}'' \alpha_n' - \alpha_{n-1}' \alpha_n'') \end{array}\right\} \equiv P_1(\alpha_1', \alpha_1''; \ \cdots \alpha_n', \alpha_n'')$$

is a necessary and sufficient condition for this. P_1 is not symmetric in the pairs of α's, since an interchange of two subscripts changes P_1 into $-P_1$.   If, however, we consider P_1^2 instead of P_1, we have a binary symmetric function which can be expressed as a polynomial in the a's

$$[P_1(\alpha_1', \alpha_1''\,;\; \cdots \alpha_n', \alpha_n'')]^2 \equiv F(a_0, \cdots a_n).$$

Moreover, F vanishes when, and only when, P_1 does.   Accordingly $F = 0$ *is a necessary and sufficient condition for* $f(x_1,\, x_2)$ *to have a multiple linear factor.*

But the vanishing of the discriminant Δ (cf. § 82) of $f(x_1,\, x_2)$ is also a necessary and sufficient condition for this.

THEOREM 2.   *F and Δ differ only by a constant factor, and are irreducible.*

The proof of this theorem, which is practically the same as that of Theorem 1, is left to the reader.

If we subject the two binary forms f and ϕ, which we may suppose written in the factored form, to the linear transformation

$$(1) \qquad \begin{cases} x_1 = c_{11}x_1' + c_{12}x_2', \\ x_2 = c_{21}x_1' + c_{22}x_2', \end{cases}$$

we get two new binary forms

$$(A_1''x_1' - A_1'x_2')(A_2''x_1' - A_2'x_2') \cdots (A_n''x_1' - A_n'x_2'),$$
$$(B_1''x_1' - B_1'x_2')(B_2''x_1' - B_2'x_2') \cdots (B_m''x_1' - B_m'x_2'),$$

where
$$A_i'' = \quad \alpha_i''c_{11} - \alpha_i'c_{21}, \qquad B_j'' = \quad \beta_j''c_{11} - \beta_j'c_{21},$$
$$A_i' = -\,\alpha_i''c_{12} + \alpha_i'c_{22}, \qquad B_j' = -\,\beta_j''c_{12} + \beta_j'c_{22},$$

so that
$$A_i''B_j' - A_i'B_j'' \equiv c(\alpha_i''\beta_j' - \alpha_i'\beta_j''),$$

where c is the determinant of the transformation (1).

Since the linear transformation (1) may be regarded as carrying over the α's and β's into the A's and B's, the last written identity shows us that $\alpha_i''\beta_j' - \alpha_i'\beta_j''$ is, in a certain sense, an invariant of weight 1.   It can, however, not be expressed rationally in terms of the a's and b's.   Such an expression is called an *irrational invariant.*

Since the resultant of f and ϕ is the product of mn such irrational invariants of weight 1, it is evident that the resultant itself is an invariant of weight mn.   Thus we get a new proof of this fact, independent of the proof given in § 82.

A similar proof can be used in the case of the discriminant of a binary form.

EXERCISES

Develop the theory of the invariants of the binary biquadratic

$$f(x_1, x_2) \equiv a_0 x_1^4 + 4\,a_1 x_1^3 x_2 + 6\,a_2 x_1^2 x_2^2 + 4\,a_3 x_1 x_2^3 + a_4 x_2^4$$
$$\equiv (\alpha_1'' x_1 - \alpha_1' x_2)(\alpha_2'' x_1 - \alpha_2' x_2)(\alpha_3'' x_1 - \alpha_3' x_2)(\alpha_4'' x_1 - \alpha_4' x_2)$$

along the following lines:

1. Start from the irrational invariants of weight 2,

$$A = (\alpha_1'' \alpha_2' - \alpha_1' \alpha_2'')(\alpha_3'' \alpha_4' - \alpha_3' \alpha_4''),$$
$$B = (\alpha_1'' \alpha_3' - \alpha_1' \alpha_3'')(\alpha_4'' \alpha_2' - \alpha_4' \alpha_2''),$$
$$C = (\alpha_1'' \alpha_4' - \alpha_1' \alpha_4'')(\alpha_2'' \alpha_3' - \alpha_2' \alpha_3''),$$

whose sum is zero, and the negatives of whose ratios are the cross-ratios of the four points (α_1', α_1''), (α_2', α_2''), (α_3', α_3''), (α_4', α_4'').

2. Form the further irrational invariants of weight 2

$$E_1 \equiv B - C, \qquad E_2 \equiv C - A, \qquad E_3 \equiv A - B\,;$$

and prove that every homogeneous symmetric polynomial in E_1, E_2, E_3 is a binary symmetric function of the four points (α_i', α_i''), and therefore an integral rational invariant of f.

3. In particular

$$G_2 \equiv E_1 E_2 + E_2 E_3 + E_3 E_1, \qquad G_3 \equiv E_1 E_2 E_3$$

are homogeneous integral rational invariants of weights 4 and 6, and of degrees 2 and 3 respectively. Prove that

$$G_2 \equiv -36 g_2, \qquad G_3 \equiv 432 g_3,$$

where

$$g_2 \equiv a_0 a_4 - 4\,a_1 a_3 + 3\,a_2^2,$$
$$g_3 \equiv a_0 a_2 a_4 + 2\,a_1 a_2 a_3 - a_0 a_3^2 - a_1^2 a_4 - a_2^3.$$

These expressions g_2 and g_3 are the simplest invariants of f.*

4. Prove that the discriminant Δ of f is given by the formula

$$\Delta \equiv g_2^3 - 27 g_3^2.$$

5. If $\Delta \neq 0$, prove that $g_3 = 0$ is a necessary and sufficient condition that the four points $f = 0$ form a harmonic range; and that $g_2 = 0$ is a necessary and sufficient condition that they form an equianharmonic range. (Cf. Exercise 3, § 33.)

6. Prove that $g_2 = g_3 = 0$ is a necessary and sufficient condition that f have at least a threefold linear factor.†

* They are among the oldest examples of invariants, having been found by Cayley and Boole in 1845.

† Notice that we here have a projective property of the locus $f = 0$ expressed by the vanishing of two integral rational invariants; cf. the closing paragraph of § 81.

7. If λ is the absolute irrational invariant

$$\lambda = -\frac{A}{B},$$

i.e. one of the cross-ratios of the points $f = 0$, prove that the absolute rational invariant

$$I = \frac{g_2^3}{\Delta}$$

can be expressed in the form

$$I = \frac{4}{27}\frac{(\lambda^2 - \lambda + 1)^3}{(\lambda - 1)^2 \lambda^2}.$$

8. Prove that a necessary and sufficient condition for the equivalence of two biquadratic binary forms neither of whose discriminants is zero is that the invariant I have the same value for the two forms.

9. Prove that a necessary and sufficient condition for the equivalence with regard to linear transformations with determinant $+1$ of two biquadratic binary forms for which g_2 and g_3 are both different from zero is that the values of g_2 and g_3 be the same for one form as for the other.

10. Prove that if the discriminant of a biquadratic binary form is not zero, the form can be reduced by means of a linear transformation of determinant $+1$ to the normal form

$$4\,x_1^3 x_2 - g_2 x_1 x_2^3 - g_3 x_2^4.$$

11. Prove that every integral rational invariant of a biquadratic binary form is a polynomial in g_2 and g_3.

------

12. Develop the theory of the invariants of a pair of binary quadratic forms along the same lines as those just sketched for a single biquadratic form.

13. Prove that every integral rational invariant of a pair of quadratic forms in n variables is an integral rational function of the invariants $\Theta_0, \cdots \Theta_n$ of § 57.

[SUGGESTION. Show first that, provided a certain integral rational function of the coefficients of the quadratic form does not vanish, there exists a linear transformation of determinant $+1$ which reduces the pair of forms to

$$\alpha_1 x_1^2 + \alpha_2 x_2^2 + \cdots + \alpha_n x_n^2,$$
$$\beta_1 x_1^2 + \beta_2 x_2^2 + \cdots + \beta_n x_n^2.$$

Then show that every integral rational invariant of the pair of quadratic forms can be expressed as a binary symmetric function of (α_1, β_1), (α_2, β_2), $\cdots (\alpha_n, \beta_n)$, and that the Θ's are precisely the elementary binary symmetric functions.]

CHAPTER XX

ELEMENTARY DIVISORS AND THE EQUIVALENCE OF λ-MATRICES

91. λ-Matrices and their Elementary Transformations. The theory of elementary divisors, invented by Sylvester, H. J. S. Smith, and, more particularly, Weierstrass, and perfected in important respects by Kronecker, Frobenius, and others, has, in the form in which we will present it,[*] for its immediate purpose the study of matrices (which without loss of generality we assume to be square) whose elements are polynomials in a single variable λ. Such matrices we will call λ-matrices.[†] The determinant of a λ-matrix is a polynomial in λ, and if this determinant vanishes identically, we will call the matrix a singular λ-matrix. By the rank of a λ-matrix we understand the order of the largest determinant of the matrix which is not identically zero.

We have occasion here, as in § 19, to consider certain *elementary transformations* which we define as follows:

DEFINITION 1. *By an elementary transformation of a λ-matrix we understand a transformation of any one of the following forms:*

(a) The interchange of two rows or of two columns.

(b) The multiplication of each element of a row (or of a column) by the same constant not zero.

(c) The addition to the elements of a row (or column) of the products of the corresponding elements of another row (or column) by one and the same polynomial in λ.

---

[*] Various modifications of the point of view here adopted are possible and important. First, we may consider matrices whose elements are polynomials in any number of variables. Secondly, we may confine ourselves to polynomials whose coefficients lie in a certain domain of rationality. Thirdly, we may approach the subject from the side of the theory of numbers, assuming that the coefficients of the polynomials are integers. The simplest case here would be that in which the elements of the matrix are themselves integers; see Exercise 2, § 91, Exercise 3, § 92, and Exercise 2, § 94.

[†] The matrix of a pencil of quadratic forms is an important example of a λ-matrix to which the general theory will be applied in Chapter XXII.

If we pass from a first matrix to a second by an elementary transformation, it is clear that we can pass back from the second to the first by an elementary transformation. Thus the following definition is justified:

DEFINITION 2. *Two λ-matrices are said to be equivalent if it is possible to pass from one to the other by means of a finite number of elementary transformations.*

We see here that all λ-matrices equivalent to a given matrix are equivalent to each other; and, as in §19, that two equivalent λ-matrices always have the same rank.

The rank of a λ-matrix is not, however, the only thing which is left unchanged by every elementary transformation. In order to show this we begin with

LEMMA 1. *If the polynomial $\phi(\lambda)$ is a factor of all the i-rowed determinants of a λ-matrix a, it will be a factor of all the i-rowed determinants of every λ-matrix obtained from a by means of an elementary transformation.*

If the transformation is of the type (a) or (b) of Definition 1, this lemma is obviously true, since these transformations have no effect on the i-rowed determinants of a except to multiply them by constants which are not zero. If it is of the type (c), let us suppose it consists in adding to the elements of the pth column of a the corresponding elements of the qth column, each multiplied by the polynomial $\psi(\lambda)$. Any i-rowed determinant of a which either does not involve the pth column, or involves both the pth and the qth, will be unaffected by this transformation. An i-rowed determinant which involves the pth column but not the qth may be written after the transformation in the form $A \pm \psi(\lambda)B$, where A and B are i-rowed determinants of a; so that here also our lemma is true.

THEOREM 1. *If a and b are equivalent λ-matrices of rank r, and $D_i(\lambda)$ is the greatest common divisor of the i-rowed determinants $(i \leq r)$ of a, then it is also the greatest common divisor of the i-rowed determinants of b.*

For by our lemma, $D_i(\lambda)$ is a factor of all the i-rowed determinants of b ; and if these determinants had a common factor of higher degree, this factor would, by our lemma, be a factor of all the i-rowed determinants of a; which is contrary to hypothesis.

The theorem just proved shows that the greatest common divisors $D_1(\lambda), \cdots D_r(\lambda)$ are invariants with regard to elementary transformations, or, more generally, that they are invariants with regard to all transformations which can be built up from a finite number of elementary transformations. In point of fact they form, along with the rank r, a complete system of invariants. To prove this we now proceed to show how, by means of elementary transformations, a λ-matrix may be reduced to a very simple normal form.

LEMMA 2. *If the first element* [*] $f(\lambda)$ *of a λ-matrix is not identically zero and is not a factor of all the other elements, then an equivalent matrix can be formed whose first element is not identically zero and is of lower degree than f.*

Suppose first there is an element $f_1(\lambda)$ in the first row which is not divisible by $f(\lambda)$ and let j denote the number of the column in which it lies. Dividing f_1 by f and calling the quotient q and the remainder r, we have
$$f_1(\lambda) \equiv q(\lambda)f(\lambda) + r(\lambda).$$

Accordingly, if to the elements of the jth column we add those of the first, each multiplied by $-q(\lambda)$, we get an equivalent matrix in which the first element of the jth column is $r(\lambda)$, which is a polynomial of degree lower than $f(\lambda)$. If now we interchange the first and jth columns, the truth of our lemma is established in the case we are considering.

A similar proof obviously applies if there is an element in the first column which is not divisible by $f(\lambda)$.

Finally, suppose every element of the first row and column is divisible by $f(\lambda)$, but that there is an element, say in the ith row and jth column, which is not divisible by $f(\lambda)$. Let us suppose the element in the first row and jth column is $\psi(\lambda)f(\lambda)$, and form an equivalent matrix by adding to the elements of the jth column $-\psi(\lambda)$ times the corresponding elements of the first column. In this matrix, $f(\lambda)$ still stands in the upper left-hand corner, the first element of the jth column is zero; the first element of the ith row has not been changed and is therefore divisible by $f(\lambda)$; and the element in the ith row and jth column is still not divisible by f. Now form another equivalent matrix by adding to the elements of the first column the corresponding elements of the jth column. The upper left-hand element is still $f(\lambda)$, while the first element of the

---

[*] By the first element of a matrix we will understand the element in the upper left-hand corner

ith row is not divisible by $f(\lambda)$. This matrix, therefore, comes under the case already treated in which there is an element in the first column which is not divisible by $f(\lambda)$, and our lemma is established.

LEMMA 3. *If we have a λ-matrix whose elements are not all identically zero, an equivalent matrix can be formed which has the following three properties :*

(a) The first element $f(\lambda)$ is not identically zero.

(b) All the other elements of the first row and of the first column are identically zero.

(c) Every element neither in the first row nor in the first column is divisible by $f(\lambda)$.

For we may first, by an interchange of rows and of columns, bring into the first place an element which is not identically zero. If this is not a factor of all the other elements, we can, by Lemma 2, find an equivalent matrix whose first element is of lower degree and is not identically zero. If this element is not a factor of all the others, we may repeat the process. Since at each step we lower the degree of the first element, there must, after a finite number of steps, come a point where the process stops, that is, where the first element is a factor of all the others. We can then, by using transformations of type (c) (Definition 1), reduce all the elements in the first row and in the first column except this first one to zero, while the other elements remain divisible by the first one. Thus our lemma is established.

Finally, we note that since $f(\lambda)$ in the lemma just proved is a factor of all the other elements of the simplified matrix, it must, by Theorem 1, be the greatest common divisor of all the elements of the original matrix.

The lemma just proved tells us that the λ-matrix of the nth order of rank $r > 0$

$$(1) \qquad \left\| \begin{array}{ccc} a_{11} & \cdots & a_{1n} \\ \cdot & \cdot & \cdot \\ \cdot & \cdot & \cdot \\ a_{n1} & \cdots & a_{nn} \end{array} \right\|$$

can be reduced by means of elementary transformations to the form

$$(2) \qquad \left\| \begin{array}{cccc} f_1(\lambda) & 0 & \cdots & 0 \\ 0 & b_{11} & \cdots & b_{1,\,n-1} \\ \cdot & \cdot & \cdot & \cdot \\ \cdot & \cdot & \cdot & \cdot \\ 0 & b_{n-1,\,1} & \cdots & b_{n-1,\,n-1} \end{array} \right\|,$$

where $f_1(\lambda) \not\equiv 0$ and where $f_1(\lambda)$ is a factor of all the b's. The last written matrix being necessarily of rank r, the matrix of the $(n-1)$th order

$$(3) \qquad \left\| \begin{array}{ccc} b_{11} & \cdots & b_{1,\,n-1} \\ \cdot & \cdots & \cdot \\ \cdot & \cdots & \cdot \\ b_{n-1,\,1} & \cdots & b_{n-1,\,n-1} \end{array} \right\|$$

is of rank $r-1$. Consequently, if $r > 1$, (3) may be reduced by means of elementary transformations to the form

$$(4) \qquad \left\| \begin{array}{cccc} f_2(\lambda) & 0 & \cdots & 0 \\ 0 & c_{11} & \cdots & c_{1,\,n-2} \\ \cdot & \cdot & \cdots & \cdot \\ \cdot & \cdot & \cdots & \cdot \\ 0 & c_{n-2,\,1} & \cdots & c_{n-2,\,n-2} \end{array} \right\|,$$

where $f_2(\lambda) \not\equiv 0$ and where $f_2(\lambda)$ is a factor of all the c's. By Theorem 1, $f_2(\lambda)$, being the greatest common divisor of all the elements of (4), is also the greatest common divisor of all the b's, and is therefore divisible by $f_1(\lambda)$.

Now it is important to notice that the elementary transformations which carry over (3) into (4) may be regarded as elementary transformations of (2) which leave the first row and column of this matrix unchanged. Thus by a succession of elementary transformations, we have reduced (1) to the form

$$(5) \qquad \left\| \begin{array}{ccccc} f_1(\lambda) & 0 & 0 & \cdots & 0 \\ 0 & f_2(\lambda) & 0 & \cdots & 0 \\ 0 & 0 & c_{11} & \cdots & c_{1,\,n-2} \\ \cdot & \cdot & \cdot & \cdots & \cdot \\ \cdot & \cdot & \cdot & \cdots & \cdot \\ 0 & 0 & c_{n-2,\,1} & \cdots & c_{n-2,\,n-2} \end{array} \right\|,$$

where neither f_1 nor f_2 vanishes identically, f_1 is a factor of f_2, and f_2 is a factor of all the c's.

If $r > 2$, we may treat the $(n-2)$-rowed matrix of the c's, which is clearly of rank $r-2$, in a similar manner. Proceeding in this way, we finally reduce our matrix (1) to the form

$$(6) \qquad \left\| \begin{array}{ccccccc} f_1(\lambda) & 0 & \cdots & 0 & 0 & \cdots & 0 \\ 0 & f_2(\lambda) & \cdots & 0 & 0 & \cdots & 0 \\ \cdot & \cdot & \cdots & \cdot & \cdot & \cdots & \cdot \\ 0 & 0 & \cdots & f_r(\lambda) & 0 & \cdots & 0 \\ 0 & 0 & \cdots & 0 & 0 & \cdots & 0 \\ \cdot & \cdot & \cdots & \cdot & \cdot & \cdots & \cdot \\ \cdot & \cdot & \cdots & \cdot & \cdot & \cdots & \cdot \\ 0 & 0 & \cdots & 0 & 0 & \cdots & 0 \end{array} \right\|,$$

where none of the f's is identically zero, and each is a factor of the next following one.

So far we have used merely elementary transformations of the forms (a) and (c), Definition 1. By means of transformations of the form (b) we can simplify (6) still further by reducing the coefficient of the highest power of λ in each of the polynomials $f_i(\lambda)$ to unity. We have thus proved the theorem:

THEOREM 2. *Every λ-matrix of the nth order and of rank r can be reduced by elementary transformations to the normal form*

$$(7) \quad \begin{Vmatrix} E_1(\lambda) & 0 & \cdots & 0 & 0 & \cdots & 0 \\ 0 & E_2(\lambda) & \cdots & 0 & 0 & \cdots & 0 \\ \cdot & \cdot & \cdot & \cdot & \cdot & \cdot & \cdot \\ 0 & 0 & \cdots & E_r(\lambda) & 0 & \cdots & 0 \\ \cdot & \cdot & \cdot & \cdot & \cdot & \cdot & \cdot \\ \cdot & \cdot & \cdot & \cdot & \cdot & \cdot & \cdot \\ 0 & 0 & \cdots & 0 & 0 & \cdots & 0 \end{Vmatrix},$$

where the coefficient of the highest power of λ in each of the polynomials $E_i(\lambda)$ is unity, and $E_i(\lambda)$ is a factor of $E_{i+1}(\lambda)$ for $i = 1, 2, \cdots r - 1$.

By Theorem 1, the greatest common divisor of the i-rowed determinants $(i \leqq r)$ of the original matrix is the same as the greatest common divisor of the i-rowed determinants of the normal form (7) to which it is reduced. These last mentioned i-rowed determinants are, however, all identically zero except those which are the product of i of the E's. Let

$$(8) \quad E_{k_1}(\lambda) E_{k_2}(\lambda) \cdots E_{k_i}(\lambda)$$

be any one of these, and suppose the integers $k_1, k_2, \cdots k_i$ to have been arranged in order of increasing magnitude. We obviously have $k_1 \geq 1, k_2 \geq 2, \cdots k_i \geq i$. Consequently E_1 is a factor of E_{k_1}, E_2 of E_{k_2}, etc. Thus

$$E_1(\lambda) E_2(\lambda) \cdots E_i(\lambda)$$

is seen to be a factor of (8), and, being itself one of the i-rowed determinants of (7), it is their greatest common divisor. That is,

THEOREM 3. *The greatest common divisor of the i-rowed determinants of a λ-matrix of rank r, when $i \leqq r$, is*

$$D_i(\lambda) \equiv E_1(\lambda) E_2(\lambda) \cdots E_i(\lambda),$$

where the E's are the elements of the normal form (7) to which the given matrix is equivalent.

It may be noticed that this greatest common divisor is so determined that the coefficient of the highest power of λ in it is unity.

We come now to the fundamental theorem:

THEOREM 4. *A necessary and sufficient condition for the equivalence of two λ-matrices of the nth order is that they have the same rank r, and that for every value of i from 1 to r inclusive, the i-rowed determinants of one matrix have the same greatest common divisor as the i-rowed determinants of the other.*

To say that this is a necessary condition is merely to restate Theorem 1. To prove it sufficient, suppose both matrices to be reduced to the normal form (7), where we will distinguish the normal form for the second matrix by attaching accents to the E's in it. If the conditions of our theorem are fulfilled, we have, by Theorem 3,

$$E_1'(\lambda) \equiv E_1(\lambda),$$
$$E_1'(\lambda)E_2'(\lambda) \equiv E_1(\lambda)E_2(\lambda),$$
$$E_1'(\lambda)E_2'(\lambda)E_3'(\lambda) \equiv E_1(\lambda)E_2(\lambda)E_3(\lambda),$$

$$\cdot \quad \cdot \quad \cdot \quad \cdot \quad \cdot \quad \cdot \quad \cdot \quad \cdot \quad \cdot \quad \cdot \quad \cdot \quad \cdot \quad \cdot$$

$$\cdot \quad \cdot \quad \cdot \quad \cdot \quad \cdot \quad \cdot \quad \cdot \quad \cdot \quad \cdot \quad \cdot \quad \cdot \quad \cdot$$

and, since none of these E's are identically zero, it follows that

$$E_i'(\lambda) \equiv E_i(\lambda) \qquad (i = 1, 2, \cdots r).$$

Thus the normal forms to which the two λ-matrices can be reduced are identical, and hence the matrices are equivalent, since two λ-matrices equivalent to a third are equivalent to each other.

EXERCISES

1. Reduce the matrix

$$\begin{Vmatrix} \lambda & 1 & 0 & 0 & 0 \\ 0 & \lambda & 0 & 0 & 0 \\ 0 & 0 & \lambda & 0 & 0 \\ 0 & 0 & 0 & \lambda-1 & 0 \\ 0 & 0 & 0 & 0 & \lambda-1 \end{Vmatrix}$$

by means of elementary transformations to the normal form of Theorem 2.

Verify the result by finding the greatest common divisors $D_i(\lambda)$ first directly, and secondly from the normal form.

2. By an elementary transformation of a matrix all of whose elements are integers is understood a transformation of any one of the following forms:

(a) The interchange of two rows or of two columns.

(b) The change of sign of all the elements of any row or column.

(c) The addition to the elements of one row (or column) of the products of the corresponding elements of another row (or column) by one and the same integer.

Starting from this definition, develop the theory of matrices whose elements are integers along the same lines as the theory of λ-matrices was developed in this section.

92. Invariant Factors and Elementary Divisors. In place of the invariants $D_i(\lambda)$ of the last section, it is, for most purposes, more convenient to introduce certain other invariants to which we will give the technical name *invariant factors*. As a basis for the definition of these invariants we state the following theorem, which is merely an immediate consequence of Theorem 3, § 91:

THEOREM 1. *The greatest common divisor of the i-rowed determinants $(i = 2, 3, \cdots r)$ of a λ-matrix of rank r is divisible by the greatest common divisor of the $(i-1)$-rowed determinants of this matrix.*

DEFINITION 1. *If* **a** *is a λ-matrix of rank r, and*

$$D_i(\lambda) \qquad (i = 1, 2, \cdots r)$$

the greatest common divisor of its i-rowed determinants so determined that the coefficient of the highest power of λ is unity; and if $D_0(\lambda) \equiv 1$; then the polynomial

$$(1) \qquad E_i(\lambda) \equiv \frac{D_i(\lambda)}{D_{i-1}(\lambda)} \qquad (i = 1, 2, \cdots r)$$

is called the ith invariant factor of **a**.

This definition shows that these E's are really invariants since they are completely determined by the D's which we proved to be invariants in § 91. Moreover, by multiplying together the first i of the relations (1), we get the formula

$$(2) \qquad D_i(\lambda) \equiv E_1(\lambda)E_2(\lambda) \cdots E_i(\lambda) \qquad (i = 1, 2, \cdots r).$$

This shows us that the E's completely determine the D's, and since these latter were seen in § 91 to form, together with the **rank**, a complete system of invariants, the same is true of the E's. That is,

THEOREM 2. *A necessary and sufficient condition that two λ-matrices be equivalent is that they have the same rank r, and that the invariant factors of one be identical respectively with the corresponding invariant factors of the other.*

Since, in the case of a non-singular matrix of the nth order, $D_n(\lambda)$ differs from the determinant of the matrix only by a constant factor, we see that in this case the determinant of the matrix is, except for a constant factor, precisely the product of all the invariant factors. This is the case which is of by far the greatest importance, and the term *invariant factor* comes from the fact that the E's are really factors of the determinant of the matrix in this case.

A reference to Theorem 3, § 91, shows that our invariant factors are precisely the polynomials E_i which occur in the normal form of Theorem 2, § 91; and, since in that normal form each E is a factor of the next following one, we have the important result,

Theorem 3. *If $E_1(\lambda)$, $\cdots$ $E_r(\lambda)$ are the successive invariant factors of a λ-matrix of rank r, then each of these E's is a factor of the next following one.*

This theorem enables us to arrange the invariant factors of a λ-matrix in the proper order by simply arranging them in the order of increasing degree, two E's of the same degree being necessarily identical.

The invariant factors (like the D's of the last section) may be spoken of as *rational* invariants of our λ-matrix since they are formed from the elements of the λ-matrix by purely rational processes, namely the elementary transformations of § 91, which involve only the rational operations of addition, subtraction, multiplication, and divison. In distinction to these the *elementary divisors*, first introduced by Weierstrass, are, in general, irrational invariants.* These we now proceed to define.

Definition 2. *If $\mathbf{a}$ is a λ-matrix of rank r, and $D_r(\lambda)$ is the greatest common divisor of the r-rowed determinants of $\mathbf{a}$, then the linear factors*
$$\lambda - \alpha, \ \lambda - \alpha', \ \lambda - \alpha'', \ \cdots$$
of $D_r(\lambda)$ are called the linear factors of $\mathbf{a}$.†

Since, by formula (2), $D_r(\lambda)$ is the product of all the invariant factors of $\mathbf{a}$, it is clear that each invariant factor is merely the product of certain integral powers, positive or zero, of the linear factors of $\mathbf{a}$. We may therefore lay down the following definition:

---

* German writers, following Frobenius, use the term *elementary divisor* to cover both kinds of invariants. This is somewhat confusing, and necessitates the use of modifying adjectives such as *simple elementary divisors* for the elementary divisors as originally defined by Weierstrass, *composite elementary divisors* for the E's. On the other hand Bromwich (*Quadratic Forms and their Classification by Means of Invariant-factors*, Cambridge, England, 1906) proposes to substitute the term *invariant factor* for the term *elementary divisor*. Inasmuch as this latter term is wholly appropriate, it seems clear that it should be retained in English as well as in German in the sense in which Weierstrass first used it.

† It will be noticed that if $\mathbf{a}$ is non-singular, the linear factors of $\mathbf{a}$ are simply the linear factors of the determinant of $\mathbf{a}$.

DEFINITION 3. *Let* **a** *be a λ-matrix of rank r, and*

$$\lambda - \alpha, \ \lambda - \alpha', \ \lambda - \alpha'', \ \cdots$$

its distinct linear factors. Then if

$$E_i(\lambda) \equiv (\lambda - \alpha)^{e_i}(\lambda - \alpha')^{e_i'}(\lambda - \alpha'')^{e_i''} \ \cdots \qquad (i = 1, \ 2, \ \cdots \ \boldsymbol{r}),$$

are the invariant factors of **a**, *such of the factors*

$$(\lambda - \alpha)^{e_1}, \qquad (\lambda - \alpha)^{e_2}, \qquad \cdots\cdots (\lambda - \alpha)^{e_r},$$
$$(\lambda - \alpha')^{e_1'}, \qquad (\lambda - \alpha')^{e_2'}, \qquad \cdots\cdots (\lambda - \alpha')^{e_r'},$$
$$(\lambda - \alpha'')^{e_1''}, \qquad (\lambda - \alpha'')^{e_2''}, \qquad \cdots\cdots (\lambda - \alpha'')^{e_r''},$$

$$\cdot \quad \cdot \quad \cdot \quad \cdot \quad \cdot \quad \cdot \quad \cdot \quad \cdot \quad \cdot \quad \cdot \quad \cdot \quad \cdot \quad \cdot \quad \cdot$$
$$\cdot \quad \cdot \quad \cdot \quad \cdot \quad \cdot \quad \cdot \quad \cdot \quad \cdot \quad \cdot \quad \cdot \quad \cdot \quad \cdot \quad \cdot$$

as are not mere constants are called the elementary divisors of **a**, *each elementary divisor being said to correspond to the linear factor of which it is a power.*[*]

Since the invariant factors completely determine the elementary divisors and *vice versa*, it is clear that the elementary divisors are not merely invariants, but that, together with the rank, they form a complete system of invariants. That is,

THEOREM 4. *A necessary and sufficient condition that two λ-matrices be equivalent is that they have the same rank and that the elementary divisors of one be identical respectively with the corresponding elementary divisors of the other.*

By means of Theorem 3 we infer the important result:

THEOREM 5. *The degrees e_i of the elementary divisors corresponding to any particular linear factor satisfy the inequalities*

$$e_i \geqq e_{i-1} \qquad\qquad (i = 2, \ 3, \ \cdots \ r).$$

By means of this theorem we can arrange the elementary divisors corresponding to any given linear factor in the proper order by simply noticing their degrees.

---

[*] It will be seen that the definition just given is equivalent to the following one, in which the conception of invariant factors is not introduced:

DEFINITION. *Let $\lambda - \alpha$ be a linear factor of the λ-matrix* **a** *of rank r, and let l_i be the exponent of the highest power of $\lambda - \alpha$ which is a factor of all the i-rowed determinants ($i \leqq r$) of* **a**. *If the integers e_i (which are necessarily positive or zero) are defined by the formula*

$$e_i = l_i - l_{i-1} \qquad\qquad (i = 1, \ 2, \ \cdots \ r),$$

then such of the expressions

$$(\lambda - \alpha)^{e_1}, \ (\lambda - \alpha)^{e_2}, \ \cdots \ (\lambda - \alpha)^{e_r}$$

as are not constants are called the elementary divisors of **a** *which correspond to the linear factor $\lambda - \alpha$.*

EXERCISES

1. If $\phi = 0$ and $\psi = 0$ are two conics of which the second is non-singular, show how the number and kind of singular conics contained in the pencil $\phi - \lambda\psi = 0$ depends on the nature of the elementary divisors of the matrix of the quadratic form $\phi - \lambda\psi$.

2. Extend Exercise 1 to the case of three dimensions.

3. Apply the considerations of this section to matrices whose elements are integers. (Cf. Exercise 2, §91).

93. The Practical Determination of Invariant Factors and Elementary Divisors. The easiest general method for determining the invariant factors of a particular λ-matrix is to reduce it by means of elementary transformations to the normal form of Theorem 2, §91, following out step by step the reduction used in the proof of that theorem. From this normal form the invariant factors may be read off; and from these the elementary divisors may be computed, although only, in general, by the solution of equations of more or less high degree.

There are, however, many cases of great importance in which the elementary divisors may more easily be obtained by other methods. The most obvious of these is to apply the definition of elementary divisors directly to the case in hand. As an illustration, we mention a matrix of the nth order which has $\alpha - \lambda$ as the element in each place of the principal diagonal, while all the other elements are zero except those which lie immediately to the right of or above the elements of the principal diagonal, these being all constants different from zero:

$$(1) \qquad \begin{Vmatrix} \alpha - \lambda & c_1 & 0 & \cdots & 0 & 0 \\ 0 & \alpha - \lambda & c_2 & \cdots & 0 & 0 \\ \cdot & \cdot & \cdot & \cdot & \cdot & \cdot \\ \cdot & \cdot & \cdot & \cdot & \cdot & \cdot \\ 0 & 0 & 0 & \cdots & \alpha - \lambda & c_{n-1} \\ 0 & 0 & 0 & \cdots & 0 & \alpha - \lambda \end{Vmatrix} \qquad (c_1 c_2 \cdots c_{n-1} \neq 0).$$

The determinant of this matrix is $(\alpha - \lambda)^n$. The determinant obtained by striking out the first column and the last row is $c_1 c_2 \cdots c_{n-1}$. Accordingly

$$D_n(\lambda) \equiv (\lambda - \alpha)^n, \quad D_{n-1}(\lambda) \equiv 1, \quad E_n(\lambda) \equiv (\lambda - \alpha)^n.$$

Thus we see that $(\lambda - \alpha)^n$ is the only elementary divisor of this matrix, while the invariant factors are $(\lambda - \alpha)^n$ and $n - 1$ 1's.

This direct method may sometimes be employed to advantage in conjunction with the method of reduction by elementary transformations.   Cf. Exercise 1 at the end of this section.

A further means of recognizing the elementary divisors in some special cases is furnished by the following theorems whose proofs, which present no difficulty, we leave to the reader:

THEOREM 1.   *If all the elements of a λ-matrix are zeros except those in the principal diagonal, and if each element of this diagonal which is not a constant is resolved into the product of a constant by powers of distinct linear factors of the form* $\lambda - \alpha$, $\lambda - \alpha'$, $\cdots$, *then these powers of linear factors will be precisely the elementary divisors of the matrix.*

THEOREM 2.   *If all the elements of a λ-matrix are zeros except those which lie in a certain number of non-overlapping principal minors, then the elementary divisors of the matrix may be found by taking the elementary divisors of all these principal minors.*

The proof of this theorem consists in reducing the given matrix to the form referred to in Theorem 1 by means of elementary transformations each of which may be regarded as an elementary transformation of one of the principal minors in question.

It should be noticed that this theorem would not be true if the words *invariant factors* were substituted in it for *elementary divisors;* cf. Exercise 3 below.   The invariant factors may, however, be computed from the elementary divisors when these have been found.

EXERCISES

1. Prove that the matrix

$$\left\|\begin{array}{ccc:ccc} \lambda - \alpha & 0 & 0 & -1 & 0 & 0 \\ 0 & \lambda - \alpha & 0 & 0 & -1 & 0 \\ 0 & 0 & \lambda - \alpha & 0 & 0 & -1 \\ \hdashline \beta^2 & 1 & 0 & \lambda - \alpha & 0 & 0 \\ 0 & \beta^2 & 1 & 0 & \lambda - \alpha & 0 \\ 0 & 0 & \beta^2 & 0 & 0 & \lambda - \alpha \end{array}\right\|$$

is equivalent to

$$\left\|\begin{array}{ccc:ccc} -1 & 0 & 0 & 0 & 0 & 0 \\ 0 & -1 & 0 & 0 & 0 & 0 \\ 0 & 0 & -1 & 0 & 0 & 0 \\ \hdashline 0 & 0 & 0 & (\lambda - \alpha)^2 + \beta^2 & 1 & 0 \\ 0 & 0 & 0 & 0 & (\lambda - \alpha)^2 + \beta^2 & 1 \\ 0 & 0 & 0 & 0 & 0 & (\lambda - \alpha)^2 + \beta^2 \end{array}\right\|,$$

T

and hence that its elementary divisors are

$$[\lambda - (\alpha + \beta i)]^3, \quad [\lambda - (\alpha - \beta i)]^3.$$

2. Generalize Exercise 1 to matrices of order $2n$.

3. Find (a) the elementary divisors, and (b) the invariant factors of the matrix

$$\left\|\begin{array}{cccc} \lambda^2(\lambda - 1)^2 & 0 & 0 & 0 \\ 0 & \lambda(\lambda - 1)^3 & 0 & 0 \\ 0 & 0 & \lambda - 1 & 0 \\ 0 & 0 & 0 & \lambda \end{array}\right\|.$$

4. Determine the invariant factors and the elementary divisors of the matrix

$$\left\|\begin{array}{ccccc} 2\lambda & 3 & 0 & 1 & \lambda \\ 4\lambda & 3(\lambda + 2) & 0 & \lambda + 2 & 2\lambda \\ 0 & 6\lambda & \lambda & 2\lambda & 0 \\ \lambda - 1 & 0 & \lambda - 1 & 0 & 0 \\ 3(\lambda - 1) & 1 - \lambda & 2(\lambda - 1) & 0 & 0 \end{array}\right\|.$$

Is this matrix equivalent to the matrix in the exercise at the end of § 91?

5. Devise a convenient rational process for computing the invariant factors of matrices of the kinds considered in Theorems 1 and 2.

94. A Second Definition of the Equivalence of λ-Matrices. The definition of equivalence of λ-matrices which we have used so far rests on the elementary transformations. These transformations are of such a special character that this definition is not convenient for most purposes. We now give a new definition which we will prove to be coextensive with the old one.

DEFINITION. *Two n-rowed λ-matrices* a *and* b *are said to be equivalent if there exist two non-singular n-rowed λ-matrices* c *and* d *whose determinants are independent of* λ, *and such that*

$$(1) \qquad\qquad b \equiv cad.*$$

Since the matrices c and d have, by hypothesis, constant determinants, the inverse matrices c^{-1} and d^{-1} will also be λ-matrices, and not matrices whose coefficients are fractional rational functions of λ as would in general be the case for the inverse of λ-matrices. Consequently, if we write (1) in the form

$$(2) \qquad\qquad a \equiv c^{-1}bd^{-1},$$

we see that the relation established by our definition between the matrices a and b is a reciprocal one, as is implied in the wording of the definition.

---

* We use here and in what follows the sign $\equiv$ between two λ-matrices to denote that every element of one matrix is identically equal to the corresponding element of the other.

In order to justify the definition just given, we begin by establishing the

LEMMA. *If* **a** *and* **b** *are n-rowed λ-matrices, and the polynomial* $\phi(\lambda)$ *is a factor of all the i-rowed determinants of* **a**, *it is a factor of all the i-rowed determinants of* **ab** *and also of* **ba**.

For, by Theorem 5, § 25, every i-rowed determinant of **ab** and also of **ba** is a homogeneous linear combination of certain i-rowed determinants of **a**.

THEOREM 1. *If* **a** *and* **b** *are equivalent according to the definition of this section, they are also equivalent according to the definition of* § 91.

For in this case there exist two non-singular λ-matrices, **c** and **d**, whose determinants are constants, such that relation (1) holds. Consequently, by Theorem 7, § 25,* **a** and **b** have the same rank r. Let $D_i(\lambda)$ be the greatest common divisor of the i-rowed determinants of **a**, where $i \leq r$. By our lemma, $D_i(\lambda)$ is a factor of all the i-rowed determinants of **ca**, and therefore, applying the lemma again, it is a factor of all the i-rowed determinants of **cad**, that is, of **b**.

We can infer further that $D_i(\lambda)$ is the *greatest* common divisor of the i-rowed determinants of **b**. For applying to relation (2) the reasoning just used, we see that the greatest common divisor of the i-rowed determinants of **b** is a factor of all the i-rowed determinants of **a**, and cannot therefore be of higher degree than $D_i(\lambda)$.

A reference to Theorem 4, § 91, now shows us that **a** and **b** are equivalent according to the definition of that section.

THEOREM 2. *If* **a** *and* **b** *are equivalent according to the definition of* § 91, *they are also equivalent according to the definition of the present section.*

We begin by showing that if we can pass from a matrix **a** to a matrix $\mathbf{a}_1$ by means of an elementary transformation, one of the following relations always holds:

$$(3) \qquad\qquad \mathbf{a}_1 \equiv \mathbf{ca} \quad \text{or} \quad \mathbf{a}_1 \equiv \mathbf{ad}$$

where **c** and **d** are non-singular matrices whose determinants are independent of λ. To prove this we consider in succession the elementary transformations of the forms which were called (a), (b), (c), in Definition 1, § 91.

* How is it that we have a right to apply this theorem to λ-matrices ?

(*a*) Suppose we interchange the pth and qth rows. This can be effected by forming the product $\mathbf{ca}$ where the matrix $\mathbf{c}$ may be obtained by interchanging the pth and qth rows (or columns) in the unit matrix

$$\begin{Vmatrix} 1 & 0 & 0 & \cdots & 0 \\ 0 & 1 & 0 & \cdots & 0 \\ \cdot & \cdot & \cdot & \cdot & \cdot \\ \cdot & \cdot & \cdot & \cdot & \cdot \\ 0 & 0 & 0 & \cdots & 1 \end{Vmatrix} .$$

Similarly the interchange of the pth and qth columns of $\mathbf{a}$ may be effected by forming the product $\mathbf{ac}$, where $\mathbf{c}$ has the same meaning as before.

In each of these cases, $\mathbf{c}$ may be regarded as a non-singular λ-matrix with constant determinant, since its elements are constants and its determinant is -1.

(*b*) To multiply the pth row of $\mathbf{a}$ by a constant k, we may form the product $\mathbf{ca}$, where $\mathbf{c}$ differs from the unit matrix only in having k instead of 1 as the pth element of the principal diagonal.

Similarly, we multiply the pth column of $\mathbf{a}$ by k, by forming the product $\mathbf{ac}$, where $\mathbf{c}$ has the same meaning as before.

If we take the constant k different from zero, $\mathbf{c}$ may be regarded as a non-singular λ-matrix with constant determinant.

(*c*) We can add to the pth row of $\mathbf{a}$ $\phi(\lambda)$ times the qth row by forming the product $\mathbf{ca}$, where $\mathbf{c}$ differs from the unit matrix only in having $\phi(\lambda)$ instead of zero as the element in the pth row and qth column.

Similarly we add to the qth column $\phi(\lambda)$ times the pth column by forming the product $\mathbf{ac}$ where $\mathbf{c}$ has the same meaning as before.

The matrix $\mathbf{c}$, whose determinant is 1, is a non-singular λ-matrix.

It being thus established that one of the relations (3) holds between any two λ-matrices which can be obtained from one another by an elementary transformation, it follows that two matrices $\mathbf{a}$ and $\mathbf{b}$ which are equivalent according to the definition of § 91 will satisfy a relation of the form

$$\mathbf{b} \equiv \mathbf{c}_p \mathbf{c}_{p-1} \cdots \mathbf{c}_1 \mathbf{a} \mathbf{d}_1 \mathbf{d}_2 \cdots \mathbf{d}_q$$

where each of the $\mathbf{c}$'s and $\mathbf{d}$'s is a non-singular λ-matrix of constant determinant which corresponds to one of the elementary transformations we use in passing from **a** to **b**. This last relation being of the form

$$\mathbf{b} \equiv \mathbf{cad},$$

where **c** and **d** are non-singular λ-matrices with constant determinants, our theorem is proved.

We have now completed the proof that our two definitions of the equivalence of λ-matrices are coextensive.

EXERCISES

1. If **a** denotes the matrix in Exercise 1, § 91, and **b** the normal form of Theorem 2, § 91, for this matrix, determine two λ-matrices, **c** and **d**, such that relation (1) holds.

Verify your result by showing that the determinants of **c** and **d** are constants.

2. Apply the considerations of this section to matrices whose elements are integers. Cf. Exercise 2, § 91, and Exercise 3, § 92.

95. Multiplication and Division of λ-Matrices. We close this chapter by giving a few developments of what might be called the elementary algebra of λ-matrices.

DEFINITION. *By the degree of a λ-matrix is understood the highest degree in λ of any one of its elements.*

For a λ-matrix of the kth degree, the element in the ith row and jth column may be written $\quad a_{ij}\lambda^k + a'_{ij}\lambda^{k-1} + \cdots + a_{ij}^{[k]},$

and at least one of the coefficients of λ^k (*i.e.* one of the a_{ij}'s) must be different from zero. If, then, we denote by $\mathbf{a}_p$ the matrix of which $a_{ij}^{[p]}$ is the element which stands in the ith row and jth column, we get the theorem

THEOREM 1. *Every λ matrix of the kth degree may be written in the form*

$$(1) \qquad \mathbf{a}_0\lambda^k + \mathbf{a}_1\lambda^{k-1} + \cdots + \mathbf{a}_k \qquad (\mathbf{a}_0 \neq 0)$$

where $\mathbf{a}_0, \cdots \mathbf{a}_k$ are matrices with constant elements; and conversely, every expression (1) is a λ-matrix of degree k.

THEOREM 2. *The product of two λ-matrices of degrees k and l*

$$\mathbf{a}_0\lambda^k + \mathbf{a}_1\lambda^{k-1} + \cdots + \mathbf{a}_k \qquad (\mathbf{a}_0 \neq 0)$$

$$\mathbf{b}_0\lambda^l + \mathbf{b}_1\lambda^{l-1} + \cdots + \mathbf{b}_l \qquad (\mathbf{b}_0 \neq 0)$$

is a λ-matrix of degree $k + l$ provided at least one of the matrices $\mathbf{a}_0$ and $\mathbf{b}_0$ is non-singular.

For this product is a λ-matrix of the form

$$c_0 \lambda^{k+l} + c_1 \lambda^{k+l-1} + \cdots + c_{k+l}$$

where c_0 has the value $a_0 b_0$ or $b_0 a_0$ according to the order in which the two given matrices are multiplied together. By Theorem 7, § 25, neither $a_0 b_0$ nor $b_0 a_0$ is zero if a_0 and b_0 are not both singular.

The next theorem relates to what we may call the division of λ-matrices.

THEOREM 3. *If* a *and* b *are two* λ*-matrices and if* b*, when written in the form* (1), *has as the coefficient of the highest power of* λ *a nonsingular matrix, then there exists one, and only one, pair of* λ*-matrices* q_1 *and* r_1 *for which*

$$a \equiv q_1 b + r_1$$

and such that either $r_1 \equiv 0$, *or* r_1 *is a* λ*-matrix of lower degree than* b; *and also one and only one pair of* λ*-matrices* q_2 *and* r_2 *for which*

$$a \equiv b q_2 + r_2$$

and such that either $r_2 \equiv 0$, *or* r_2 *is a* λ*-matrix of lower degree than* b.

The proof of this theorem is practically identical with the proof of Theorem 1, § 63.

EXERCISE

DEFINITION. *By a real matrix is understood a matrix whose elements are real; by a real* λ*-matrix, a matrix whose elements are real polynomials in* λ; *and by a real elementary transformation, an elementary transformation in which the constant in* (b) *and the polynomial in* (c), *Definition 1, § 91, are real.*

Show that all the results of this chapter still hold if we interpret the words *matrix*, λ*-matrix*, and *elementary transformation* to mean *real matrix*, *real* λ*-matrix*, and *real elementary transformation*, respectively.

CHAPTER XXI

THE EQUIVALENCE AND CLASSIFICATION OF PAIRS OF BILINEAR FORMS AND OF COLLINEATIONS

96. The Equivalence of Pairs of Matrices. The applications of the theory of elementary divisors with which we shall be concerned in this chapter and the next have reference to problems in which λ-matrices occur only indirectly. A typical problem is the theory of a pair of bilinear forms. The matrices $\mathbf{a}$ and $\mathbf{b}$ of these two forms have constant elements, and we get our λ-matrix only by considering the matrix $\mathbf{a} - \lambda\mathbf{b}$ of the pencil of forms determined by the two given forms. It will be noticed that this matrix is of the first degree, and in fact we shall deal, from now on, exclusively with λ-matrices of the first degree.

By the side of this simplification, a new difficulty is introduced, as will be clear from the following considerations. We shall subject the two sets of variables in the bilinear forms to two non-singular linear transformations whose coefficients we naturally assume to be constants, that is, independent of λ. These transformations have the effect of multiplying the λ-matrix, $\mathbf{a} - \lambda\mathbf{b}$, by certain non-singular matrices whose elements are constants (cf. § 36) and therefore, by § 94, carry it over into an equivalent λ-matrix which is evidently of the first degree. The transformations of § 94, however, were far more general than those just referred to, so that it is not at all obvious whether every λ-matrix of the first degree equivalent to the given one can be obtained by transformations of the sort just referred to or not.

These considerations show the importance of the following theorem :

THEOREM 1. *If $\mathbf{a}_1$, $\mathbf{a}_2$, $\mathbf{b}_1$, $\mathbf{b}_2$ are matrices with constant elements of which the last two are non-singular, and if the λ-matrices of the first degree*

$$\mathbf{m}_1 \equiv \mathbf{a}_1 - \lambda\mathbf{b}_1, \qquad \mathbf{m}_2 \equiv \mathbf{a}_2 - \lambda\mathbf{b}_2,$$

are equivalent, then there exist two non-singular matrices, $\mathbf{p}$ and $\mathbf{q}$ whose elements are independent of λ, and such that

$$(1) \qquad \mathbf{m}_2 \equiv \mathbf{p}\mathbf{m}_1\mathbf{q}.$$

Since $\mathbf{m}_1$ and $\mathbf{m}_2$ are equivalent, there exist two non-singular λ-matrices, $\mathbf{p}_0$ and $\mathbf{q}_0$, whose determinants are constants and such that

$$(2) \qquad \mathbf{m}_2 \equiv \mathbf{p}_0\mathbf{m}_1\mathbf{q}_0.$$

The matrix $\mathbf{q}_0$ has, therefore, an inverse, $\mathbf{q}_0^{-1}$, which is also a λ-matrix.

Let us now divide $\mathbf{p}_0$ by $\mathbf{m}_2$ and $\mathbf{q}_0^{-1}$ by $\mathbf{m}_1$ by means of Theorem 3, § 95, in such a way as to get matrices $\mathbf{p}_1$, $\mathbf{p}$, $\mathbf{s}_1$, $\mathbf{s}$ which satisfy the relations

$$(3) \qquad \mathbf{p}_0 \equiv \mathbf{m}_2\mathbf{p}_1 + \mathbf{p}, \qquad \mathbf{q}_0^{-1} \equiv \mathbf{s}_1\mathbf{m}_1 + \mathbf{s},$$

$\mathbf{p}$ and $\mathbf{s}$ being matrices whose elements are independent of λ. From (2) we get

$$\mathbf{p}_0\mathbf{m}_1 \equiv \mathbf{m}_2\mathbf{q}_0^{-1}.$$

Substituting here from (3), we have

$$\mathbf{m}_2\mathbf{p}_1\mathbf{m}_1 + \mathbf{p}\mathbf{m}_1 \equiv \mathbf{m}_2\mathbf{s}_1\mathbf{m}_1 + \mathbf{m}_2\mathbf{s},$$

or

$$(4) \qquad \mathbf{m}_2(\mathbf{p}_1 - \mathbf{s}_1)\mathbf{m}_1 \equiv \mathbf{m}_2\mathbf{s} - \mathbf{p}\mathbf{m}_1.$$

From this identity we may infer that $\mathbf{p}_1 \equiv \mathbf{s}_1$ and therefore

$$(5) \qquad \mathbf{m}_2\mathbf{s} \equiv \mathbf{p}\mathbf{m}_1.$$

For if $\mathbf{p}_1 - \mathbf{s}_1$ were not identically zero, $\mathbf{m}_2(\mathbf{p}_1 - \mathbf{s}_1)$ would be a λ-matrix of at least the first degree (cf. Theorem 2, §95), and hence the left-hand side of (4) would be a λ-matrix of at least the second degree. But this is impossible, since the right-hand side of (4) is a λ-matrix of at most the first degree.

If we knew that $\mathbf{p}$ and $\mathbf{s}$ were both non-singular, our theorem would follow at once from (5); for we could write (5) in the form

$$(6) \qquad \mathbf{m}_2 \equiv \mathbf{p}\mathbf{m}_1\mathbf{s}^{-1}$$

and $\mathbf{p}$ and $\mathbf{s}^{-1}$ would be non-singular matrices with constant elements. Moreover, we see from (5) that $\mathbf{p}$ and $\mathbf{s}$ are either both singular or both non-singular. Our theorem will thus be proved if we can show that $\mathbf{s}$ is non-singular.

For this purpose let us substitute in the identity

$$I \equiv q_0 q_0^{-1}$$

for q_0^{-1} its value from (3),

(7) $$I \equiv q_0 s_1 m_1 + q_0 s.$$

Now divide q_0 by m_2 by means of Theorem 3, §95, in such a way as to get

(8) $$q_0 \equiv q_1 m_2 + q$$

where q is a matrix with constant elements.

Substituting this value in (7), we have

$$I \equiv q_0 s_1 m_1 + q_1 m_2 s + qs.$$

Referring to (5), we see that this may be written

(9) $$I - qs \equiv (q_0 s_1 + q_1 p) m_1.$$

From this we infer that $q_0 s_1 + q_1 p$ must be identically zero, and therefore

(10) $$I = qs.$$

For if $q_0 s_1 + q_1 p$ were not identically zero, the right-hand side of (9) would be a λ-matrix of at least the first degree, while the left-hand side of (9) does not involve λ.

Equation (10) shows that s is non-singular, and thus our theorem is proved. It shows us, however, also that q is non-singular, and that $q = s^{-1}$, so that equation (6) becomes $m_2 \equiv pm_1 q$.

We may, therefore, add the following

COROLLARY. *The matrices* p *and* q *whose existence is stated in the above theorem may be obtained as the remainders in the division of* p_0 *and* q_0 *in* (2) *by* m_2 *by means of the formulæ:*

$$p_0 \equiv m_2 p_1 + p, \qquad q_0 \equiv q_1 m_2 + q.$$

From this theorem concerning λ-matrices of the first degree we can now deduce the following theorem concerning pairs of matrices with constant elements. It is this theorem which forms the main foundation for such applications of the theory of elementary divisors as we shall give.

We shall naturally speak of two pairs of matrices with constant elements a_1, b_1 and a_2, b_2 as equivalent if two non-singular matrices p and q exist for which

(11) $$a_2 = pa_1 q, \qquad b_2 = pb_1 q.$$

THEOREM 2. *If a_1, b_1 and a_2, b_2 are two pairs of matrices independent of λ, and if b_1 and b_2 are non-singular, a necessary and sufficient condition that these two pairs of matrices be equivalent is that the two λ-matrices*

$$m_1 \equiv a_1 - \lambda b_1, \qquad m_2 \equiv a_2 - \lambda b_2$$

have the same invariant factors, — or, if we prefer, the same elementary divisors.

For if the pairs of matrices are equivalent, equations (11) hold; hence, multiplying the second of these equations by λ and subtracting it from the first, we have

$$(12) \qquad\qquad m_2 \equiv pm_1q,$$

that is the λ-matrices m_1 and m_2 are equivalent, and therefore have the same invariant factors, and the same elementary divisors. On the other hand, it follows at once from the assumption that b_1 and b_2 are non-singular, that m_1 and m_2 are non-singular, and hence have the same rank. Consequently if m_1 and m_2 have the same invariant factors, or the same elementary divisors, they are equivalent. Since they are of the first degree, there must, by Theorem 1, exist two non-singular matrices p and q, whose elements are independent of λ, which satisfy the identity (12). From this identity, the two equations (11) follow at once; and the two pairs of matrices are equivalent. Thus the proof of our theorem is complete.

A case of considerable importance is that in which the matrices b_1 and b_2 both reduce to the unit matrix I. In this case m_1 and m_2 reduce to what are known as the *characteristic matrices* of a_1 and a_2 respectively, according to the following definition:

DEFINITION. *If a is a matrix of the nth order with constant elements and I the unit matrix of the nth order, the λ-matrix*

$$A \equiv a - \lambda I$$

is called the characteristic matrix of a; the determinant of A is called the characteristic function of a; and the equation of the nth degree in λ formed by setting this determinant equal to zero is called the characteristic equation of a.

We can now deduce from Theorem 2 the following more special result:

THEOREM 3. *If a_1 and a_2 are two matrices independent of λ, a necessary and sufficient condition that a non-singular matrix p exist such that* [*]

$$(13) \qquad\qquad a_2 = pa_1p^{-1}$$

is that the characteristic matrices A_1 and A_2 of a_1 and a_2 have the same invariant factors, — or, if we prefer, the same elementary divisors.

For if A_1 and A_2 have the same invariant factors (or elementary divisors), there exist, by Theorem 2, two non-singular matrices p and q such that

$$a_2 = pa_1q, \qquad I = pIq.$$

The second of these equations shows us that $q = p^{-1}$; and this value being substituted in the first, we see that p is the matrix whose existence our theorem asserts.

That, on the other hand, A_1 and A_2 have the same invariant factors and elementary divisors if equation (13) is fulfilled, is at once obvious.

97. The Equivalence of Pairs of Bilinear Forms. Suppose we have a pair of bilinear forms in $2n$ variables

$$\phi_1 \equiv \sum_1^n a'_{ij} x_i y_j, \qquad\qquad \psi_1 \equiv \sum_1^n b'_{ij} x_i y_j,$$

and also a second pair

$$\phi_2 \equiv \sum_1^n a''_{ij} x_i y_j, \qquad\qquad \psi_2 \equiv \sum_1^n b''_{ij} x_i y_j,$$

and let us assume that ψ_1 and ψ_2 are non-singular. We will inquire under what conditions the two pairs of forms are equivalent, that is, under what conditions a first non-singular linear transformation for the x's and a second for the y's,

$$c\begin{cases} x_1 = c_{11}x'_1 + \cdots + c_{1n}x'_n \\ \;\cdot\quad\cdot\quad\cdot\quad\cdot\quad\cdot \\ \;\cdot\quad\cdot\quad\cdot\quad\cdot\quad\cdot \\ x_n = c_{n1}x'_1 + \cdots + c_{nn}x'_n \end{cases} \qquad d\begin{cases} y_1 = d_{11}y'_1 + \cdots + d_{1n}y'_n \\ \;\cdot\quad\cdot\quad\cdot\quad\cdot\quad\cdot \\ \;\cdot\quad\cdot\quad\cdot\quad\cdot\quad\cdot \\ y_n = d_{n1}y'_1 + \cdots + d_{nn}y'_n \end{cases}$$

can be found which together carry over ϕ_1 into ϕ_2 and ψ_1 into ψ_2.

---

[*] Two matrices connected by a relation of the form (13) are sometimes called *similar* matrices. This conception of similarity is evidently merely a special case of the general conception of equivalence as defined in § 29, the transformations considered being of the form (13) instead of the more general form usually considered in this chapter and the last.

If we denote the conjugate of the matrix $\mathbf{c}$ by $\mathbf{c}'$ and the matrices of $\phi_1, \psi_1, \phi_2, \psi_2$ by $\mathbf{a}_1, \mathbf{b}_1, \mathbf{a}_2, \mathbf{b}_2$ respectively, we know, by Theorem 1, § 36, that the transformations $\mathbf{c}, \mathbf{d}$ carry over ϕ_1 and ψ_1 into forms with matrices

$$\mathbf{c}'\mathbf{a}_1\mathbf{d}, \qquad \mathbf{c}'\mathbf{b}_1\mathbf{d}$$

respectively; so that, if these are the forms ϕ_2 and ψ_2, we have

(1) $$\mathbf{a}_2 = \mathbf{c}'\mathbf{a}_1\mathbf{d}, \qquad \mathbf{b}_2 = \mathbf{c}'\mathbf{b}_1\mathbf{d}.$$

Consequently, by Theorem 2, § 96, the two λ-matrices

$$\mathbf{a}_1 - \lambda\mathbf{b}_1, \qquad \mathbf{a}_2 - \lambda\mathbf{b}_2$$

have the same invariant factors and elementary divisors.

Conversely, by the same theorem, if these two λ-matrices have the same invariant factors (or elementary divisors), two constant matrices $\mathbf{c}'$ and $\mathbf{d}$ exist which satisfy both equations (1); and hence there exists a linear transformation of the x's and another of the y's which together carry over ϕ_1 into ϕ_2 and ψ_1 into ψ_2. Thus we have proved the

THEOREM. *If ϕ_1, ψ_1 and ϕ_2, ψ_2 are two pairs of bilinear forms in $2\,n$ variables of which ψ_1 and ψ_2 are non-singular, a necessary and sufficient condition that these two pairs of forms be equivalent is that the matrices of the two pencils*

$$\phi_1 - \lambda\psi_1, \qquad \phi_2 - \lambda\psi_2$$

*have the same invariant factors,— or, if we prefer, the same elementary divisors.**

EXERCISE

Prove that the theorem of this section remains true if the bilinear forms $\phi_1, \psi_1, \phi_2, \psi_2$ are real and the term *equivalent* is understood to mean *equivalent with regard to real non-singular linear transformations.*

98. The Equivalence of Collineations. A second important application of the theory of elementary divisors is to the theory of collineations. For the sake of simplicity we will consider the case of two dimensions

$$\mathbf{a} \begin{cases} x_1' = a_{11}x_1 + a_{12}x_2 + a_{13}x_3, \\ x_2' = a_{21}x_1 + a_{22}x_2 + a_{23}x_3, \\ x_3' = a_{31}x_1 + a_{32}x_2 + a_{33}x_3, \end{cases}$$

although the reasoning will be seen to be perfectly general.

---

* For the sake of brevity, we shall, in future, speak of these invariant factors and elementary divisors as the invariant factors and elementary divisors of the *pairs of forms* ϕ_1, ψ_1 and ϕ_2, ψ_2 respectively.

We have so far regarded a collineation merely as a means of transforming certain geometric figures. It is possible to adopt another point of view, and to study the collineation in itself with special reference to the relative position of points before and after the transformation. Thus suppose we have a figure consisting of the points A_1, A_2, $\cdots$, finite or infinite in number, and suppose these points are carried over by the collineation **a** into the points A_1', A_2', $\cdots$. These two sets of points together form a geometric figure. It is the properties of such figures as this that we call the properties of the collineation. Such properties may be either projective or metrical. Thus it would be a metrical property of a collineation if it carried over some particular pair of perpendicular lines into a pair of perpendicular lines; it would be a projective property of the collineation if it carried over some particular triangle into itself. We shall be concerned only with the projective properties of collineations.

As an example, let us consider the *fixed points* of the collineation, that is points whose initial and final position is the same. In order that (x_1, x_2, x_3) be a fixed point it is necessary and sufficient that

$$x_1' = \lambda x_1, \qquad x_2' = \lambda x_2, \qquad x_3' = \lambda x_3,$$

that is, substituting in **a**, that a constant λ exist such that,

$$
\begin{aligned}
(a_{11} - \lambda)x_1 + & \quad a_{12}x_2 + & \quad a_{13}x_3 = 0, \\
a_{21}x_1 + & \quad (a_{22} - \lambda)x_2 + & \quad a_{23}x_3 = 0, \\
a_{31}x_1 + & \quad a_{32}x_2 + & \quad (a_{33} - \lambda)x_3 = 0.
\end{aligned}
$$

(1)

The matrix of this system of equations is precisely what we have called the characteristic matrix of the matrix **a** of the linear transformation. The characteristic function is a polynomial of the third degree in λ which, when equated to zero, has one, two, or three distinct roots. Let λ_1 be one of these roots. When this is substituted in (1), these equations are satisfied by the coördinates of one or more points,—the fixed points of the collineation **a**. The number and distribution of these fixed points give an important example of a projective property of a collineation; and it is readily seen that collineations may have wholly different properties in this respect, one having three fixed points, another two, and still another an infinite number.

Coming back now to the two sets of points A_1, A_2, $\cdots$ and A_1', A_2', $\cdots$ which correspond to one another by means of the **collinea-**

tion $\mathbf{a}$ (which may be singular or non-singular), let us subject all these points to a non-singular collineation $\mathbf{c}$, which carries over $A_1, A_2, \cdots$ into $B_1, B_2, \cdots$ and $A_1', A_2', \cdots$ into $B_1', B_2', \cdots$ respectively. The figure formed by the B's will have the same projective properties as that formed by the A's; and consequently if we can find a collineation $\mathbf{b}$ which carries over $B_1, B_2, \cdots$ into $B_1', B_2', \cdots$, this collineation will have the same projective properties as the collineation $\mathbf{a}$. Such a collineation is clearly given by the formula

$$(2) \qquad \qquad \mathbf{b} = \mathbf{cac}^{-1}$$

since $\mathbf{c}^{-1}$ carries over the points B_i into the points A_i, $\mathbf{a}$ then carries over these into A_i', and $\mathbf{c}$ carries over the points A_i' into the points B_i'.

Since two collineations $\mathbf{a}$ and $\mathbf{b}$ related by formula (2) are indistinguishable so far as their projective properties go (though they may have very different metrical properties), we will call them equivalent according to the following

DEFINITION. *Two collineations $\mathbf{a}$ and $\mathbf{b}$ shall be called equivalent if a non-singular collineation $\mathbf{c}$ exists such that relation (2) is fulfilled.*

A reference to Theorem 3, § 96, now gives us the fundamental theorem:

THEOREM. *A necessary and sufficient condition that two collineations be equivalent is that their characteristic matrices have the same invariant factors, — or, if we prefer, the same elementary divisors.*

EXERCISES

1. If $P_1, P_2, \cdots P_k$ are fixed points of a non-singular collineation in space of $n - 1$ dimensions which correspond to k distinct roots of the characteristic equation, prove that these points are linearly independent.

2. Discuss the distribution of the fixed points of a collineation
      (a) in two dimensions,
      (b) in three dimensions,
for all possible cases of non-singular collineations.

3. Discuss the distribution of
      (a) the fixed lines of a collineation in two dimensions,
      (b) the fixed planes of a collineation in three dimensions,
for all possible cases of non-singular collineations; paying special attention to their relation to the fixed points.

4. Two real collineations, $\mathbf{a}$ and $\mathbf{b}$, may be said to be equivalent if there exists a real non-singular collineation $\mathbf{c}$ such that $\mathbf{b} = \mathbf{cac}^{-1}$.

With this understanding of the term *equivalence*, show that the theorem of the present section holds for real collineations.

99. Classification of Pairs of Bilinear Forms. We consider again the pair of bilinear forms

$$\phi \equiv \sum_1^n a_{ij}x_iy_j, \qquad \psi \equiv \sum_1^n b_{ij}x_iy_j,$$

of which we assume the second to be non-singular, and form the λ-matrix.

$$(1) \qquad\qquad a - \lambda b.$$

Using a slightly different notation from that employed in § 92, we will denote the elementary divisors of (1) by

$$(\lambda - \lambda_1)^{e_1}, \quad (\lambda - \lambda_2)^{e_2}, \quad \cdots \cdots \quad (\lambda - \lambda_k)^{e_k}, \qquad (e_1 + e_2 + \cdots + e_k = n),$$

so that the linear factors $\lambda - \lambda_i$ need not all be distinct from one another. The most important thing concerning these elementary divisors is, for many purposes, their *degrees*, $e_1, e_2, \cdots e_k$. When we wish to indicate these degrees without writing out the elementary divisors in full, we will use the symbol $[e_1 \, e_2 \cdots e_k]$, called the *characteristic* of the λ-matrix (1), or of the pair of forms ϕ, ψ. It will be seen that this characteristic is a sort of arithmetical invariant of the pair of bilinear forms, since two pairs of bilinear forms which are equivalent necessarily have the same characteristic. The converse of this, however, is not true, since for the equivalence of two pairs of bilinear forms the identity of the elementary divisors themselves, not merely the equality of their degrees, is necessary.

All pairs of bilinear forms which have the same characteristic are said to form a *category*. Thus, for example, in the case of pairs of bilinear forms in six variables we should distinguish between three categories corresponding to the three characteristics,

$$[1 \quad 1 \quad 1], \qquad [2 \quad 1], \qquad [3],$$

which are obviously the only possible ones in this case. In fact, we must inquire whether these three categories really all exist. This question we answer in the affirmative by writing down the following pairs of bilinear forms in six variables which represent these three categories:

$$\text{I. } [1\,1\,1] \quad \begin{cases} \lambda_1 x_1 y_1 + \lambda_2 x_2 y_2 + \lambda_3 x_3 y_3, \\ x_1 y_1 + x_2 y_2 + x_3 y_3, \end{cases}$$

$$\begin{Vmatrix} \lambda_1 - \lambda & 0 & 0 \\ 0 & \lambda_2 - \lambda & 0 \\ 0 & 0 & \lambda_3 - \lambda \end{Vmatrix}.$$

II. [2 1]
$$\begin{cases} \lambda_1 x_1 y_1 + \lambda_1 x_2 y_2 + x_1 y_2 + \lambda_2 x_3 y_3, \\ x_1 y_1 + x_2 y_2 + x_3 y_3, \end{cases}$$

$$\left\| \begin{array}{ccc} \lambda_1 - \lambda & 1 & 0 \\ 0 & \lambda_1 - \lambda & 0 \\ 0 & 0 & \lambda_2 - \lambda \end{array} \right\|.$$

III. [3]
$$\begin{cases} \lambda_1 x_1 y_1 + \lambda_1 x_2 y_2 + \lambda_1 x_3 y_3 + x_1 y_2 + x_2 y_3, \\ x_1 y_1 + x_2 y_2 + x_3 y_3, \end{cases}$$

$$\left\| \begin{array}{ccc} \lambda_1 - \lambda & 1 & 0 \\ 0 & \lambda_1 - \lambda & 1 \\ 0 & 0 & \lambda_1 - \lambda \end{array} \right\|.$$

The pairs of bilinear forms we have just written down do more than merely establish the existence of our three categories. They establish the fact that not only the degrees of the elementary divisors are arbitrary (subject merely to the condition that their sum be three), but that, subject to this restriction, the elementary divisors themselves may be arbitrarily chosen. They are, moreover, *normal forms* to one or the other of which every pair of bilinear forms in six variables, of which the first is non-singular, may be reduced by non-singular linear transformations.

The general theorem here is this:

THEOREM. *If $\lambda_1, \lambda_2, \cdots \lambda_k$ are any constants, equal or unequal, and $e_1, e_2, \cdots e_k$ are any positive integers whose sum is n, there exist pairs of bilinear forms in $2n$ variables, the second form in each pair being non-singular, which have the elementary divisors*

$$(2) \qquad (\lambda - \lambda_1)^{e_1}, \quad (\lambda - \lambda_2)^{e_2}, \quad \cdots\cdots \quad (\lambda - \lambda_k)^{e_k}.$$

The proof of this theorem consists in considering the pair of bilinear forms

$$(3) \qquad \begin{cases} \phi \equiv \left(\overset{e_1}{\underset{1}{\Sigma}} \lambda_1 x_i y_i + \overset{e_1}{\underset{2}{\Sigma}} x_{i-1} y_i \right) + \left(\overset{e_1+e_2}{\underset{e_1+1}{\Sigma}} \lambda_2 x_i y_i + \overset{e_1+e_2}{\underset{e_1+2}{\Sigma}} x_{i-1} y_i \right) \\ \qquad\qquad + \cdots\cdots + \left(\overset{n}{\underset{n-e_k+1}{\Sigma}} \lambda_k x_i y_i + \overset{n}{\underset{n-e_k+2}{\Sigma}} x_{i-1} y_i \right), \\ \psi \equiv x_1 y_1 + x_2 y_2 + \cdots + x_n y_n, \end{cases}$$

of which the second is non-singular. These forms have a λ-matrix which may be indicated, for brevity, as

$$(4) \qquad \left\|\begin{array}{ccccc} \mathbf{M}_1 & & & & \\ & \mathbf{M}_2 & & & \\ & & \cdot\ \cdot\ \cdot & & \\ & & \cdot\ \cdot\ \cdot & & \\ & & \cdot\ \cdot\ \cdot & & \\ & & & & \mathbf{M}_k \end{array}\right\|,$$

where the letters $\mathbf{M}_1, \cdots \mathbf{M}_k$ represent not single terms but blocks of terms ; $\mathbf{M}_i$ standing for the matrix of order e_i

$$\mathbf{M}_i = \left\|\begin{array}{ccccc} \lambda_i - \lambda & 1 & 0 & \cdots & 0 \\ 0 & \lambda_i - \lambda & 1 & \cdots & 0 \\ \cdot & \cdot & \cdot & \cdots & \cdot \\ \cdot & \cdot & \cdot & \cdots & \cdot \\ 0 & 0 & 0 & \cdots & \lambda_i - \lambda \end{array}\right\| ;$$

while all the terms of the matrix (4) are zero which do not stand in one of the blocks of terms $\mathbf{M}_i$. The elementary divisors of (4) are, as we see by a reference to § 93 (Formula (1) and Theorem 2), precisely the expressions (2). Thus our theorem is proved.

A reference to § 97 shows that formula (3) is a normal form to which every pair of bilinear forms in $2n$ variables with the elementary divisors (2) can be reduced.*

***

* Many other normal forms might be chosen in place of (3). Thus, for instance, we might have used in place of (3) the form

$$(3') \quad \left\{\begin{aligned} \phi &\equiv \left(\sum_{1}^{e_1}\lambda_1 c_1 x_i y_{e_1-i+1} + \sum_{1}^{e_1-1} d_1 x_i y_{e_1-i}\right) + \left(\sum_{e_1+1}^{e_1+e_2}\lambda_2 c_2 x_i y_{2e_1+e_2-i+1} + \sum_{e_1+1}^{e_1+e_2-1} d_2 x_i y_{2e_1+e_2-i}\right) \\ &\qquad + \cdots + \left(\sum_{n-e_k+1}^{n}\lambda_k c_k x_i y_{2n-e_k-i+1} + \sum_{n-e_k+1}^{n-1} d_k x_i y_{2n-e_k-i}\right), \\ \psi &\equiv \sum_{1}^{e_1} c_1 x_i y_{e_1-i+1} + \sum_{e_1+1}^{e_1+e_2} c_2 x_i y_{2e_1+e_2-i+1} + \sum_{e_1+e_2+1}^{e_1+e_2+e_3} c_3 x_i y_{2e_1+2e_2+e_3-i+1} \\ &\qquad + \cdots + \sum_{n-e_k+1}^{n} c_k x_i y_{2n-e_k-i+1}, \end{aligned}\right.$$

where the constants $c_1, \cdots c_k, d_1, \cdots d_k$ may be chosen at pleasure provided, merely that none of them are zero. For instance, they may all be assigned the value 1.

Let us now return to the classification of pairs of bilinear forms. For a given number, $2n$, of variables we have obviously only a finite number of categories. We may subdivide these categories into *classes* by noticing which, if any, of the elementary divisors correspond to the same linear factor. This we can indicate in the characteristic by connecting by parentheses those integers which are the degrees of elementary divisors corresponding to one and the same linear factor. Thus, in the case $n = 8$, the characteristic

$$[(2\,1)(1\,1\,1)2]$$

would indicate that the λ-matrix has just three distinct linear factors; that to one of these there correspond two elementary divisors of degrees two and one respectively, to another three elementary divisors of the first degree, and to the last a single elementary divisor of degree two.

Two pairs of bilinear forms which are equivalent belong necessarily to the same class, but two pairs of bilinear forms which belong to the same class are not necessarily equivalent.

To illustrate what has just been said, let us again consider the case $n = 3$. Here we have now, instead of three categories, six classes, which are exhibited in the following table:

	a	b	c
I.	$[1\,1\,1]$	$[(1\,1)\,1]$	$[(1\,1\,1)]$
II.	$[2\,1]$	$[(2\,1)]$	
III.	$[3]$		

The λ-matrix of this pair of forms may be written in the form (4), where, however, $\mathbf{M}_i$ now stands for the matrix of order e_i:

$$\mathbf{M}_i = \begin{Vmatrix} 0 & \cdots & 0 & d_i & c_i(\lambda_i - \lambda) \\ 0 & \cdots & d_i & c_i(\lambda_i - \lambda) & 0 \\ \cdot & \cdot & \cdot & \cdot & \cdot \\ \cdot & \cdot & \cdot & \cdot & \cdot \\ c_i(\lambda_i - \lambda) & \cdots & 0 & 0 & 0 \end{Vmatrix}.$$

It will be noticed that the matrices $\mathbf{M}_i$, and therefore also the bilinear forms (3'), are symmetrical, a fact which will make this normal form important when we come to the subject of quadratic forms in the next chapter.

Constants similar to the constants c_i and d_i which we have introduced in (3') might also have been introduced in (3).

The three classes Ia, Ib, Ic form together the category I, and are all represented by the normal form given for that category above, the only difference being that in class Ia the three quantities λ_1, λ_2, λ_3 are all distinct, in class Ib two, and only two, of them are equal, while in class Ic they are all equal. Similarly category II is now divided into two classes, IIa and IIb, for both of which the normal form of category II holds good, λ_1 and λ_2 being, however, different in that normal form for class IIa and equal for class IIb. Finally category III consists of only a single class.

For some purposes it is desirable to carry this subdivision still farther. The second of our two bilinear forms, ψ, has been assumed throughout to be non-singular. The first, ϕ, may be singular or non-singular; and it is readily seen that a necessary and sufficient condition that ϕ be singular is that one, at least, of the constants λ, which enter into the linear factors of the λ-matrix be zero. Thus it will be seen that in a single class we shall have pairs of bilinear forms both of which are non-singular and others one of which is singular, and we may wish to separate into different sub-classes the pairs of forms which belong to one or the other of these two cases.

Let us go a step farther in this same direction, and inquire how the rank of ϕ is connected with the values of the constants λ_i. We notice that the matrix of ϕ is equal to the matrix of the pencil $\phi - \lambda\psi$ when $\lambda = 0$. Accordingly, if ϕ is of rank r, every $(r+1)$-rowed determinant of the matrix of $\phi - \lambda\psi$ will be divisible by λ, while at least one r-rowed determinant of this matrix is not divisible by λ. It is then necessary, as we see by a reference to the definition of elementary divisors (cf. the footnote to Definition 3, § 92), that just $n - r$ of the constants λ_i which enter into the elementary divisors should be zero. Since the converse of these statements is also true, we may say that *a necessary and sufficient condition that the form ϕ be of rank r is that just $n - r$ of the elementary divisors be of the form λ^{e_i}*. Let us, in the characteristic $[e_1\ e_2\ \cdots\ e_k]$, place a small zero above each of the integers e_i which is the degree of such an elementary divisor; and regard two pairs of bilinear forms as belonging to a single class when, and only when, their characteristics coincide in the distribution of these zeros as well as in other respects. Here again two equivalent pairs of forms will always belong to the same class, but the converse will not be true.

As an illustration, let us again take the case $n = 3$. We have now fourteen classes instead of six.

$$[1\ 1\ 1],\ [(1\ 1)1],\ [(1\ 1\ 1)],\ [2\ 1],\ [(2\ 1)],\ [3], \qquad (r = 3),$$

$$[1\ 1\ \overset{0}{1}],\ [(1\ 1)\ \overset{0}{1}],\ [2\ \overset{0}{1}],\ [\overset{0}{2}\ 1],\ [\overset{0}{3}], \qquad (r = 2),$$

$$[(\overset{0}{1}\ \overset{0}{1})\ 1],\ [(\overset{0}{2}\ \overset{0}{1})], \qquad (r = 1),$$

$$[(\overset{0}{1}\ \overset{0}{1}\ \overset{0}{1})], \qquad (r = 0).$$

We have indicated, in each case, the rank r of the form ϕ. Thus in the first six cases ϕ is non-singular; in the next five it is of rank 2, etc.

EXERCISES

1. Prove that there exist pairs of real bilinear forms in $2\,n$ variables of which the second is non-singular, and which have the elementary divisors

$$(\lambda - \lambda_1)^{e_1}, \quad (\lambda - \lambda_2)^{e_2}, \quad \cdots\cdots \quad (\lambda - \lambda_k)^{e_k} \qquad (e_1 + e_2 + \cdots + e_k = n),$$

provided that such of these elementary divisors as are not real admit of arrangement in conjugate imaginary pairs. (Cf. Exercises 1, 2, § 93.)

2. Classify pairs of real bilinear forms in six variables (the second form in each pair being non-singular), distinguishing between real and imaginary elementary divisors.

100. Classification of Collineations. The classification of pairs of bilinear forms which we gave in the last section may obviously be regarded, from a more general point of view, as a classification of pairs of matrices, the second matrix of each pair being assumed to be non-singular. From this point of view it admits of application to the classification of collineations, since, as we saw in § 98, to every collineation corresponds a pair of matrices of which one is non-singular, namely the unit matrix **I** and the matrix of the linear transformation. Moreover, the normal form (3) of § 99 is precisely adapted to the treatment of the more special kind of equivalence which we have to consider here, since the matrix of the form ψ is precisely the unit matrix. We may therefore state at once the fundamental theorem:

THEOREM 1. *If* $\lambda_1, \lambda_2, \cdots \lambda_k$ *are any constants, equal or unequal, and* $e_1, e_2, \cdots e_k$ *any positive integers whose sum is* n, *there exists a collineation in space of* $n - 1$ *dimensions whose characteristic matrix has the elementary divisors*

$$(\lambda - \lambda_1)^{e_1}, \quad (\lambda - \lambda_2)^{e_2}, \quad \cdots\cdots (\lambda - \lambda_k)^{e_k}.$$

To this we may add

THEOREM 2. *Every collineation of the kind mentioned in Theorem 1 is equivalent to the collineation whose matrix is*

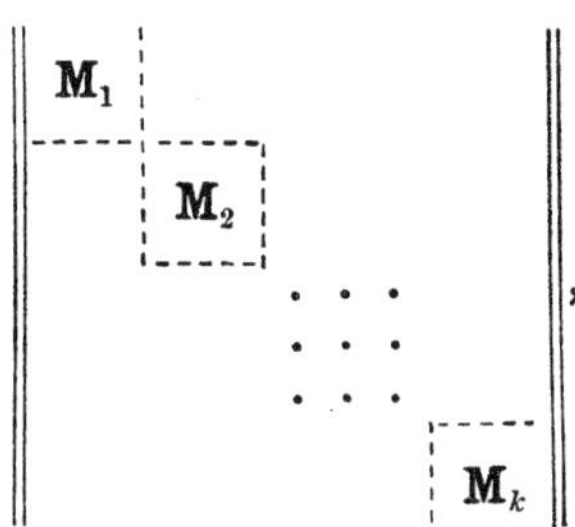

$$\left\|\begin{array}{ccccc} \mathbf{M}_1 & & & & \\ & \mathbf{M}_2 & & & \\ & & \cdot\ \cdot\ \cdot & & \\ & & \cdot\ \cdot\ \cdot & & \\ & & \cdot\ \cdot\ \cdot & & \\ & & & & \mathbf{M}_k \end{array}\right\|,$$

where $\mathbf{M}_i$ stands for the matrix of order e_i,

$$\mathbf{M}_i = \left\|\begin{array}{ccccc} \lambda_i & 1 & 0 & \cdots & 0 \\ 0 & \lambda_i & 1 & \cdots & 0 \\ \cdot & \cdot & \cdot & \cdot & \cdot \\ \cdot & \cdot & \cdot & \cdot & \cdot \\ 0 & 0 & 0 & \cdots & \lambda_i \end{array}\right\|.$$

We thus get a classification of collineations into categories and a subdivision of these categories into classes precisely as in § 99. For instance, in the case $n = 3$ (collineations in the plane), we have three categories whose characteristics and representative normal forms we give:

I. [1 1 1]
$$\begin{cases} x_1' = \lambda_1 x_1 \\ x_2' = \qquad \lambda_2 x_2 \\ x_3' = \qquad\qquad \lambda_3 x_3, \end{cases}$$

II. [2 1]
$$\begin{cases} x_1' = \lambda_1 x_1 + x_2 \\ x_2' = \qquad \lambda_1 x_2 \\ x_3' = \qquad\qquad \lambda_2 x_3, \end{cases}$$

III. [3]
$$\begin{cases} x_1' = \lambda_1 x_1 + x_2 \\ x_2' = \qquad \lambda_1 x_2 + x_3 \\ x_3' = \qquad\qquad \lambda_1 x_3. \end{cases}$$

These categories we should then subdivide either into six classes as on page 290 or into fourteen classes as on page 292. This latter classification is the desirable one in this case. We proceed to give a list of these fourteen classes with a characteristic property of each.

That the normal forms of the collineations have these properties will be at once evident, and from this it follows that all the collineations of the class have the property in question, since the properties mentioned are obviously all projective. That the properties mentioned are really *characteristic* properties, that is, serve to distinguish one class from another, can only be seen *a posteriori*, by noticing that no one of the properties mentioned is shared by two classes.

[1 1 1]   Three distinct non-collinear fixed points.*

[(1 1) 1]   Every point of a certain line and one point not on this line are fixed.

[(1 1 1)]   The identical collineation.

[2 1]   Two distinct fixed points.

[(2 1)]   Every point of a certain line is fixed.

[3]   One fixed point.

In all these cases the collineation is non-singular. The remaining collineations are singular. In the next three, one point P of the plane is not transformed at all, while all other points go over on to a line p which does not pass through P, and every one of whose points corresponds to an infinite number of points.

$[1\,1\,\overset{0}{1}]$ There are two fixed points on p.

$[(1\,1)\,\overset{0}{1}]$ Every point on p is fixed.

$[2\,\overset{0}{1}]$ One fixed point on p.

In the next two cases one point P is not transformed at all, while all other points go over on to a line p which passes through P, and every one of whose points corresponds to an infinite number of points.

$[\overset{0}{2}\,1]$ One fixed point.

$[\overset{0}{3}]$ No fixed point.

The remaining collineations are so simple that they are not merely characterized, but completely described, by the property we mention.

$[(\overset{0}{1}\,\overset{0}{1})\,1]$ The points on a certain line are not transformed. All other points go over into a single point which does not lie on this line.

---

* It should be understood here and in what follows that the fixed points which are mentioned are the only fixed points of the collineation in question.

$[(\overset{0}{2}\,\overset{0}{1})]$ The points on a certain line are not transformed.   All other points go over into a single point on this line.

$[(\overset{0}{1}\,\overset{0}{1}\,\overset{0}{1})]$ No point in the plane is transformed.

This last case is of course not a transformation at all.

EXERCISES

1. Classify, in a similar manner, the projective transformations in one dimension.

2. Classify the collineations in space of three dimensions.

3. Classify the real projective transformations in space of one, two, and three dimensions.   (Cf. Exercises 1, 2, § 99.)

CHAPTER XXII

THE EQUIVALENCE AND CLASSIFICATION OF PAIRS OF QUADRATIC FORMS

101. Two Theorems in the Theory of Matrices. In order to justify the applications we wish to make of the theory of elementary divisors to the subject of quadratic forms, it will be necessary for us to turn back for a moment to the general theory of matrices.

DEFINITION. *If $\phi(x)$ is a polynomial:*

$$\phi(x) \equiv a_0 x^m + a_1 x^{m-1} + \cdots + a_{m-1} x + a_m,$$

then
$$a_0 \mathbf{x}^m + a_1 \mathbf{x}^{m-1} + \cdots + a_{m-1} \mathbf{x} + a_m \mathbf{I}$$

is called a polynomial in the matrix $\mathbf{x}$ and is denoted by $\phi(\mathbf{x})$.[*]

We come now to one of the most fundamental theorems in the whole theory of matrices:

THEOREM 1. *If $\mathbf{a}$ is a matrix, and $\phi(\lambda)$ its characteristic function, then*
$$\phi(\mathbf{a}) = 0.$$

This equation is called *the Hamilton-Cayley equation.*

Let $\mathbf{c}$ be the characteristic matrix of $\mathbf{a}$:

$$\mathbf{c} = \mathbf{a} - \lambda \mathbf{I}.$$

This being a λ-matrix of the first degree, its adjoint $\mathbf{C}$ will be a λ-matrix of degree not higher than $n-1$, if n is the order of the matrix $\mathbf{a}$:

(1) $$\mathbf{C} \equiv \mathbf{C}_{n-1}\lambda^{n-1} + \mathbf{C}_{n-2}\lambda^{n-2} + \cdots + \mathbf{C}_0.$$

We may also write

(2) $$\phi(\lambda) \equiv k_n \lambda^n + k_{n-1} \lambda^{n-1} + \cdots + k_0.$$

Now referring to formula (5), § 25, we see that

$$\mathbf{aC} - \lambda \mathbf{C} \equiv \phi(\lambda) \mathbf{I}.$$

---

[*] It should be noticed that, according to this definition, the coefficients of a polynomial in $\mathbf{x}$ are scalars. Contrast this with a λ-matrix, in which the coefficients are matrices and the variable a scalar. Both of these conceptions would be included in expressions of the form:

$$a_0\mathbf{x}^m \mathbf{b}_0 + a_1\mathbf{x}^{m-1}\mathbf{b}_1 + \cdots + a_{m-1}\mathbf{x}\mathbf{b}_{m-1} + a_m.$$

Substituting here from (1) and (2), we have, on equating corresponding powers of λ,

$$aC_0 \qquad\qquad = k_0 I,$$
$$aC_1 - \quad C_0 = k_1 I,$$
$$aC_2 - \quad C_1 = k_2 I,$$
$$\cdots \cdots \cdots$$
$$\cdots \cdots \cdots$$
$$aC_{n-1} - C_{n-2} = k_{n-1} I,$$
$$- C_{n-1} = k_n I.$$

If we multiply these equations in succession by I, a, a^2, $\cdots$ a^n,
and add, the first members cancel out, and we get

$$k_0 I + k_1 a + k_2 a^2 + \cdots + k_n a^n = 0.$$

This is precisely the equation

$$\phi(a) = 0$$

which we wished to establish.

As a means of deducing our second theorem, we next establish a
lemma which relates merely to scalar quantities.

LEMMA. *If $\psi(x)$ is a polynomial of the nth degree $(n > 0)$ whose
constant term is not zero, there exists a polynomial $\chi(x)$ of degree less
than n such that*
$$(\chi(x))^2 - x$$
is divisible by $\psi(x)$.

Let $x - a$, $x - b$, $x - c$, $\cdots$ be the distinct linear factors of $\psi(x)$,
so that we may write

$$\psi(x) \equiv k(x - a)^\alpha (x - b)^\beta (x - c)^\gamma \cdots \qquad (\alpha + \beta + \gamma + \cdots = n).$$

None of the constants a, b, c, $\cdots$ are zero, since, by hypothesis, the
constant term of ψ is not zero. Let us, further, denote by $\psi_1(x)$ the
polynomial obtained from ψ by omitting the factor $(x - a)^\alpha$, by
$\psi_2(x)$ the polynomial obtained from ψ by omitting the factor $(x - b)^\beta$,
etc., and finally let us form, with undetermined coefficients, the
polynomials

$$A(x) \equiv A_0 + A_1(x - a) + A_2(x - a)^2 + \cdots + A_{\alpha-1}(x - a)^{\alpha-1},$$
$$B(x) \equiv B_0 + B_1(x - b) + B_2(x - b)^2 + \cdots + B_{\beta-1}(x - b)^{\beta-1},$$
$$C(x) \equiv C_0 + C_1(x - c) + C_2(x - c)^2 + \cdots + C_{\gamma-1}(x - c)^{\gamma-1},$$
$$\cdots \cdots \cdots \cdots \cdots$$
$$\cdots \cdots \cdots \cdots \cdots$$

From these polynomials we now form the polynomial

$$\chi(x) \equiv A(x)\psi_1(x) + B(x)\psi_2(x) + C(x)\psi_3(x) + \cdots$$

whose degree can obviously not exceed $n-1$. We wish to show that the coefficients A_i, B_i, $\cdots$ can be so determined that this polynomial $\chi(x)$ satisfies the conditions of our lemma.

Since ψ_2, ψ_3, $\cdots$ are all divisible by $(x-a)^\alpha$, a necessary and sufficient condition that $(\chi(x))^2 - x$ be divisible by this factor is that the polynomial

$$\phi(x) \equiv (A(x))^2(\psi_1(x))^2 - x$$

be divisible by $(x-a)^\alpha$. We have

$$\phi(a) = A_0^2 k^2 (a-b)^{2\beta}(a-c)^{2\gamma} \cdots - a.$$

In order that $\phi(x)$ be divisible by $x-a$ it is therefore necessary and sufficient that

(3)
$$A_0^2 = \frac{a}{k^2(a-b)^{2\beta}(a-c)^{2\gamma} \cdots}.$$

Neither numerator nor denominator here being zero, we thus obtain two distinct values for A_0, both different from zero. If we give to A_0 one of these values, $\phi(x)$ is divisible by $x-a$. A necessary and sufficient condition that it be also divisible by $(x-a)^2$ is that $\phi'(a) = 0$, accents here, and in what follows, denoting differentiation. We shall see in a moment that this condition can be imposed in one, and only one, way by a suitable choice of A_1. The condition that $\phi(x)$ be divisible by $(x-a)^3$ is then simply $\phi''(a) = 0$. We wish to show that this process can be continued until we have finally imposed the condition that $\phi(x)$ be divisible by $(x-a)^\alpha$. For this purpose we use the method of mathematical induction, and assume that A_0, $\cdots A_{s-1}$ have been so determined that $\phi(a) = \phi'(a) = \cdots = \phi^{[s-1]}(a) = 0$. It remains then merely to show that A_s can be so determined that $\phi^{[s]}(a) = 0$. For this purpose we notice that

(4)
$$\phi^{[s]}(x) \equiv 2A^{[s]}(x)A(x)(\psi_1(x))^2 + R_s(x)$$

where $R_s(x)$ is an integral rational function with numerical coefficients of ψ_1, ψ_1', $\cdots \psi_1^{[s]}$, A, A', $\cdots A^{[s-1]}$. Since

$$A(a) = A_0,\ A'(a) = A_1,\ A''(a) = 2!A_2,\ \cdots\ A^{[s-1]}(a) = (s-1)!A_{s-1},$$

it follows that $R(a)$ is a known constant, that is, that it does not depend on any of the still undetermined constants $A_s, A_{s+1}, \cdots A_{\alpha-1}$, nor on the B's, C's, etc. Consequently we see from (4) that a necessary and sufficient condition that $\phi^{[s]}(a) = 0$ is that A_s have the value

$$(5) \qquad A_s = \frac{-R_s(a)}{2\,s!\,A_0\,(\psi_1(a))^2}.$$

Determining the coefficients A_1, A_2, $\cdots$ A_{a-1} in succession by means of this formula, we finally determine the polynomial $A(x)$ in such a way that $\phi(x)$ is divisible by $(x-a)^a$. For this determination, $(\chi(x))^2 - x$ will, as we saw above, be divisible by $(x-a)^a$.

In the same way we can now determine the coefficients of $B(x)$ so that $(\chi(x))^2 - x$ is divisible by $(x-b)^\beta$; then we determine the coefficients of $C(x)$ so that $(\chi(x))^2 - x$ is divisible by $(x-c)^\gamma$; etc. When all the polynomials A, B, C, $\cdots$ are thus determined, $(\chi(x))^2 - x$ is divisible by $\psi(x)$, and our lemma is proved.

THEOREM 2. *If* $\mathbf{a}$ *is a non-singular matrix of order n, there exist matrices* $\mathbf{b}$ *of order n (necessarily non-singular) with the following properties:*

$$\mathbf{b}^2 = \mathbf{a},$$

$\mathbf{b}$ *is a polynomial in* $\mathbf{a}$ *of degree less than n.*

Since $\mathbf{a}$ is non-singular, its characteristic function $\phi(\lambda)$ is a polynomial of the nth degree whose constant term is not zero. Hence, by the preceding lemma, a polynomial $\chi(\lambda)$ of degree less than n can be determined such that

$$(\chi(\lambda))^2 - \lambda \equiv \phi(\lambda) f(\lambda)$$

where $f(\lambda)$ is also a polynomial. From this identity it follows that

$$(\chi(\mathbf{a}))^2 - \mathbf{a} = \phi(\mathbf{a}) f(\mathbf{a}).$$

Since, by Theorem 1, $\phi(\mathbf{a}) = 0$, the last equation may be written

$$(\chi(\mathbf{a}))^2 = \mathbf{a},$$

so that $\mathbf{b} = \chi(\mathbf{a})$ is a matrix satisfying the conditions of our theorem, which is thus proved.

102. Symmetric Matrices. The application of the theory of elementary divisors to the subject of quadratic forms rests on the following proposition:

THEOREM 1. *If* $\mathbf{a}_1$ *and* $\mathbf{a}_2$ *are symmetric matrices and if there exist two non-singular matrices* $\mathbf{p}$ *and* $\mathbf{q}$ *such that*

$$(1) \qquad \mathbf{a}_2 = \mathbf{p}\mathbf{a}_1\mathbf{q},$$

then there also exists a non-singular matrix $\mathbf{P}$ *such that*

$$(2) \qquad\qquad \mathbf{a}_2 = \mathbf{P}'\mathbf{a}_1\mathbf{P}$$

where $\mathbf{P}'$ *is the conjugate of* $\mathbf{P}$.[*]

Let us denote by $\mathbf{p}'$ and $\mathbf{q}'$ the conjugates of $\mathbf{p}$ and $\mathbf{q}$ respectively. Taking the conjugates of both sides of (1), and remembering that $\mathbf{a}_1$ and $\mathbf{a}_2$, being symmetric, are their own conjugates, we get, by Theorem 6, § 22,

$$(3) \qquad\qquad \mathbf{a}_2 = \mathbf{q}'\mathbf{a}_1\mathbf{p}'.$$

By equating the values of $\mathbf{a}_2$ in (1) and (3), we readily deduce the further relation

$$(4) \qquad\qquad (\mathbf{q}')^{-1}\mathbf{p}\mathbf{a}_1 = \mathbf{a}_1\mathbf{p}'\mathbf{q}^{-1}.$$

For brevity we will let

$$(5) \qquad\qquad \mathbf{U} = (\mathbf{q}')^{-1}\mathbf{p}, \qquad \mathbf{U}' = \mathbf{p}'\mathbf{q}^{-1},$$

and note that $\mathbf{U}'$ is the conjugate of $\mathbf{U}$; cf. Exercise 6, § 25.   Equation (4) may then be written

$$(6) \qquad\qquad \mathbf{U}\mathbf{a}_1 = \mathbf{a}_1\mathbf{U}'.$$

From this equation we infer at once the following further ones :

$$(7) \qquad \begin{cases} \mathbf{U}^2\mathbf{a}_1 = \mathbf{U}\mathbf{a}_1\mathbf{U}' = \mathbf{a}_1\mathbf{U}'^2, \\[6pt] \mathbf{U}^3\mathbf{a}_1 = \mathbf{U}\mathbf{a}_1\mathbf{U}'^2 = \mathbf{a}_1\mathbf{U}'^3, \\[6pt] \quad \cdot \quad \cdot \quad \cdot \quad \cdot \quad \cdot \quad \cdot \\[2pt] \quad \cdot \quad \cdot \quad \cdot \quad \cdot \quad \cdot \quad \cdot \\[6pt] \mathbf{U}^k\mathbf{a}_1 = \mathbf{U}\mathbf{a}_1\mathbf{U}'^{k-1} = \mathbf{a}_1\mathbf{U}'^k. \end{cases}$$

Let us now multiply the equations (6) and (7) and also the equation $\mathbf{a}_1 = \mathbf{a}_1$ by any set of scalar constants and add them together. We see in this way that if $\chi(\mathbf{U})$ is any polynomial in $\mathbf{U}$,

$$(8) \qquad\qquad \chi(\mathbf{U})\mathbf{a}_1 = \mathbf{a}_1\chi(\mathbf{U}').$$

---

[*] A proof of this theorem much simpler than that given in the text is the following:

From (1) we infer at once that $\mathbf{a}_1$ and $\mathbf{a}_2$ have the same rank.   Hence the quadratic forms of which $\mathbf{a}_1$ and $\mathbf{a}_2$ are the matrices are equivalent to each other by Theorem 4, § 46.   If we denote by $\mathbf{P}$ the matrix of the linear transformation which carries over the quadratic form $\mathbf{a}_1$ into the form $\mathbf{a}_2$, we see, from Theorem 1, § 43, that equation (2) holds.

This proof would not enable us to infer that $\mathbf{P}$ can be expressed in terms of $\mathbf{p}$ and $\mathbf{q}$ alone, and this is essential for our purposes.

We will choose the polynomial

$$V = \chi(U)$$

so that V is non-singular and

$$V^2 = U,$$

as is seen to be possible by Theorem 2, § 101. Denoting by V' the conjugate of V, we evidently have

$$V' = \chi(U'),$$

so that we may write (8) in the form

$$Va_1 = a_1V',$$

or

$$a_1 = V^{-1}a_1V'.$$

We now substitute this value in (1) and get

(9)
$$a_2 = pV^{-1}a_1V'q.$$

From the first equation (5) we infer the formula

$$pV^{-1} = q'V.$$

Consequently pV^{-1} is the conjugate of $V'q$, so that if we let

$$P = V'q,$$

equation (9) may be written

$$a_2 = P'a_1P,$$

and our theorem is proved.

The proof just given enables us to add the

COROLLARY. *As the matrix P of the foregoing theorem may be taken the matrix $V'q$ where V' is the conjugate of any one of the square roots, determined by Theorem 2, § 101, of $(q')^{-1}p$.*

In particular it will be seen that P depends on p and q but *not* on a_1 or a_2. Hence if a_1, a_2, b_1, b_2 are symmetric matrices, and there exist two non-singular matrices p and q such that

$$a_2 = pa_1q, \qquad b_2 = pb_1q,$$

then there exists a non-singular matrix P such that

$$a_2 = P'a_1P, \qquad b_2 = P'b_1P.$$

From this and Theorem 2, § 96, we infer

THEOREM 2. *If* $a_1, a_2, b_1, b_2,$ *are symmetric matrices of which* b_1, b_2 *are non-singular, a necessary and sufficient condition that a non-singular matrix* P *exist such that*

$$(10) \qquad\qquad a_2 = P'a_1P, \qquad b_2 = P'b_1P,$$

where P' *is the conjugate of* P, *is that the matrices*

$$a_1 - \lambda b_1, \qquad a_2 - \lambda b_2$$

have the same invariant factors, — or, if we prefer, the same elementary divisors.

If, in particular, $b_1 = b_2 = I$, where I is the unit matrix, we have, from the second equation (10), the formula

$$I = P'P.$$

Such a matrix P we call an *orthogonal matrix* according to the definition, which will readily be seen to be equivalent to the one given in the first footnote on page 154 :

DEFINITION. *By an orthogonal matrix we understand a non-singular matrix whose inverse is equal to its conjugate.*

In the special case just referred to, Theorem 2 may be stated in the following form:

THEOREM 3. *If* a_1 *and* a_2 *are two symmetric matrices, a necessary and sufficient condition that an orthogonal matrix* P *exist such that*

$$a_2 = P'a_1P$$

is that the characteristic matrices of a_1 *and* a_2 *have the same invariant factors, — or, if we prefer, the same elementary divisors.*

If this theorem is compared with Theorem 3, § 96, it will be seen that it differs from it only in two respects, first that a_1 and a_2 are assumed to be symmetric, and secondly that P is required to be orthogonal.

103. The Equivalence of Pairs of Quadratic Forms. Let us consider the two pairs of quadratic forms

$$\phi_1 \equiv \sum_1^n a'_{ij} x_i x_j, \qquad \psi_1 \equiv \sum_1^n b'_{ij} x_i x_j,$$

and
$$\phi_2 \equiv \sum_1^n a''_{ij} x_i x_j, \qquad \psi_2 \equiv \sum_1^n b''_{ij} x_i x_j,$$

of which the two forms ψ_1 and ψ_2 are assumed to be non-singular. We will inquire under what conditions these two pairs of forms are equivalent ; that is, under what conditions a linear transformation

$$c \begin{cases} x_1 = c_{11}x_1' + \cdots + c_{1n}x_n' \\ \cdot \quad \cdot \quad \cdot \quad \cdot \quad \cdot \quad \cdot \quad \cdot \quad \cdot \quad \cdot \\ \cdot \quad \cdot \quad \cdot \quad \cdot \quad \cdot \quad \cdot \quad \cdot \quad \cdot \quad \cdot \\ x_n = c_{n1}x_1' + \cdots + c_{nn}x_n' \end{cases}$$

exists which carries over ϕ_1 into ϕ_2 and, at the same time, ψ_1 into ψ_2.

If we denote the conjugate of the matrix c by c', and the matrices of the forms ϕ_1, ψ_1, ϕ_2, ψ_2 by a_1, b_1, a_2, b_2 respectively, we know, by Theorem 1, § 43, that the transformation c carries over ϕ_1 and ψ_1 into forms with the matrices

$$c'a_1c, \qquad c'b_1c$$

respectively; so that, if these are the forms ϕ_2 and ψ_2, we have

(1) $$a_2 = c'a_1c, \qquad b_2 = c'b_1c.$$

Consequently, by Theorem 2, § 102, the two λ-matrices

$$a_1 - \lambda b_1, \qquad a_2 - \lambda b_2$$

have the same invariant factors and elementary divisors.

Conversely, by the same theorem, if these two λ-matrices have the same invariant factors (or elementary divisors), a matrix c, independent of λ, exists which satisfies both equations (1) ; and hence the two pairs of quadratic forms are equivalent. Thus we have proved

THEOREM 1. *If ϕ_1, ψ_1 and ϕ_2, ψ_2 are two pairs of quadratic forms in n variables, in which ψ_1 and ψ_2 are non-singular, a necessary and sufficient condition that these two pairs of forms be equivalent is that the matrices of the two pencils*

$$\phi_1 - \lambda\psi_1, \qquad \phi_2 - \lambda\psi_2$$

*have the same invariant factors, — or, if we prefer, the same elementary divisors.**

A special case of this theorem which is of considerable importance is that in which both of the forms ψ_1 and ψ_2 reduce to

$$x_1^2 + x_2^2 + \cdots + x_n^2.$$

---

*For brevity, we shall speak of these invariant factors and elementary divisors as the invariant factors and elementary divisors of the *pairs of forms* ϕ_1, ψ_1 and ϕ_2, ψ_2 respectively.

In this case we have to deal with orthogonal transformations (cf. the Definition in Exercise 1, § 52), and our theorem may be stated in the form *

THEOREM 2. *If $\mathbf{a}_1$ and $\mathbf{a}_2$ are the matrices of two quadratic forms, a necessary and sufficient condition that there exist an orthogonal transformation which carries over one of these forms into the other is that the characteristic matrices of $\mathbf{a}_1$ and $\mathbf{a}_2$ have the same invariant factors, — or, if we prefer, the same elementary divisors.*

To illustrate the meaning of the theorems of this section, let us consider again briefly the problem of the simultaneous reduction of two quadratic forms to sums of squares. In Chapter XIII we became acquainted with two cases in which this reduction is possible; cf. Theorem 2, § 58, and Theorem 2, § 59. We are in a position now to state a necessary and sufficient condition for the possibility of this reduction, provided that one of the two forms is non-singular.

For this purpose, consider the two quadratic forms

$$\phi \equiv k_1 x_1^2 + k_2 x_2^2 + \cdots + k_n x_n^2,$$
$$\psi \equiv c_1 x_1^2 + c_2 x_2^2 + \cdots + c_n x_n^2,$$

where we assume, in order that the second form may be non-singular, that none of the c's vanish. The matrix of the pencil $\phi - \lambda\psi$ is

$$\left\|\begin{array}{ccccc} k_1 - c_1\lambda & 0 & 0 & \cdots & 0 \\ 0 & k_2 - c_2\lambda & 0 & \cdots & 0 \\ \cdot & \cdot & \cdot & \cdots & \cdot \\ \cdot & \cdot & \cdot & \cdots & \cdot \\ 0 & 0 & 0 & \cdots & k_n - c_n\lambda \end{array}\right\|,$$

and the elementary divisors of this matrix are

$$\lambda - \frac{k_1}{c_1}, \qquad \lambda - \frac{k_2}{c_2}, \qquad \cdots\cdots \qquad \lambda - \frac{k_n}{c_n},$$

all of the first degree. Consequently, any pair of quadratic forms equivalent to the pair just considered must have a λ-matrix whose elementary divisors are all of the first degree.

Conversely, if we have a pair of quadratic forms, of which the first is non-singular, whose λ-matrix has elementary divisors all of

---

* This theorem is, of course, essentially equivalent to Theorem 5, § 102, of which it may be regarded as an immediate consequence.

the first degree, we can obviously choose the constants k and c in such a way that the λ-matrix of the forms ϕ and ψ just considered has these same elementary divisors, and therefore the given forms are equivalent to these special forms ϕ and ψ. Thus we have proved the theorem :

THEOREM 3. *If ϕ and ψ are quadratic forms and ψ is non-singular, a necessary and sufficient condition that it be possible to reduce ϕ and ψ simultaneously by a non-singular linear transformation to forms into which only the square terms enter is that all the elementary divisors of the pair of forms be of the first degree.*

This theorem obviously includes as a special case Theorem 2 of § 58, since the elementary divisors are necessarily of the first degree when the λ-equation has no multiple roots.

Comparing the theorem just proved with Theorem 2, § 59, we see that under the conditions of that theorem the elementary divisors must be of the first degree. Hence

THEOREM 4. *If ψ is a non-singular, definite, quadratic form, and ϕ is a real quadratic form, all the elementary divisors of this pair of forms are necessarily of the first degree.*

104. Classification of Pairs of Quadratic Forms. We consider the pair of quadratic forms

$$(1) \qquad \phi \equiv \sum_1^n a_{ij} x_i x_j, \qquad \psi \equiv \sum_1^n b_{ij} x_i x_j,$$

and assume, as before, that ψ is non-singular. We denote the elementary divisors of these forms, as in § 99, by

$$(\lambda - \lambda_1)^{e_1}, \qquad (\lambda - \lambda_2)^{e_2}, \quad \cdots \quad (\lambda - \lambda_k)^{e_k} \qquad (e_1 + e_2 + \cdots + e_k = n).$$

The symbol $[e_1 \, e_2 \, \cdots e_k]$ we call the *characteristic* of the pair of quadratic forms; and all pairs of quadratic forms which have the same characteristic we speak of as forming a *category*.*

We have here, precisely as in the case of bilinear forms, the theorem:

THEOREM. *If $\lambda_1, \lambda_2, \cdots \lambda_k$ are any constants, equal or unequal, and $e_1, e_2, \cdots e_k$ are any positive integers whose sum is n, there exist pairs*

---

* Thus, for instance, all pairs of forms of which the second is non-singular and which admit of simultaneous reduction to sums of squares, form a category whose characteristic is $[1 \, 1 \, \cdots 1]$. Cf. Theorem 3, § 103.

of quadratic forms in n variables, the second form in each pair being non-singular, which have the elementary divisors

$$(2) \qquad (\lambda - \lambda_1)^{e_1}, \qquad (\lambda - \lambda_2)^{e_2}, \quad \cdots \cdots (\lambda - \lambda_k)^{e_k}.$$

The proof of this theorem consists in considering the following pair of quadratic forms, analogous to the normal form $(3')$ of § 99:

$$(3) \begin{cases} \phi \equiv \left(\sum_1^{e_1} \lambda_1 c_1 x_i x_{e_1-i+1} + \sum_1^{e_1-1} d_1 x_i x_{e_1-i} \right) + \left(\sum_{e_1+1}^{e_1+e_2} \lambda_2 c_2 x_i x_{2e_1+e_2-i+1} + \sum_{e_1+1}^{e_1+e_2-1} d_2 x_i x_{2e_1+e_2-i} \right) \\[2ex] \qquad\qquad + \cdots + \left(\sum_{n-e_k+1}^{n-1} \lambda_k c_k x_i x_{2n-e_k-i+1} + \sum_{n-e_k+1}^{n-1} d_k x_i x_{2n-e_k-i} \right), \\[3ex] \psi \equiv \sum_1^{e_1} c_1 x_i x_{e_1-i+1} + \sum_{e_1+1}^{e_1+e_2} c_2 x_i x_{2e_1+e_2-i+1} + \sum_{e_1+e_2+1}^{e_1+e_2+e_3} c_3 x_i x_{2e_1+2e_2+e_3-i+1} \\[2ex] \qquad\qquad\qquad\qquad + \cdots + \sum_{n-e_k+1}^{n} c_k x_i x_{2n-e_k-i+1}, \end{cases}$$

where $c_1, \cdots c_k, d_1, \cdots d_k$ are constants which may be chosen at pleasure, provided none of them are zero.

The λ-matrix of this pair of forms is the same as the λ-matrix of the pair of bilinear forms $(3')$ of § 99, and therefore has the desired elementary divisors.

A reference to Theorem 1, § 103, shows that formula (3) yields a normal form to which every pair of quadratic forms, of which the second is non-singular and whose elementary divisors are given by (2), can be reduced.

The categories, of which we have so far spoken, may be divided into classes by the same methods we used in § 99 in the case of bilinear forms. This may be done, as before, either by simply noting which of the λ_i's are equal to each other, or by further distinguishing between the cases where some of the λ_i's are zero.

We are now in a position to see exactly in what way our elementary divisors give us a more powerful instrument than we had in the invariants Θ_i of § 57. These invariants Θ_i, being the coefficients of the λ-equation of our pair of forms, determine the constants λ_i, which are the roots of this equation, as well as the multiplicities of these roots. They do *not* determine the degrees e_i of the elementary divisors, and the use of the Θ_i's alone does not, in all cases, enable us to determine whether two pairs of forms are equivalent or not. Thus, for instance, we may have two pairs of forms with exactly the same invariants Θ_i but with characteristics $[(1\,1)\,1\,1\,\cdots\,1]$ and $[2\,1\,1\,\cdots\,1]$ respectively.* It will be seen, therefore, that the Θ_i's form in only a very technical sense a complete system of invariants.

EXERCISES

1. Form a numerical example in the case $n = 3$ to illustrate the statement made in the next to the last sentence of this section.

2. Prove that if two equivalent pairs of quadratic forms have two elementary divisors of the first degree which correspond to the same linear factor, there exist an infinite number of linear transformations which carry over one pair of forms into the other.

3. Prove the general theorem, of which Exercise 2 is a special case, namely, that if two equivalent pairs of quadratic forms have a characteristic in which one or more parentheses appear, there exist an infinite number of linear transformations which carry over one pair of forms into the other.

4. Prove that if two equivalent pairs of quadratic forms have a characteristic in which no parentheses appear, only a finite number of linear transformations exist which carry over one pair of forms into the other. †

How are these transformations related to each other?

105. Pairs of Quadratic Equations, and Pencils of Forms or Equations.‡ In dealing with quadratic forms, the questions of equivalence and classification do not always present themselves to us in precisely the form in which we have considered them in the last two sections. We frequently have to deal not with the quadratic forms themselves but with the *equations* obtained by setting the forms equal to zero. Two such pairs of equations we shall regard as equivalent, not merely if the forms in them are equivalent, but also if one pair of forms can be obtained from the other by multiplication by constants different from zero.

Let us consider two quadratic forms ϕ, ψ, of which we assume, as before, that the second is non-singular, and inquire what the effect on the elementary divisors,

$$(1) \qquad (\lambda - \lambda_1)^{e_1}, \qquad (\lambda - \lambda_2)^{e_2}, \;\cdots\cdots\; (\lambda - \lambda_k)^{e_k}$$

---

* We may, in the case $n = 3$, put the same thing geometrically (cf. the next section) by saying that it is impossible to distinguish between the case of two conics having double contact and that of two conics having simple contact at a single point by the use of the invariants Θ_i alone, whereas these two cases are at once distinguished by the use of elementary divisors.

† The exercise in § 58 is practically a special case of this.

‡ Questions similar to those treated in this section might have been taken up in the last chapter for the case of bilinear forms.

of these forms will be if the forms are multiplied respectively by the constants p, q which are both assumed to be different from zero. Let us write

$$\phi_1 \equiv p\phi, \qquad \psi_1 \equiv q\psi.$$

Then

(2)
$$\phi_1 - \lambda\psi_1 \equiv p\,(\phi - \lambda'\psi)$$

where
$$\lambda' = \frac{q}{p}\lambda.$$

Let $\lambda - \alpha$ be any one of the linear factors of the matrix of $\phi - \lambda\psi$, so that α is any one of the constants λ_1, λ_2, $\cdots \lambda_k$; and let us denote, as in the footnote to Definition 3, § 92, by l_i the exponent of the highest power of $\lambda - \alpha$ which is a factor of all the i-rowed determinants of this matrix. Then it is clear, from (2), that l_i is the exponent of the highest power of $\lambda' - \alpha$ which is a factor of all the i-rowed determinants of the matrix of $\phi_1 - \lambda\psi_1$. In other words,

$$\left(\lambda - \frac{p\alpha}{q}\right)^{l_i}$$

is the highest power of the linear factor $\lambda - p\alpha/q$ which is a factor of all the i-rowed determinants of the matrix of $\phi_1 - \lambda\psi_1$. Turning now to the definition of elementary divisors as given in the footnote to Definition 3, § 92, we see that the elementary divisors of the matrix of $\phi_1 - \lambda\psi_1$ differ from those of the matrix of $\phi - \lambda\psi$ only in having the constants λ_i replaced by the constants $p\lambda_i/q$. We thus have the result :

THEOREM 1. *If the pair of quadratic forms ϕ, ψ, of which the second is assumed to be non-singular, has the elementary divisors*

$$(\lambda - \lambda_1)^{e_1}, \quad (\lambda - \lambda_2)^{e_2}, \quad \cdots\cdots (\lambda - \lambda_k)^{e_k},$$

and if p, q are constants different from zero, then the pair of quadratic forms $p\phi$, $q\psi$ has the elementary divisors

$$(\lambda - \lambda_1')^{e_1}, \quad (\lambda - \lambda_2')^{e_2}, \quad \cdots\cdots (\lambda - \lambda_k')^{e_k}$$

where
$$\lambda_i' = \frac{p}{q}\lambda_i.$$

In particular, it will be seen that these two pairs of forms have the same characteristic, even when the conception of the characteristic is refined not merely by inserting parentheses but also by the use of the small zeros.

The theorem just proved shows that pairs of homogeneous quadratic equations, of which the second equation in each pair is non-singular, may be classified by the use of their characteristics precisely as was done in the last section for pairs of quadratic forms. We proceed to illustrate this in the case $n = 3$, where we may consider that we have to deal with the classification of pairs of conics in a plane, one of the conics being non-singular.

We have here three categories represented by the following normal forms :*

I.   [1 1 1] $\begin{cases} \phi \equiv \lambda_1 x_1^2 + \lambda_2 x_2^2 - \lambda_3 x_3^2 \\ \psi \equiv x_1^2 + x_2^2 - x_3^2. \end{cases}$

II.   [2 1] $\begin{cases} \phi \equiv 2\lambda_1 x_1 x_2 + x_1^2 + \lambda_2 x_3^2 \\ \psi \equiv 2 x_1 x_2 + x_3^2. \end{cases}$

III. [3] $\begin{cases} \phi \equiv 2\lambda_1 x_1 x_3 + \lambda_1 x_2^2 + 2x_1 x_2 \\ \psi \equiv 2 x_1 x_3 + x_2^2. \end{cases}$

We next subdivide these categories into classes, and, by an examination of the normal form in each case, we are enabled at once to characterize each class by certain projective properties which it has, and which are shared by no other class.† Since the conic ψ is non-singular in all cases, this fact need not be explicitly stated.

[1 1 1] ϕ and ψ intersect in four distinct points.

[(1 1) 1] ϕ and ψ have double contact.

[(1 1 1)] ϕ and ψ coincide.

[2 1] ϕ and ψ meet in three distinct points at one of which they touch.

[(2 1)] ϕ and ψ have contact of the third order.

[3] ϕ and ψ have contact of the second order.

In all of the above cases ϕ, as well as ψ, is non-singular.

In the next five cases, ϕ consists of a pair of distinct straight lines.

* We assign to the constants c_i and k_i, in formula (3) of the last section, values so chosen that the loci $\phi = 0$, $\psi = 0$ are real when the constants λ_i are real. This is, of course, not essential, since we are not concerned with questions of reality.

† In order to verify the statements made below, the reader should have some knowledge of the theory of the contact of conics; cf. for instance Salmon's *Conic Sections*, Chapter XIV., pages 232–238.

$[1\ 1\ \overset{0}{1}]$ ϕ and ψ intersect in four distinct points.

$[(1\ 1)\ \overset{0}{1}]$ Both of the lines of which ϕ consists touch ψ.

$[\overset{0}{2}\ 1]$ One of the lines of which ϕ consists touches ψ, while the other cuts it in two points distinct from the point of contact of the first.

$[\overset{0}{2}\ 1]$ The two lines of which ϕ consists intersect on ψ, and neither of them touches ψ.

$[\overset{0}{3}]$ The two lines of which ϕ consists intersect on ψ, and one of them touches ψ.

In the next two cases, ϕ consists of a single line.

$[(\overset{0}{1}\ \overset{0}{1})\ 1\ ]$ The line ϕ meets ψ in two distinct points.

$[(\overset{0}{2}\ \overset{0}{1})]$ The line ϕ touches ψ.

Finally we have the case:

$[(\overset{0}{1}\ \overset{0}{1}\ \overset{0}{1}])$ Here $\phi \equiv 0$, and we have no conic other than ψ.

Suppose finally that we wish to classify not *pairs* of quadratic forms or equations but *pencils* of quadratic forms or equations. Consider the pencil of quadratic forms

$$\phi - \lambda\psi$$

where ϕ and ψ are quadratic forms, and ψ is non-singular, and suppose that the elementary divisors of the pair of forms ϕ, ψ are given by formula (1) above. The question presents itself whether, if, in place of the forms ϕ, ψ, we take any other two forms of the pencil

$$\phi_1 \equiv \phi - \mu\psi, \qquad \psi_1 \equiv \phi - \nu\psi,$$

the constants μ, ν being so chosen that $\mu \neq \nu$ and that ψ_1 is non-singular, the pair of forms ϕ_1, ψ_1, will have these same elementary divisors (1). If this were the case, we could properly speak of (1) as the elementary divisors of the pencil. This, however, is *not* the case, and *the pencil of quadratic forms cannot properly be said to have elementary divisors.**

---

* We here regard the pencil as merely an aggregate of an infinite number of quadratic forms, namely, all the forms which can be obtained from the expression $\phi - \lambda\psi$ by giving to λ different values. In this sense we cannot speak of the elementary divisors of the pencil. If, however, we wish to regard the polynomial in the x's and λ, $\phi - \lambda\psi$, as the pencil, we may speak of its elementary divisors, meaning thereby simply what we have called the elementary divisors of the pair of forms ϕ, ψ.

There is, however, a simple relation between the elementary divisors of two pairs of forms taken from the same pencil. In order to show this, let us determine the elementary divisors of the pair of forms ϕ_1, ψ_1, above. For this purpose consider the expression $\phi_1 - \lambda\psi_1$, which, when $\lambda \neq 1$, may be written

$$(3) \qquad \phi_1 - \lambda\psi_1 \equiv (1 - \lambda)[\phi - \lambda'\psi]$$

where
$$\lambda' = \frac{\mu - \nu\lambda}{1 - \lambda}.$$

Now suppose, as above, that $\lambda - \alpha$ is any one of the linear factors of the matrix of $\phi - \lambda\psi$, and that l_i is the exponent of the highest power of $\lambda - \alpha$ which is a factor of all the i-rowed determinants of this matrix. Then any one of the i-rowed determinants of the matrix of $\phi - \lambda'\psi$ may, when $\lambda \neq 1$, be written in the form

$$(\lambda' - \alpha)^{l_i} f(\lambda')$$

where f is a polynomial in λ' of degree not greater than $i - l_i$. Accordingly, by (3), the corresponding i-rowed determinant of the matrix of $\phi_1 - \lambda\psi_1$ may be written

$$[\mu - \nu\lambda - \alpha(1 - \lambda)]^{l_i} f_1(\lambda)$$

where f_1 is a polynomial in λ. Thus we see that

$$\left[\lambda - \frac{\alpha - \mu}{\alpha - \nu}\right]^{l_i}$$

is a factor of every i-rowed determinant of the matrix of $\phi_1 - \lambda\psi_1$. Similar reasoning, carried through in the reverse order, shows that this is the highest power of $\lambda - \dfrac{\alpha - \mu}{\alpha - \nu}$

which is a factor of all these i-rowed determinants. Hence

THEOREM 2. *If the pair of quadratic forms ϕ, ψ, of which the second is non-singular, have the elementary divisors*

$$(\lambda - \lambda_1)^{e_1}, \qquad (\lambda - \lambda_2)^{e_2}, \quad \cdots\cdots \quad (\lambda - \lambda_k)^{e_k},$$

and if μ, ν are any two constants distinct from each other and such that ν is distinct from all the constants λ_1, λ_2, $\cdots$ λ_k, then the two forms

$$\phi_1 \equiv \phi - \mu\psi, \qquad \psi_1 \equiv \phi - \nu\psi,$$

of which the second will then be non-singular, will have the elementary divisors

$$(\lambda - \lambda_1')^{e_1}, \qquad (\lambda - \lambda_2')^{e_2}, \quad \cdots\cdots \quad (\lambda - \lambda_k')^{e_k}$$

where
$$\lambda_i' = \frac{\lambda_i - \mu}{\lambda_i - \nu} \qquad\qquad (i = 1, 2, \cdots k).$$

In particular, it will be seen that the two pairs of forms ϕ, ψ and ϕ_1, ψ_1 have the same characteristic $[e_1\ e_2 \cdots e_k]$ even if we put in parentheses to indicate which of the e's correspond to equal λ_i's. The characteristics will not, however, necessarily be the same if we put in small zeros to indicate which of the e's correspond to vanishing λ_i's, since λ_i and λ_i' do not usually vanish together. Accordingly, in classifying pencils of quadratic forms, we may use the characteristic of any pair of distinct forms of the pencil, the second of which is non-singular, *but we must not introduce the small zeros into these characteristics.* This classification, of course, applies only to what may be called *non-singular pencils*, that is, pencils whose forms are not all singular.

It will readily be seen that what has just been said applies without essential change to the case of pencils of homogeneous quadratic equations. We may therefore illustrate it by the classification of non-singular pencils of conics.* We have here six classes of pencils which we characterize as follows:

[1 1 1]   The conics all pass through four distinct points.

[(1 1) 1]   The conics all pass through two points at which they have double contact with each other.

[(1 1 1)]   The conics all coincide.

[2 1]   The conics all pass through three points at one of which they touch one another.

[(2 1)]   The conics all pass through one point at which they have contact of the third order.

[3]   The conics all pass through two points, at one of which they have contact of the second order.

EXERCISES

1. Determine, by the use of elementary divisors, the nature of each of the following pairs of conics:

$$(a)\quad \begin{cases} 3\,x_1^2 + 7\,x_2^2 \qquad\ + 8\,x_1 x_2 - 10\,x_2 x_3 + 4\,x_1 x_3 = 0 \\ 2\,x_1^2 + 3\,x_2^2 - x_3^2 + 4\,x_1 x_2 -\ 6\,x_2 x_3 + 6\,x_1 x_3 = 0. \end{cases}$$

$$(b)\quad \begin{cases} 3\,x_1^2 - x_2^2 - 3\,x_3^2 - 3\,x_1 x_2 + 3\,x_2 x_3 + x_1 x_3 = 0 \\ 2\,x_1^2 + x_2^2 -\ x_3^2 - 2\,x_1 x_2 - 2\,x_2 x_3 + 2\,x_1 x_3 = 0. \end{cases}$$

2. Give a classification of pairs of binary quadratic equations, the second equation of each pair being non-singular, and interpret the work geometrically.

* For a similar classification of pencils of quadrics we refer to p. 46 of Bromwich's book : *Quadratic Forms and their Classification by Means of Invariant Factors.*

106. Conclusion. We wish, in this section, to point out some of the important questions connected with the subject of elementary divisors, which, in order to keep our treatment within proper limits, we have been obliged to leave out of consideration.

If ϕ_1, ψ_1 and ϕ_2, ψ_2 are two pairs of bilinear or quadratic forms of which ψ_1, ψ_2 are non-singular, we have found a method of determining whether these two pairs of forms are equivalent or not. If we use the invariant factors instead of the elementary divisors, our method involves only the use of the rational operations (addition, subtraction, multiplication, and division), and can, therefore, be actually carried through in any concrete case. In fact we have explained in § 93 some really practical methods of determining the invariant factors of a λ-matrix, so that the problem of determining whether or not two pairs of bilinear or quadratic forms, the second form in each pair being non-singular, are equivalent, may be regarded as solved, not merely from the theoretical, but also from the practical point of view.

There is, however, another question here, which we have not treated, namely, if the two pairs of forms turn out to be equivalent, to find a linear transformation which carries over one into the other. This problem, too, we may consider that we have solved from a theoretical point of view; for the proof we have given that if two pairs of forms have the same elementary divisors there exists a linear transformation which carries over one pair of forms into the other, consisted, as will be seen on examination, in actually giving a method whereby such a linear transformation could be determined. In fact, in the case of bilinear forms, the processes involved are, here again, merely the rational processes; so that, given two equivalent pairs of bilinear forms, the second form of each pair being non-singular, we are in a position to find, in any concrete case, linear transformations of the x's and y's which carry over one pair of forms into the other. Even here the arrangement of the work in a practical manner might require further consideration.

In the case of quadratic forms the problem becomes a much more difficult one, inasmuch as the processes involved in the determination of the required linear transformation are no longer rational; cf. the Lemma of § 101. That this is not merely a defect of the method we have used, but is inherent in the problem itself, will be seen by a consideration of simple numerical examples.   Let, for instance,

$$\phi_1 \equiv 2x_1^2 + 3x_2^2, \qquad \phi_2 \equiv 2x_1^2 - 3x_2^2,$$
$$\psi_1 \equiv \ x_1^2 + \ x_2^2, \qquad \psi_2 \equiv \ x_1^2 - \ x_2^2.$$

Here the pairs of forms ϕ_1, ψ_1 and ϕ_2, ψ_2 both have the elementary divisors

$$\lambda - 2, \qquad \lambda - 3,$$

and are therefore equivalent. The linear transformation which carries over one pair of forms into the other cannot, however, be real (and therefore its coefficients cannot be determined rationally from the coefficients of the given forms) since ϕ_1 and ψ_1 are definite, ϕ_2 and ψ_2 indefinite.

We have, therefore, here the problem of devising a practical method of determining a linear transformation which carries over a first pair of quadratic forms into a second given equivalent pair.   A method of this sort, which is a practical one when once the elementary divisors have been determined, will be found in Bromwich's book on quadratic forms referred to in the footnote on p. 312.

Another point at which our treatment is incomplete is in the restriction we have always made in assuming that, in the pair of bilinear or quadratic forms ϕ, ψ, the form ψ is non-singular. Although this is the case in many of the most important problems to which one wishes to apply the method of elementary divisors, it is still a restriction which it is desirable to remove. This may be done in part by making use not, as we have done, of the pencil $\phi - \lambda\psi$, but of the more general pencil $\mu\phi - \lambda\psi$, μ and λ being variable parameters. The determinants of the matrix of this pencil are binary forms in (μ, λ), and the whole subject of elementary divisors admits an easy extension to this case, the elementary divisors being now integral powers of linear binary forms. The only case which cannot be treated in this way is that in which not only ϕ and ψ are both singular, but every form of the pencil $\mu\phi - \lambda\psi$ is singular. This *singular case*, which was explicitly excluded by Weierstrass in his original paper, requires a special treatment which has been given by Kronecker. Cf., for the case of quadratic forms, the book of Bromwich already referred to.

Still another question is the application of the method of elementary divisors to the case in which the two forms ϕ, ψ are real, and only real linear transformations are admitted. In the case of bilinear forms, this question presents no serious difficulty; cf. the exercises of §§ 97, 99. In the case of quadratic forms, however, the irrational processes involved in the proof of the Lemma of § 101 introduce an essential difficulty, since they are capable of introducing imaginary quantities. Moreover, this difficulty does not lie merely in the method of treatment. The theorems themselves which we have established do not remain true, as is seen by a reference to the numerical example given earlier in this section for another purpose, where we have two pairs of real quadratic forms which, although they have the same elementary divisors, are not equivalent with regard to real linear transformations.

We must content ourselves with merely mentioning this important subject, and referring, for one of the fundamental theorems, to p. 69 of the book of Bromwich.

For further information concerning the subject of elementary divisors the reader is referred to Muth's *Theorie und Anwendung der Elementartheiler*, Leipzig, Teubner, 1899. In English, the book of Bromwich already referred to and some sections in Mathews' revision of Scott's *Determinants* will be found useful.

INDEX

(The numbers refer to the pages.)

317

A series of hardcover reprints of major works in mathematics, science and engineering.
All editions are 5⅝ × 8½ unless otherwise noted.

Mathematics

Theory of Approximation, N. I Achieser. Unabridged republication of the 1956 edition. 320pp. 49543-4

The Origins of the Infinitesimal Calculus, Margaret E. Baron. Unabridged republication of the 1969 edition. 320pp. 49544-2

A Treatise on the Calculus of Finite Differences, George Boole. Unabridged republication of the 2nd and last revised edition. 352pp. 49523-X

Space and Time, Emile Borel. Unabridged republication of the 1926 edition. 15 figures. 256pp. 49545-0

An Elementary Treatise on Fourier's Series, William Elwood Byerly. Unabridged republication of the 1893 edition. 304pp. 49546-9

Substance and Function & Einstein's Theory of Relativity, Ernst Cassirer. Unabridged republication of the 1923 double volume. 480pp. 49547-7

A History of Geometrical Methods, Julian Lowell Coolidge. Unabridged republication of the 1940 first edition. 13 figures. 480pp. 49524-8

Linear Groups with an Exposition of Galois Field Theory, Leonard Eugene Dickson. Unabridged republication of the 1901 edition. 336pp. 49548-5

Continuous Groups of Transformations, Luther Pfahler Eisenhart. Unabridged republication of the 1933 first edition. 320pp. 49525-6

Transcendental and Algebraic Numbers, A. O. Gelfond. Unabridged republication of the 1960 edition. 208pp. 49526-4

Lectures on Cauchy's Problem in Linear Partial Differential Equations, Jacques Hadamard. Unabridged reprint of the 1923 edition. 320pp. 49549-3

The Theory of Branching Processes, Theodore E. Harris. Unabridged, corrected republication of the 1963 edition. xiv+230pp. 49508-6

The Continuum, Edward V. Huntington. Unabridged republication of the 1917 edition. 4 figures. 96pp. 49550-7

Lectures on Ordinary Differential Equations, Witold Hurewicz. Unabridged republication of the 1958 edition. xvii+122pp. 49510-8

Mathematical Methods and Theory in Games, Programming, and Economics: Two Volumes Bound as One, Samuel Karlin. Unabridged republication of the 1959 edition. 848pp. 49527-2

Famous Problems of Elementary Geometry, Felix Klein. Unabridged reprint of the 1930 second edition, revised and enlarged. 112pp. 49551-5

Lectures on the Icosahedron, Felix Klein. Unabridged republication of the 2nd revised edition, 1913. 304pp. 49528-0

On Riemann's Theory of Algebraic Functions, Felix Klein. Unabridged republication of the 1893 edition. 43 figures. 96pp. 49552-3

A Treatise on the Theory of Determinants, Thomas Muir. Unabridged republication of the revised 1933 edition. 784pp. 49553-1

A Survey of Minimal Surfaces, Robert Osserman. Corrected and enlarged republication of the work first published in 1969. 224pp. 49514-0

The Variational Theory of Geodesics, M. M. Postnikov. Unabridged republication of the 1967 edition. 208pp. 49529-9

DOVER PHOENIX EDITIONS

An Introduction to the Approximation of Functions, Theodore J. Rivlin. Unabridged republication of the 1969 edition. 160pp. 49554-X

An Essay on the Foundations of Geometry, Bertrand Russell. Unabridged republication of the 1897 edition. 224pp. 49555-8

Elements of Number Theory, I. M. Vinogradov. Unabridged republication of the first edition, 1954. 240pp. 49530-2

Asymptotic Expansions for Ordinary Differential Equations, Wolfgang Wasow. Unabridged republication of the 1976 corrected, slightly enlarged reprint of the original 1965 edition. 384pp. 49518-3

Physics

Semiconductor Statistics, J. S. Blakemore. Unabridged, corrected, and slightly enlarged republication of the 1962 edition. 141 illustrations. xviii+318pp. 49502-7

Wave Propagation in Periodic Structures, L. Brillouin. Unabridged republication of the 1946 edition. 131 illustrations. 272pp. 49556-6

The Conceptual Foundations of the Statistical Approach in Mechanics, Paul and Tatiana Ehrenfest. Unabridged republication of the 1959 edition. 128pp. 49504-3

The Analytical Theory of Heat, Joseph Fourier. Unabridged republication of the 1878 edition. 20 figures. 496pp. 49531-0

States of Matter, David L. Goodstein. Unabridged republication of the 1975 edition. 154 figures. 4 tables. 512pp. 49506-X

The Principles of Mechanics, Heinrich Hertz. Unabridged republication of the 1900 edition. 320pp. 49557-4

Thermodynamics of Small Systems, Terrell L. Hill. Unabridged and corrected republication in one volume of the two-volume edition published in 1963–1964. 32 illustrations. 408pp. $6\frac{1}{2}$ x $9\frac{1}{4}$. 49509-4

Theoretical Physics, A. S. Kompaneyets. Unabridged republication of the 1961 edition. 56 figures. 592pp. 49532-9

Quantum Mechanics, H. A. Kramers. Unabridged republication of the 1957 edition. 14 figures. 512pp. 49533-7

The Theory of Electrons, H. A. Lorentz. Unabridged reproduction of the 1915 edition. 9 figures. 352pp. 49558-2

The Principles of Physical Optics, Ernst Mach. Unabridged republication of the 1926 edition. 279 figures. 10 portraits. 336pp. 49559-0

The Scientific Papers of James Clerk Maxwell, James Clerk Maxwell. Unabridged republication of the 1890 edition. 197 figures. 39 tables. Total of 1,456pp. *Volume I* (640pp.) 49560-4; *Volume II* (816pp.) 49561-2

Vectors and Tensors in Crystallography, Donald E. Sands. Unabridged and corrected republication of the 1982 edition. xviii+228pp. 49516-7

Principles of Mechanics and Dynamics, Sir William (Lord Kelvin) Thompson and Peter Guthrie Tait. Unabridged republication of the 1912 edition. 168 diagrams. Total of 1,088pp. *Volume I* (528pp.) 49562-0; *Volume II* (560pp.) 49563-9

Treatise on Irreversible and Statistical Thermophysics: An Introduction to Nonclassical Thermodynamics, Wolfgang Yourgrau, Alwyn van der Merwe, and Gough Raw. Unabridged, corrected republication of the 1966 edition. xx+268pp. 49519-1

Engineering

Principles of Aeroelasticity, Raymond L. Bisplinghoff and Holt Ashley. Unabridged, corrected republication of the original 1962 edition. xi+527pp. 49500-0

Statics of Deformable Solids, Raymond L. Bisplinghoff, James W. Mar, and Theodore H. H. Pian. Unabridged and corrected Dover republication of the edition published in 1965. 376 illustrations. xii+322pp. $6\frac{1}{2}$ x $9\frac{1}{4}$. 49501-9